?itman Research Notes in Mathematics ;eries

;ubmission of proposals for consideration

;uggestions for publication, in the form of outlines and representative samples, are invited ›y the Editorial Board for assessment. Intending authors should approach one of the main :ditors or another member of the Editorial Board, citing the relevant AMS subject :lassifications. Alternatively, outlines may be sent directly to the publisher's offices. Refereeing is by members of the board and other mathematical authorities in the topic :oncerned, throughout the world.

?reparation of accepted manuscripts

)n acceptance of a proposal, the publisher will supply full instructions for the preparation of nanuscripts in a form suitable for direct photo-lithographic reproduction. Specially printed ;rid sheets can be provided and a contribution is offered by the publisher towards the cost of yping. Word processor output, subject to the publisher's approval, is also acceptable.

llustrations should be prepared by the authors, ready for direct reproduction without further mprovement. The use of hand-drawn symbols should be avoided wherever possible, in ›rder to maintain maximum clarity of the text.

[he publisher will be pleased to give any guidance necessary during the preparation of a ypescript, and will be happy to answer any queries.

mportant note

n order to avoid later retyping, intending authors are strongly urged not to begin final ›reparation of a typescript before receiving the publisher's guidelines. In this way it is hoped o preserve the uniform appearance of the series.

.ongman Scientific & Technical
.ongman House
3urnt Mill
Iarlow, Essex, CM20 2JE
JK
Telephone (0279) 426721)

Titles in this series. A full list is available on request from the publisher.

100 Optimal control of variational inequalities
V Barbu
101 Partial differential equations and dynamical systems
W E Fitzgibbon III
102 Approximation of Hilbert space operators Volume II
C Apostol, L A Fialkow, D A Herrero and D Voiculescu
103 Nondiscrete induction and iterative processes
V Ptak and F-A Potra
104 Analytic functions – growth aspects
O P Juneja and G P Kapoor
105 Theory of Tikhonov regularization for Fredholm equations of the first kind
C W Groetsch
106 Nonlinear partial differential equations and free boundaries. Volume I
J I Díaz
107 Tight and taut immersions of manifolds
T E Cecil and P J Ryan
108 A layering method for viscous, incompressible L_p flows occupying R^n
A Douglis and E B Fabes
109 Nonlinear partial differential equations and their applications: Collège de France Seminar. Volume VI
H Brezis and J L Lions
110 Finite generalized quadrangles
S E Payne and J A Thas
111 Advances in nonlinear waves. Volume II
L Debnath
112 Topics in several complex variables
E Ramírez de Arellano and D Sundararaman
113 Differential equations, flow invariance and applications
N H Pavel
114 Geometrical combinatorics
F C Holroyd and R J Wilson
115 Generators of strongly continuous semigroups
J A van Casteren
116 Growth of algebras and Gelfand–Kirillov dimension
G R Krause and T H Lenagan
117 Theory of bases and cones
P K Kamthan and M Gupta
118 Linear groups and permutations
A R Camina and E A Whelan
119 General Wiener–Hopf factorization methods
F-O Speck
120 Free boundary problems: applications and theory. Volume III
A Bossavit, A Damlamian and M Fremond
121 Free boundary problems: applications and theory. Volume IV
A Bossavit, A Damlamian and M Fremond
122 Nonlinear partial differential equations and their applications: Collège de France Seminar. Volume VII
H Brezis and J L Lions
123 Geometric methods in operator algebras
H Araki and E G Effros
124 Infinite dimensional analysis–stochastic processes
S Albeverio
125 Ennio de Giorgi Colloquium
P Krée
126 Almost-periodic functions in abstract spaces
S Zaidman
127 Nonlinear variational problems
A Marino, L Modica, S Spagnolo and M Degliovanni
128 Second-order systems of partial differential equations in the plane
L K Hua, W Lin and C-Q Wu
129 Asymptotics of high-order ordinary differential equations
R B Paris and A D Wood
130 Stochastic differential equations
R Wu
131 Differential geometry
L A Cordero
132 Nonlinear differential equations
J K Hale and P Martinez-Amores
133 Approximation theory and applications
S P Singh
134 Near-rings and their links with groups
J D P Meldrum
135 Estimating eigenvalues with *a posteriori/a priori* inequalities
J R Kuttler and V G Sigillito
136 Regular semigroups as extensions
F J Pastijn and M Petrich
137 Representations of rank one Lie groups
D H Collingwood
138 Fractional calculus
G F Roach and A C McBride
139 Hamilton's principle in continuum mechanics
A Bedford
140 Numerical analysis
D F Griffiths and G A Watson
141 Semigroups, theory and applications. Volume I
H Brezis, M G Crandall and F Kappel
142 Distribution theorems of L-functions
D Joyner
143 Recent developments in structured continua
D De Kee and P Kaloni
144 Functional analysis and two-point differential operators
J Locker
145 Numerical methods for partial differential equations
S I Hariharan and T H Moulden
146 Completely bounded maps and dilations
V I Paulsen
147 Harmonic analysis on the Heisenberg nilpotent Lie group
W Schempp
148 Contributions to modern calculus of variations
L Cesari
149 Nonlinear parabolic equations: qualitative properties of solutions
L Boccardo and A Tesei
150 From local times to global geometry, control and physics
K D Elworthy

151 A stochastic maximum principle for optimal control of diffusions
U G Haussmann
152 Semigroups, theory and applications. Volume II
H Brezis, M G Crandall and F Kappel
153 A general theory of integration in function spaces
P Muldowney
154 Oakland Conference on partial differential equations and applied mathematics
L R Bragg and J W Dettman
155 Contributions to nonlinear partial differential equations. Volume II
J I Díaz and P L Lions
156 Semigroups of linear operators: an introduction
A C McBride
157 Ordinary and partial differential equations
B D Sleeman and R J Jarvis
158 Hyperbolic equations
F Colombini and M K V Murthy
159 Linear topologies on a ring: an overview
J S Golan
160 Dynamical systems and bifurcation theory
M I Camacho, M J Pacifico and F Takens
161 Branched coverings and algebraic functions
M Namba
162 Perturbation bounds for matrix eigenvalues
R Bhatia
163 Defect minimization in operator equations: theory and applications
R Reemtsen
164 Multidimensional Brownian excursions and potential theory
K Burdzy
165 Viscosity solutions and optimal control
R J Elliott
166 Nonlinear partial differential equations and their applications: Collège de France Seminar. Volume VIII
H Brezis and J L Lions
167 Theory and applications of inverse problems
H Haario
168 Energy stability and convection
G P Galdi and B Straughan
169 Additive groups of rings. Volume II
S Feigelstock
170 Numerical analysis 1987
D F Griffiths and G A Watson
171 Surveys of some recent results in operator theory. Volume I
J B Conway and B B Morrel
172 Amenable Banach algebras
J-P Pier
173 Pseudo-orbits of contact forms
A Bahri
174 Poisson algebras and Poisson manifolds
K H Bhaskara and K Viswanath
175 Maximum principles and eigenvalue problems in partial differential equations
P W Schaefer
176 Mathematical analysis of nonlinear, dynamic processes
K U Grusa
177 Cordes' two-parameter spectral representation theory
D F McGhee and R H Picard
178 Equivariant K-theory for proper actions
N C Phillips
179 Elliptic operators, topology and asymptotic methods
J Roe
180 Nonlinear evolution equations
J K Engelbrecht, V E Fridman and E N Pelinovski
181 Nonlinear partial differential equations and their applications: Collège de France Seminar. Volume IX
H Brezis and J L Lions
182 Critical points at infinity in some variational problems
A Bahri
183 Recent developments in hyperbolic equations
L Cattabriga, F Colombini, M K V Murthy and S Spagnolo
184 Optimization and identification of systems governed by evolution equations on Banach space
N U Ahmed
185 Free boundary problems: theory and applications. Volume I
K H Hoffmann and J Sprekels
186 Free boundary problems: theory and applications. Volume II
K H Hoffmann and J Sprekels
187 An introduction to intersection homology theory
F Kirwan
188 Derivatives, nuclei and dimensions on the frame of torsion theories
J S Golan and H Simmons
189 Theory of reproducing kernels and its applications
S Saitoh
190 Volterra integrodifferential equations in Banach spaces and applications
G Da Prato and M Iannelli
191 Nest algebras
K R Davidson
192 Surveys of some recent results in operator theory. Volume II
J B Conway and B B Morrel
193 Nonlinear variational problems. Volume II
A Marino and M K V Murthy
194 Stochastic processes with multidimensional parameter
M E Dozzi
195 Prestressed bodies
D Iesan
196 Hilbert space approach to some classical transforms
R H Picard
197 Stochastic calculus in application
J R Norris
198 Radical theory
B J Gardner
199 The C^*-algebras of a class of solvable Lie groups
X Wang
200 Stochastic analysis, path integration and dynamics
K D Elworthy and J C Zambrini

201 Riemannian geometry and holonomy groups
S Salamon
202 Strong asymptotics for extremal errors and polynomials associated with Erdös type weights
D S Lubinsky
203 Optimal control of diffusion processes
V S Borkar
204 Rings, modules and radicals
B J Gardner
205 Two-parameter eigenvalue problems in ordinary differential equations
M Faierman
206 Distributions and analytic functions
R D Carmichael and D Mitrovic
207 Semicontinuity, relaxation and integral representation in the calculus of variations
G Buttazzo
208 Recent advances in nonlinear elliptic and parabolic problems
P Bénilan, M Chipot, L Evans and M Pierre
209 Model completions, ring representations and the topology of the Pierce sheaf
A Carson
210 Retarded dynamical systems
G Stepan
211 Function spaces, differential operators and nonlinear analysis
L Paivarinta
212 Analytic function theory of one complex variable
C C Yang, Y Komatu and K Niino
213 Elements of stability of visco-elastic fluids
J Dunwoody
214 Jordan decomposition of generalized vector measures
K D Schmidt
215 A mathematical analysis of bending of plates with transverse shear deformation
C Constanda
216 Ordinary and partial differential equations. Volume II
B D Sleeman and R J Jarvis
217 Hilbert modules over function algebras
R G Douglas and V I Paulsen
218 Graph colourings
R Wilson and R Nelson
219 Hardy-type inequalities
A Kufner and B Opic
220 Nonlinear partial differential equations and their applications: Collège de France Seminar. Volume X
H Brezis and J L Lions
221 Workshop on dynamical systems
E Shiels and Z Coelho
222 Geometry and analysis in nonlinear dynamics
H W Broer and F Takens
223 Fluid dynamical aspects of combustion theory
M Onofri and A Tesei
224 Approximation of Hilbert space operators. Volume I. 2nd edition
D Herrero
225 Operator theory: proceedings of the 1988 GPOTS–Wabash conference
J B Conway and B B Morrel
226 Local cohomology and localization
J L Bueso Montero, B Torrecillas Jover and A Verschoren
227 Nonlinear waves and dissipative effects
D Fusco and A Jeffrey
228 Numerical analysis 1989
D F Griffiths and G A Watson
229 Recent developments in structured continua. Volume II
D De Kee and P Kaloni
230 Boolean methods in interpolation and approximation
F J Delvos and W Schempp
231 Further advances in twistor theory. Volume I
L J Mason and L P Hughston
232 Further advances in twistor theory. Volume II
L J Mason and L P Hughston
233 Geometry in the neighborhood of invariant manifolds of maps and flows and linearization
U Kirchgraber and K Palmer
234 Quantales and their applications
K I Rosenthal
235 Integral equations and inverse problems
V Petkov and R Lazarov
236 Pseudo-differential operators
S R Simanca
237 A functional analytic approach to statistical experiments
I M Bomze
238 Quantum mechanics, algebras and distributions
D Dubin and M Hennings
239 Hamilton flows and evolution semigroups
J Gzyl
240 Topics in controlled Markov chains
V S Borkar
241 Invariant manifold theory for hydrodynamic transition
S Sritharan
242 Lectures on the spectrum of $L^2(\Gamma\backslash G)$
F L Williams
243 Progress in variational methods in Hamiltonian systems and elliptic equations
M Girardi, M Matzeu and F Pacella
244 Optimization and nonlinear analysis
A Ioffe, M Marcus and S Reich
245 Inverse problems and imaging
G F Roach
246 Semigroup theory with applications to systems and control
N U Ahmed
247 Periodic-parabolic boundary value problems and positivity
P Hess
248 Distributions and pseudo-differential operators
S Zaidman
249 Progress in partial differential equations: the Metz surveys
M Chipot and J Saint Jean Paulin
250 Differential equations and control theory
V Barbu

251 Stability of stochastic differential equations with respect to semimartingales
X Mao
252 Fixed point theory and applications
J Baillon and M Théra
253 Nonlinear hyperbolic equations and field theory
M K V Murthy and S Spagnolo
254 Ordinary and partial differential equations. Volume III
B D Sleeman and R J Jarvis
255 Harmonic maps into homogeneous spaces
M Black
256 Boundary value and initial value problems in complex analysis: studies in complex analysis and its applications to PDEs 1
R Kühnau and W Tutschke
257 Geometric function theory and applications of complex analysis in mechanics: studies in complex analysis and its applications to PDEs 2
R Kühnau and W Tutschke
258 The development of statistics: recent contributions from China
X R Chen, K T Fang and C C Yang
259 Multiplication of distributions and applications to partial differential equations
M Oberguggenberger
260 Numerical analysis 1991
D F Griffiths and G A Watson
261 Schur's algorithm and several applications
M Bakonyi and T Constantinescu
262 Partial differential equations with complex analysis
H Begehr and A Jeffrey
263 Partial differential equations with real analysis
H Begehr and A Jeffrey
264 Solvability and bifurcations of nonlinear equations
P Drábek
265 Orientational averaging in mechanics of solids
A Lagzdins, V Tamuzs, G Teters and A Kregers
266 Progress in partial differential equations: elliptic and parabolic problems
C Bandle, J Bemelmans, M Chipot, M Grüter and J Saint Jean Paulin
267 Progress in partial differential equations: calculus of variations, applications
C Bandle, J Bemelmans, M Chipot, M Grüter and J Saint Jean Paulin
268 Stochastic partial differential equations and applications
G Da Prato and L Tubaro
269 Partial differential equations and related subjects
M Miranda
270 Operator algebras and topology
W B Arveson, A S Mishchenko, M Putinar, M A Rieffel and S Stratila
271 Operator algebras and operator theory
W B Arveson, A S Mishchenko, M Putinar, M A Rieffel and S Stratila
272 Ordinary and delay differential equations
J Wiener and J K Hale
273 Partial differential equations
J Wiener and J K Hale
274 Mathematical topics in fluid mechanics
J F Rodrigues and A Sequeira
275 Green functions for second order parabolic integro-differential problems
M G Garroni and J F Menaldi
276 Riemann waves and their applications
M W Kalinowski
277 Banach C(K)-modules and operators preserving disjointness
Y A Abramovich, E L Arenson and A K Kitover
278 Limit algebras: an introduction to subalgebras of C*-algebras
S C Power
279 Abstract evolution equations, periodic problems and applications
D Daners and P Koch Medina
280 Emerging applications in free boundary problems
J M Chadam and H Rasmussen
281 Free boundary problems involving solids
J M Chadam and H Rasmussen
282 Free boundary problems in fluid flow with applications
J M Chadam and H Rasmussen
283 Asymptotic problems in probability theory: stochastic models and diffusions on fractals
K D Elworthy and N Ikeda
284 Asymptotic problems in probability theory: Wiener functionals and asymptotics
K D Elworthy and N Ikeda
285 Dynamical systems
R Bamon, R Labarca, J Lewowicz and J Palis
286 Models of hysteresis
A Visintin
287 Moments in probability and approximation theory
G Anastassiou
288 Mathematical aspects of penetrative convection
B Straughan
289 Ordinary and partial differential equations. Vol ume IV
B D Sleeman and R J Jarvis
290 *K*-theory for real *C**-algebras
H Schröder
291 Recent developments in theoretical fluid mechanics
G P Galdi and J Necas
292 Propagation of a curved shock and nonlinear ray theory
P Prasad
293 Non-classical elastic solids
M Ciarletta and D Ieşan
294 Multigrid methods
J Bramble
295 Entropy and partial differential equations
W A Day
296 Progress in partial differential equations: the Metz surveys 2
M Chipot
297 Nonstandard methods in the calculus of variations
C Tuckey
298 Barrelledness, Baire-like- and (LF)-spaces
M Kunzinger
299 Nonlinear partial differential equations and their applications. Collège de France Seminar. Volume XI
H Brezis and J L Lions
300 Introduction to operator theory
T Yoshino

301 Generalized fractional calculus and applications
V Kiryakova

D F Griffiths and G A Watson (Editors)

University of Dundee

Numerical analysis 1993

Proceedings of the 15th Dundee Conference, June–July 1993

Copublished in the United States with
John Wiley & Sons, Inc., New York

Longman Scientific & Technical
Longman Group UK Limited
Longman House, Burnt Mill, Harlow
Essex CM20 2JE, England
and Associated companies throughout the world.

Copublished in the United States with
John Wiley & Sons Inc., 605 Third Avenue, New York, NY 10158

First published 1994

AMS Subject Classification: 65-06

ISSN 0269-3674

ISBN 0 582 22568 X

British Library Cataloguing in Publication Data

A catalogue record for this book is
available from the British Library

Library of Congress Cataloging-in-Publication Data

A catalog record for this book is available

Printed and bound in Great Britain
by Biddles Ltd, Guildford and King's Lynn

Contents

Preface

The 15th Dundee Biennial Conference on Numerical Analysis was held at the University of Dundee during the 4 days June 29– July 2, 1993. The meeting opened in the splendour of the new West Park Conference Centre, which was completed in 1992. When plans for the Conference were being drawn up in 1991, it was expected that use would be made of the new Conference Centre for all the talks. Unfortunately, it became clear when the project was completed that these intentions would be frustrated because the new building contained only one sufficiently large lecture room, and this necessitated making rather belated alternative arrangements. After holding the first morning session in the main hall of the new Centre, the meeting therefore moved for the rest of the time to the more familiar surroundings of the main University campus, and to the more traditional pattern of invited talks (although in the Ustinov Room of the Bonar Hall) interspersed with submitted ones in rooms in the University Tower Building.

The meeting was attended by around 210 people from over 25 countries, with about half the participants coming from outside the UK. The opening talk at the Conference was the A R Mitchell lecture, delivered by Professor M J D Powell. The choice of Mike Powell to give this lecture was not only motivated by a recognition of the many important contributions he has made to numerical analysis, but also by his long association with Dundee and his strong support for the Dundee meetings.

Twelve other invited speakers presented talks (unfortunately Professor Roger Temam was prevented at the last minute from attending due to illness), and in addition there were 100 submitted talks, presented in 4 parallel sessions. This volume contains full versions of all the invited papers; the titles of all contributed talks given at the meeting, together with the names and addresses of the presenters (correct at the time of the meeting), are also listed here. Also included is an updated version of the talk given by Professor Gene Golub as the A R Mitchell lecture at the 1991 Conference. Unfortunately, the text of that talk was not available when the Proceedings of the 1991 meeting went to press. However, we were given the opportunity to include it in this volume, and because of its value as a contribution to numerical analysis, and because of its place in Dundee history as the first A R Mitchell lecture, we are pleased to take the unusual step of including it here.

We would like to take this opportunity of thanking all the speakers, chairmen and participants for their contributions. We thank in particular the after dinner speaker, Professor Sean McKee, of the University of Strathclyde, who kept his audience entertained (and the chairman on the edge of his seat) during a wide-ranging and spirited presentation, which included an epic poem.

We are as always grateful for help received from members of the Department of Mathematics and Computer Science both before and during the Conference, in particular the secretarial assistance of Mrs S Fox. The Conference is also indebted to the University of Dundee for making available various University facilities throughout the week, and for the provision of a Reception for the participants in West Park Hall. Finally, we would like to thank the publishers of these Proceedings, Longman Scientific and Technical, for their co-operation during the pre-publication process.

D F Griffiths
G A Watson

November 1993

Invited Speakers

J.W. Barrett:	Mathematics Department, Imperial College, Queens Gate, London SW7 2BX, UK
I.S. Duff:	Rutherford Appleton Laboratory, Numerical Analysis Group, Central Computing Department, Atlas Centre, Didcot, Oxon OX11 0QX, UK
C.M. Elliott :	Centre for Mathematical Analysis and its Applications, University of Sussex, Brighton BN1 9QH, UK
P.E. Gill:	Department of Mathematics, University of California at San Diego, 9500 Gilman Drive, La Jolla, CA 92093-0112, USA
D.J. Higham:	Department of Mathematics & Computer Science, The University, Dundee DD1 4HN, Scotland, UK
N.K. Nichols:	Department of Mathematics, Reading University, Whiteknights, PO BOX 220, Reading, RG6 2AX, UK
M.J.D. Powell:	Department of Applied Mathematics and Theoretical Physics, University of Cambridge, Silver Street, Cambridge CB3 9EW, UK
J. M. Sanz-Serna:	Departmento Matemática Aplicada y Computación, Universidad de Valladolid, Facultad de Ciencias, Valladolid, Spain.
M.N. Spijker:	Department of Mathematics & Computer Science, University of Leiden, Niels Bohrweg 1, 2333 CA Leiden, The Netherlands
G.W. Stewart:	Department of Computer Science and Institute for Advanced Computer Studies, University of Maryland, College Park, MD 20742, USA
A.M. Stuart:	Program in Scientific Computing and Computational Mathematics, Division of Applied Mechanics, Durand 252, Stanford University, CA 94305, USA
M.J. Todd:	School of Operations Research and Industrial Engineering, Cornell University, Ithaca, New York 14853-3801, USA
P. Townsend:	Department of Computer Science, University of Wales, Singleton Park, Swansea SA2 8PP, UK

J W BARRETT AND WENBIN LIU

Finite element approximation of degenerate quasilinear elliptic and parabolic problems

Abstract In this paper we present an overview of our recent work in this area. Energy type error bounds are derived for the finite element approximation of degenerate quasilinear elliptic equations of second order, which include the p-Laplacian. A key step in the analysis is to prove abstract error bounds initially in a quasi-norm, which naturally arises in degenerate problems of this type. These error bounds improve on existing results in the literature and are often optimal for sufficiently regular solutions, which we show to be achievable for a subclass of data. We extend these results to the corresponding parabolic problems and to the corresponding obstacle problems (variational inequalities). Finally, we extend these results to a non-Newtonian flow problem.

1 Introduction

Consider the quasilinear problem: Given $k \in C^1(0,\infty), f$ and g; find u such that

$$-\nabla.(k(|\nabla u|)\nabla u) = f \quad \text{in}\, \Omega \subset \mathbf{R}^2, \qquad u = g \quad \text{on}\, \partial\Omega. \tag{1.1}$$

It follows that

$$\nabla.(k(|\nabla u|)\nabla u) \equiv k(|\nabla u|)[a u_{x_1x_1} + 2b u_{x_1x_2} + c u_{x_2x_2}]\,,$$

where

$$a \equiv 1 + \sigma(|\nabla u|)\tfrac{(u_{x_1})^2}{|\nabla u|^2}\,,\ b \equiv \sigma(|\nabla u|)\tfrac{u_{x_1}u_{x_2}}{|\nabla u|^2}\,,\ c \equiv 1 + \sigma(|\nabla u|)\tfrac{(u_{x_2})^2}{|\nabla u|^2}$$

and

$$\sigma(s) \equiv \frac{k'(s)s}{k(s)}.$$

(1.1) is elliptic if the eigenvalues of the coefficient matrix

$$k(|\nabla u|)\begin{bmatrix} a & b \\ b & c \end{bmatrix},$$

$$k(|\nabla u|) \qquad \text{and} \qquad k(|\nabla u|) + k'(|\nabla u|)|\nabla u| \equiv [k(|\nabla u|)|\nabla u|]'\,,$$

are of one sign; see for example [15]. Without loss of generality, (1.1) is said to be (i) elliptic and non-degenerate if for all $\gamma > 0$ there exist constants C_γ, $M_\gamma > 0$ such that

$$C_\gamma \ \geq \ k(s),\ [k(s)s]' \ \geq \ M_\gamma \qquad \forall s \in [0,\gamma]; \tag{1.2}$$

and (ii) elliptic and degenerate if (1.2) only holds on choosing either

$$C_\gamma \ = \ \infty \text{ or } M_\gamma \ = \ 0.$$

The p-Laplacian, $k(s) \equiv s^{p-2}$ for a given $p \in (1,2) \cup (2,\infty)$, is an example of a degenerate quasilinear elliptic equation since $C_\gamma = \infty$ if $p \in (1,2)$ and $M_\gamma = 0$ if $p \in (2,\infty)$. Whereas, $k(s) \equiv (1+s)^{p-2}$ for a given $p \in (1,2) \cup (2,\infty)$ is an example of a non-degenerate quasilinear elliptic equation. Note that both of these examples collapse to the linear Laplacian if $p = 2$.

It should be noted that degenerate elliptic equations have limited regularity for infinitely smooth data. For example consider the p-Laplacian with radial symmetry:

$$-\nabla.(|\nabla u|^{p-2}\nabla u) = f \quad \text{in } \Omega, \qquad u = 0 \quad \text{on } \partial\Omega; \tag{1.3}$$

where $\Omega \equiv r \leq 1$, $r^2 \equiv x_1^2 + x_2^2$ and $f(x_1, x_2) \equiv F(r)$. If $f \equiv 1$, then a simple calculation yields that $u(x_1, x_2) = C[1 - r^{p/(p-1)}]$ for an appropriate constant C. It follows that $u \in W^{2,s}(\Omega)$ for $p > 2$ only if $s < 2(p-1)/(p-2)$. Hence $s \to 2$ as $p \to \infty$. A similar calculation leads to limited regularity in the case $p \in (1,2)$ also.

We will return to the question of regularity for more general data $k(\cdot)$, f, g and Ω in section 3. In the above and throughout this paper we adopt the standard notation $W^{\nu,s}(\Omega)$ for Sobolev spaces with norm $\|\cdot\|_{W^{\nu,s}(\Omega)}$ and semi-norm $|\cdot|_{W^{\nu,s}(\Omega)}$. We end this section by stating our assumptions on $k(\cdot)$.

Assumptions (A) There exist constants $p \in (1,\infty)$, $\alpha \in [0,1]$ and $C, M > 0$ such that for all $s \geq r \geq 0$

$$\begin{aligned} C\,(s-r)\,[(s+r)^\alpha(1+s+r)^{1-\alpha}]^{p-2} &\geq k(s)s - k(r)r \\ &\geq M\,(s-r)\,[(s+r)^\alpha(1+s+r)^{1-\alpha}]^{p-2}. \end{aligned}$$

Many functions $k(\cdot)$ met in practical problems satisfy the above. For example, $k(s) \equiv [s^\alpha(1+s)^{1-\alpha}]^{p-2}$ for given $p \in (1,\infty)$ and $\alpha \in [0,1]$ satisfies (A). Note that choosing $\alpha = 1$ in this example yields $k(s) \equiv s^{p-2}$, the p-Laplacian. In fact choosing any $\alpha \in (0,1]$ with $p \neq 2$ gives rise to a degenerate $k(\cdot)$. Whereas, choosing $\alpha = 0$ for any p yields a non-degenerate $k(\cdot)$. Therefore we see from this example that the parameter $\alpha \in [0,1]$ in (A) for a given p measures the degree of degeneracy in $k(\cdot)$.

Finally, we note that the results in this paper carry over to $k \in C(\overline{\Omega} \times (0,\infty))$ where we denote $k(\underline{x}, t)$ by $k(t)$ and require the Assumptions (A) to hold for all $\underline{x} \in \overline{\Omega}$.

2 Elliptic Problem

Let $\Omega \subset \mathbf{R}^d$, $d = 1$ or 2, be a bounded domain with a Lipschitz boundary $\partial\Omega$ in the case $d = 2$. Given $k(\cdot)$, satisfying (A), $f \in L^2(\Omega)$ and for ease of exposition zero Dirichlet data, i.e. $g \equiv 0$; we consider the problem:

(CP) Find u such that

$$-\nabla.(k(|\nabla u|)\nabla u) = f \quad \text{in } \Omega, \qquad u = 0 \quad \text{on } \partial\Omega.$$

Such problems occur for example in the mathematical modelling of nonlinear diffusion, nonlinear elasticity and Non-Newtonian fluids. The weak formulation of (CP) is

(WP) Find $u \in V \equiv W_0^{1,p}(\Omega) \equiv \{ v \in W^{1,p}(\Omega) \;:\; v = 0 \text{ on } \partial\Omega \}$ such that

$$\int_\Omega k(|\nabla u|)\nabla u.\nabla v \;=\; \int_\Omega f\, v \qquad \forall v \in V.$$

Associated with (WP) is the minimization problem:

(MP) Find $u \in V$ such that

$$J(u) \;\leq\; J(v) \qquad \forall v \in V,$$

where

$$J(v) \equiv \int_\Omega [\, \mathcal{K}(|\nabla v|) \;-\; f\, v \,] \qquad \text{and} \qquad \mathcal{K}(s) \equiv \int_0^s k(r) r \, dr \quad \forall s \geq 0.$$

For example, for the p-Laplacian, $k(s) \equiv s^{p-2}$, it follows that

$$J(v) \;\equiv\; \frac{1}{p}\int_\Omega |\nabla v|^p \;-\; \int_\Omega f\, v.$$

The following Lemma plays a crucial role in establishing our improved error bounds.

Lemma 2.1 *Under Assumptions (A) on $k(\cdot)$ it follows that for all $\underline{x}$, $\underline{y} \in \mathbf{R}^d$, $d \geq 1$,*

$$(i) \quad |k(|\underline{x}|)\underline{x} - k(|\underline{y}|)\underline{y}| \;\leq\; C\,|\underline{x} - \underline{y}|\,[(|\underline{x}| + |\underline{y}|)^\alpha (1 + |\underline{x}| + |\underline{y}|)^{1-\alpha}]^{p-2}$$

and

$$(ii) \quad [k(|\underline{x}|)\underline{x} - k(|\underline{y}|)\underline{y}] \cdot [\underline{x} - \underline{y}] \;\geq\; M\,|\underline{x} - \underline{y}|^2\,[(|\underline{x}| + |\underline{y}|)^\alpha (1 + |\underline{x}| + |\underline{y}|)^{1-\alpha}]^{p-2}.$$

Proof: See the proof of Lemma 2.1 in [25]. □

In addition the Assumptions (A) yield that there exist constants M_1, M_2, C_1, $C_2 > 0$ such that

$$-M_1 \;+\; M_2 s^p \;\leq\; \mathcal{K}(s) \;\leq\; C_1 \;+\; C_2 s^p \qquad \forall s \geq 0.$$

Therefore it follows from the above and Lemma 2.1 that the Assumptions (A) imply that $J(\cdot)$ is coercive, strictly convex, continuous and differentiable on V; with (WP) being its Euler equation. Hence it follows that there exists a unique solution u solving $(WP) \equiv (MP)$ and $\|u\|_{W^{1,p}(\Omega)} \leq C(\|f\|_{L^2(\Omega)})$.

Let $V^h \subset V$. Then the corresponding discsretisation of $(WP) \equiv (MP)$ is

$(WP)^h$ Find $u^h \in V^h$ such that

$$\int_\Omega k(|\nabla u^h|)\nabla u^h.\nabla v^h \;=\; \int_\Omega f\, v^h \qquad \forall v^h \in V^h.$$

It follows, as above, that $(WP)^h$ is equivalent to

$(MP)^h$ Find $u^h \in V^h$ such that

$$J(u^h) \leq J(v^h) \qquad \forall v^h \in V^h.$$

Moreover, there exists a unique solution u^h solving $(WP)^h \equiv (MP)^h$ and

$$\|u^h\|_{W^{1,p}(\Omega)} \leq C(\|f\|_{L^2(\Omega)}). \tag{2.1}$$

In order to prove an abstract error bound for this approximation we introduce two crucial ingredients.

Lemma 2.2 *For all constants $p \in (1, \infty)$, $\alpha \in [0,1]$, $a, r, s \geq 0$ and $\epsilon \in (0,1]$*

$$[(a+r)^\alpha(1+a+r)^{1-\alpha}]^{p-2}\, r\, s$$
$$\leq \epsilon\,[(a+r)^\alpha(1+a+r)^{1-\alpha}]^{p-2}\, r^2 \,+\, \epsilon^{-\gamma}\,[(a+s)^\alpha(1+a+s)^{1-\alpha}]^{p-2}\, s^2,$$

where $\gamma = max\{1, p-1\}$.

Proof: See the proof of Lemma 2.2 in [26]. □

We define

$$|v|^\rho_{(p,\alpha)} \equiv \int_\Omega [(|\nabla u| + |\nabla v|)^\alpha(1 + |\nabla u| + |\nabla v|)^{1-\alpha}]^{p-2}\, |\nabla v|^2, \tag{2.2}$$

where $\rho = \max\,\{p, 2\}$ and u is the unique solution of (WP). By adapting the proof of Lemma 3.1 in [7], it is a simple matter to show that $|\cdot|_{(p,\alpha)}$ is a quasi-norm on V; that is, it satisfies all the properties of a norm except the homogeneity property. In addition by applying Holder inequalities (adapting the proof of Lemma 3.2 in [5]), it is easily deduced that

(i) for $p \in (1,2]$ and $\sigma \in [p, 2 + \alpha(p-2)]$

$$M(\|v\|_{W^{1,p}(\Omega)})\|v\|^2_{W^{1,p}(\Omega)} \leq |v|^2_{(p,\alpha)} \leq C\|v\|^\sigma_{W^{1,\sigma}(\Omega)}, \tag{2.3}$$

and (ii) for $p \in [2, \infty)$ and $\sigma \in [2 + \alpha(p-2), p]$

$$M\|v\|^\sigma_{W^{1,\sigma}(\Omega)} \leq |v|^p_{(p,\alpha)} \leq C(\|u\|_{W^{1,\infty}(\Omega)}, \|v\|_{W^{1,\infty}(\Omega)})\|v\|^2_{W^{1,2}(\Omega)}. \tag{2.4}$$

We now prove an abstract error bound.

Theorem 2.1 *The unique solutions u and u^h of (WP) and $(WP)^h$ are such that*

$$|u - u^h|_{(p,\alpha)} \leq C\,|u - v^h|_{(p,\alpha)} \qquad \forall v^h \in V^h. \tag{2.5}$$

Proof: As $V^h \subset V$, it follows from (WP) and $(WP)^h$ that

$$\int_\Omega [k(|\nabla u|)\nabla u \;-\; k(|\nabla u^h|)\nabla u^h] \,.\nabla(u - u^h) \;=\;$$

$$\int_\Omega [k(|\nabla u|)\nabla u \;-\; k(|\nabla u^h|)\nabla u^h] \,.\nabla(u - v^h) \qquad \forall v^h \in V^h.$$

Setting $e \equiv u \;-\; u^h$ and $e^A \equiv u \;-\; v^h$, where v^h is any element of V^h, it follows from Lemma 2.1, and noting that for all $\underline{x}$, $\underline{y} \in \mathbf{R}^d$

$$\frac{1}{2}\,[\,|\underline{x}| + |\underline{y}|\,] \;\leq\; |\underline{x}| \;+\; |\underline{x} - \underline{y}| \;\leq\; 2\,[\,|\underline{x}| + |\underline{y}|\,],$$

and Lemma 2.2 that

$$\begin{aligned}
M \int_\Omega & [(|\nabla u| + |\nabla e|)^\alpha (1 + |\nabla u| + |\nabla e|)^{1-\alpha}]^{p-2}\, |\nabla e|^2 \\
&\leq \int_\Omega [k(|\nabla u|)\nabla u \;-\; k(|\nabla u^h|)\nabla u^h] \,.\nabla e \\
&= \int_\Omega [k(|\nabla u|)\nabla u \;-\; k(|\nabla u^h|)\nabla u^h] \,.\nabla e^A \\
\leq C \int_\Omega & [(|\nabla u| + |\nabla e|)^\alpha (1 + |\nabla u| + |\nabla e|)^{1-\alpha}]^{p-2}\, |\nabla e|\, |\nabla e^A| \\
\leq C \int_\Omega & [(|\nabla u| + |\nabla e^A|)^\alpha (1 + |\nabla u| + |\nabla e^A|)^{1-\alpha}]^{p-2}\, |\nabla e^A|^2.
\end{aligned}$$

Hence we obtain the desired result (2.5) after noting (2.2). □

It follows from (2.5) that u^h converges to u at the optimal rate of convergence in $|\cdot|_{(p,\alpha)}$. Before deriving explicit error bounds in more familiar Sobolev spaces, we first review the previous work on quasilinear degenerate elliptic equations.

For non-degenerate and smooth $k(\cdot)$ optimal $W^{1,2}(\Omega)$, $L^2(\Omega)$ and $L^\infty(\Omega)$ error bounds have been proved by Freshe & Rannacher [14], Dobrowolski & Rannacher [12]. For the p-Laplacian; Glowinski & Marrocco [17], see also Ciarlet [10], proved that for all $v^h \in V^h$

$$\|u - u^h\|_{W^{1,p}(\Omega)} \;\leq\; C\,[\|u - v^h\|_{W^{1,p}(\Omega)}]^\sigma,$$

where

$$\sigma = \frac{1}{3-p} \;\text{ if } p \in (1,2] \quad \text{and} \quad \sigma = \frac{1}{p-1} \;\text{ if } p \in [2,\infty).$$

By exploiting the minimization property of u^h, Tyukhtin [30] improved the above to

$$\sigma = \frac{p}{2} \;\text{ if } p \in (1,2] \quad \text{and} \quad \sigma = \frac{2}{p} \;\text{ if } p \in [2,\infty).$$

Finally, Chow [9] extended Tyukhtin's result to more general $k(\cdot)$ satisfying assumptions similar to (A).

Returning to the abstract error bound (2.5), applying the relationships (2.3), (2.4) and noting (2.1) yields the following error bounds in standard Sobolev spaces:

If $p \in (1,2]$, $\sigma \in [p, 2+\alpha(p-2)]$ then for all $v^h \in V^h$

$$\|u - u^h\|_{W^{1,p}(\Omega)} \leq C[\|u - v^h\|_{W^{1,\sigma}(\Omega)}]^{\sigma/2}. \tag{2.6}$$

If $p \in [2,\infty)$, $\sigma \in [2+\alpha(p-2), p]$ then for all $v^h \in V^h$

$$\|u - u^h\|_{W^{1,\sigma}(\Omega)} \leq C(\|u\|_{W^{1,\infty}(\Omega)}, \|v^h\|_{W^{1,\infty}(\Omega)})[\|u - v^h\|_{W^{1,2}(\Omega)}]^{2/\sigma}. \tag{2.7}$$

We note that if $k(\cdot)$ is non-degenerate, i.e. $\alpha = 0$, then (2.6) and (2.7) yield optimal error bounds on choosing $\sigma = 2$. Moreover, from these error bounds one can recover and improve on all the above known error bounds for degenerate $k(\cdot)$. For example, for the p-Laplacian we have $\alpha = 1$ and hence $\sigma = p$; in which case (2.6) and (2.7) give rise to the result in [30] with the approximation error being in a weaker norm in the case $p \in (2,\infty)$. Furthermore, if $k(\cdot)$ satisfies (A) for an $\alpha \in (0,1)$ then (2.6) and (2.7) yield improved error bounds on those in [9]; as the above analysis takes into account the degree of degeneracy in $k(\cdot)$. In addition, our analysis does not exploit the minimization property of u^h and can therefore be adapted to problems with non-self-adjoint lower order terms, e.g.

$$-\nabla.(k(|\nabla u|)\nabla u) + \nabla.(\underline{b}u) = f,$$

and to the corresponding parabolic problems.

Due to the general lack of regularity of u when $k(\cdot)$ is degenerate, there is no point in using a high order approximation space V^h. Therefore we restrict ourselves to the case V^h consisting of continuous piecewise linears on a regular partitioning of Ω. This assumes that Ω is polygonal if $d = 2$. However, the above analysis is easily adapted for a general convex domain Ω, which is approximated in the standard way by a polygonal $\Omega^h \equiv \cup_{\tau \in T^h}\overline{\tau}$; where T^h is a partitioning of Ω^h into regular triangles, each of maximum diameter bounded above by h. However, for ease of exposition, throughout the rest of this paper, we will assume that $\Omega^h \equiv \Omega$; but we note that all the results developed in this paper remain true for a general convex domain Ω.

For this choice of V^h, the error bounds (2.6) and (2.7) then yield for $\sigma = 2+\alpha(p-2)$:

If $p \in (1,2]$ and $u \in W^{2,\sigma}(\Omega)$ then

$$\|u - u^h\|_{W^{1,p}(\Omega)} \leq Ch^{\sigma/2}.$$

If $p \in [2,\infty)$ and $u \in W^{2,2}(\Omega) \cap W^{1,\infty}(\Omega)$ then

$$\|u - u^h\|_{W^{1,\sigma}(\Omega)} \leq Ch^{2/\sigma}.$$

However, exploiting the abstract error bound (2.5) and the quasi-norm (2.2) in a more sophisticated way one can prove optimal error bounds on assuming sufficient regularity on u. We have, for example, the following results, see [5] in the case of the p-Laplacian and [25] for general $k(\cdot)$ satisfying (A).

If $p \in (1,2)$ and $u \in C^{2,[2-\sigma]/\sigma}(\overline{\Omega}) \cap W^{3,1}(\Omega)$ with $\sigma = 2+\alpha(p-2)$ then

$$\|u - u^h\|_{W^{1,p}(\Omega)} \leq Ch. \tag{2.8}$$

If $p \in (2,\infty)$, $u \in W^{2,2}(\Omega) \cap W^{1,\infty}(\Omega)$ and $\int_\Omega |f|^{-2} < \infty$ then

$$\|u - u^h\|_{W^{1,1}(\Omega)} \leq Ch. \tag{2.9}$$

The regularity requirement for the optimal error bound (2.8) does not seem minimal as it does not collapse to $W^{2,2}(\Omega)$ for $p = 2$, when the elliptic operator in (CP) collapses to the Laplacian. For the p-Laplacian, the regularity requirement for (2.8) is weakened in [22] to $u \in W^{(2+p)/p,p}(\Omega)$; provided f is piecewise Holder continuous and the partitioning T^h is quasi-uniform. Adapting the argument in [22] the regularity requirement in general for (2.8) to be hold can be relaxed to $u \in W^{(2+\sigma)/\sigma,\sigma}(\Omega)$ under the above additional constraints. Note that this does collapse to $W^{2,2}(\Omega)$ for a non-degenerate problem when $\sigma = 2$; that is, either $\alpha = 0$ or $p = 2$.

The above analysis can also be generalised to degenerate quasilinear elliptic systems: Given $k(\cdot)$, satisfying (A), and $\underline{f} \equiv (f_j)_{j=1}^J$, $J = 1$ or 2; find $\underline{u} \equiv (u_j)_{j=1}^J$ such that

$$-\nabla.(k(|\nabla \underline{u}|)\nabla \underline{u}) = \underline{f} \ \text{ in } \Omega, \qquad \underline{u} = \underline{0} \ \text{ on } \partial\Omega; \tag{2.10}$$

where $[\nabla \underline{u}]_{ij} \equiv \partial u_j / \partial x_i$ and $|\nabla \underline{u}|$ denotes its Euclidean norm. This system is of Uhlenbeck type, since the nonlinearity depends on $|\nabla \underline{u}|$. One can also generalise the above analysis to some non-Uhlenbeck systems. For example consider the problem: Given $k(\cdot)$, satisfying (A), and f; find u such that

$$-\sum_{i=1}^{2} \tfrac{\partial}{\partial x_i}[(k(|\tfrac{\partial u}{\partial x_i}|)\tfrac{\partial u}{\partial x_i}] = f \ \text{ in } \Omega, \qquad u = 0 \ \text{ on } \partial\Omega. \tag{2.11}$$

The finite element approximation of (2.10) and (2.11) are studied in [24], where it is seen that there are some notable differences between Uhlenbeck and non-Uhlenbeck systems when it comes to obtaining optimal error bounds.

Finally, we note in this section that in addition to the a priori error bounds above; one can produce a posteriori error indicators, which can be used in an adaptive finite element code. We define

$$|v|^{\rho}_{(p,\alpha)^h} \equiv \int_\Omega [(|\nabla u^h| + |\nabla v|)^\alpha (1 + |\nabla u^h| + |\nabla v|)^{1-\alpha}]^{p-2} |\nabla v|^2, \tag{2.12}$$

where $\rho = \max\{p, 2\}$. This is just (2.2) with u replaced by u^h, the unique solution of $(WP)^h$. Similarly, one can show that $|\cdot|_{(p,\alpha)^h}$ is a quasi-norm on V. One can then extend on the results in [2] and show, see [28] for details, that

(i) For $p \in (1,2]$ and $p' = p/(p-1)$

$$|u - u^h|_{(p,\alpha)^h} \leq C\{\sum_{\tau \in T^h} h_\tau^{p'} |f|^{p'}_{0,p',\tau} + \sum_{t \in \Gamma^h} h_t |[k(|\nabla u^h|)\nabla u^h.\underline{n}_t]_t|^{p'}_{0,p',t}\}^{1/p'}.$$

(ii) For $p \in [2,\infty)$, $\sigma = 2 + \alpha(p-2)$ and $\sigma' = \sigma/(\sigma - 1)$

$$|u - u^h|_{(p,\alpha)^h} \leq C\{\sum_{\tau \in T^h} h_\tau^{\sigma'} |f|^{\sigma'}_{0,p',\tau} + \sum_{t \in \Gamma^h} h_t |[k(|\nabla u^h|)\nabla u^h.\underline{n}_t]_t|^{\sigma'}_{0,\sigma',t}\}^{1/p}.$$

Here Γ^h is the set of all interior sides t of T^h, h_t the length of side t, $\underline{n}_t$ a unit normal to t, and $[\cdot]_t$ the jump across t. The above indicators have been successfully implemented in an adaptive finite element code, once again see [28] for details.

3 Regularity

The regularity requirements on u for the optimal a priori error bounds (2.8) and (2.9) certainly hold for some data. For example, it is a simple matter to deduce, see [5], for the radially symmetric p-Laplacian (1.3) that

$$p \in (1,2),\ f \in C^1(\overline{\Omega}) \quad \Rightarrow \quad u \in C^{2,[2-p]/p}(\overline{\Omega}) \cap W^{3,1}(\Omega);$$

$$p \in (2,\infty),\ f \text{ constant} \quad \Rightarrow \quad u \in W^{2,2}(\Omega) \cap W^{1,\infty}(\Omega) \text{ and } \int_\Omega |f|^{-2} < \infty.$$

We now study the regularity issue for more general data. From DiBenedetto [11] and Tolksdorf [29] one has the following local regularity result for (1.1): if $f \in L^\infty(\Omega)$ then $u \in C^{1,\mu}(\Omega)$ for some $\mu \in (0,1)$; and in addition if $g = 0$ and $\partial\Omega \in C^2$ then $u \in C^{1,\mu}(\overline{\Omega})$. A global regularity result for more general data is proved by Lieberman [19]: if $f \in L^\infty(\Omega)$, $g \in C^{1,\gamma}(\overline{\Omega})$ and $\partial\Omega \in C^{1,\nu}$; then $u \in C^{1,\mu}(\overline{\Omega})$ for some γ, ν, $\mu \in (0,1)$.

Sharp local regularity results have been proved in the case $f = 0$ for the p-Laplacian by Iwaniec & Manfredi [18]. As $f = 0$ it allows the use of complex function theory. In particular their analysis exploits the fact that the complex map $u_{x_1} - iu_{x_2}$ is quasi-regular. They prove that $u \in C^{m,\beta}(\Omega) \cap W^{m+2,q}_{loc}(\Omega)$; where $\beta \in (0,1]$, $q \in [1,2]$ and the integer $m \geq 1$ are determined uniquely from the relations

$$6(m+\beta) = 7 + \tfrac{1}{p-1} + [1 + \tfrac{14}{p-1} + \tfrac{1}{(p-1)^2}]^{1/2} \quad \text{and} \quad q < 2/(2-\beta). \qquad (3.1)$$

We note that this result is sharp and not true for $f \neq 0$, e.g. $f = 1$. In addition, we note that $p > 2$ implies that $m = 1$.

Unfortunately, all the regularity results in the literature are, as above, for either global first or local higher derivatives; and are therefore not adequate for the finite element error bounds in the previous section. We now report on some of our recent regularity results.

For the p-Laplacian and for $k(\cdot)$ satisfying assumptions similar to (A) we have proved in [21] that

(i) If $p \in (1,2)$ and $f \in L^q(\Omega)$ with $q > 2$, then

$$\partial\Omega \in C^2,\ g \in H^2(\Omega) \ \Rightarrow\ u \in H^2(\Omega);$$

$$\Omega \text{ convex polygonal},\ g = 0 \ \Rightarrow\ u \in H^2(\Omega) \cap C^{1,\mu}(\overline{\Omega}) \text{ for some } \mu \in (0,1).$$

(ii) If $p \in (2,\infty)$, $\partial\Omega \in C^2$, $f = 0$ and $g \in H^2(\Omega)$; then $u \in H^2(\Omega)$.

In [23] we have extended the sharp interior results for the p-Laplacian with $f = 0$ in [18] to general $k(\cdot)$ satisfying assumptions similar to (A), see [23] for the precise conditions. Assumptions (A) imply that $k(s) \approx C\, s^{\alpha(p-2)} \equiv C\, s^{\overline{p}-2}$ for s near 0, where $\overline{p} \equiv 2 + \alpha(p-2)$. We then prove that $u \in C^{m,\beta}(\Omega) \cap W^{m+2,q}_{loc}(\Omega)$; where $\beta \in (0,1]$, $q \in [1,2]$ and the integer $m \geq 1$ are determined uniquely from the relations (3.1) with p replaced by $\overline{p}$. Therefore, as one would expect, it is the degenerate part of $k(\cdot)$ which determines the regularity of u. Furthermore, in [23] we use this sharp local regularity to prove the following higher order global regularity result for a special class of boundary data g:

Let $p \in (1,2)$, $\overline{p} \equiv 2 + \alpha(p-2)$, $\partial\Omega \in C^{2,1}$, $f = 0$ and $g \in C^{2,[2-\overline{p}]/\overline{p}}(\overline{\Omega}) \cap W^{3,r}(\Omega), r \in (1,p]$. In addition let g be such that its tangential derivative along $\partial\Omega$ is only zero at the global extrema of g on $\partial\Omega$. Then we prove that $u \in C^{2,[2-\overline{p}]/\overline{p}}(\overline{\Omega}) \cap W^{3,r}(\Omega)$. Hence it follows that the optimal finite element error bound (2.8) holds for this data.

4 Parabolic Problem

We now extend the results of section 2 to the corresponding degenerate quasilinear parabolic problem. Given $k(\cdot)$, satisfying (A), $f \in C([0,T];L^2(\Omega))$ and $u_0 \in V \equiv W_0^{1,p}(\Omega)$; the problem we then wish to study is

(CP) Find u such that

$$u_t - \nabla.(k(|\nabla u|)\nabla u) = f \text{ in } \Omega \times (0,T],$$
$$u = 0 \text{ on } \partial\Omega \times (0,T], \qquad u(\underline{x},0) = u_0(\underline{x}) \ \forall \underline{x} \in \Omega.$$

The weak formulation of (CP) is

(WP) Find $u \in L^\infty(0,T;V) \cap H^1(0,T;L^2(\Omega))$ such that for a.e. $t \in (0,T)$

$$\int_\Omega u_t v + \int_\Omega k(|\nabla u|)\nabla u.\nabla v = \int_\Omega f\, v \qquad \forall v \in V,$$
$$u(\underline{x},0) = u_0(\underline{x}) \qquad \forall \underline{x} \in \Omega.$$

It is a simple matter to deduce, by adapting the argument for the parabolic p-Laplacian in [31] and [6], that there exists a unique u solving (WP) and $u \in C([0,T];L^2(\Omega))$.

Let $V^h \subseteq V$. Then the corresponding approximation of (WP), based on a backward Euler time discretisation, is

$(WP)^h$ Let $U^0 \in V^h$ approximate u_0, $\Delta t = T/N$ and for $n = 1 \to N$ $f^n(\underline{x}) \equiv f(\underline{x}, n\Delta t)$ for all $\underline{x} \in \Omega$. Then for $n = 1 \to N$ find $U^n \in V^h$ such that

$$\int_\Omega \frac{U^n - U^{n-1}}{\Delta t} v^h + \int_\Omega k(|\nabla U^n|)\nabla U^n.\nabla v^h = \int_\Omega f^n\, v^h \qquad \forall v^h \in V^h.$$

Using the results of section 2 it is easily deduced that there exists a unique solution to $(WP)^h$ and

$$\max_{n=1\to N} \|U^n\|_{W^{1,p}(\Omega)} \le C(\Delta t \sum_{n=1}^{N} \|f\|^2_{L^2(\Omega)}, \|U^0\|_{W^{1,p}(\Omega)}). \tag{4.1}$$

Introducing the corresponding quasi-norm on $L^\rho(0,T;V)$

$$\|v\|_{L^\rho(0,T;\,|\cdot|_{(p,\alpha)})} \equiv [\int_0^T |v(\cdot,t)|^\rho_{(p,\alpha)}\, dt\,]^{1/\rho},$$

where $\rho = \max\{p,2\}$ and $|\cdot|_{(p,\alpha)}$ is defined as in (2.2) with u the unique solution of (WP), we have the following abstract error bound.

Theorem 4.1 *Let $f \in C^{0,1}([0,T];L^2(\Omega))$ then the unique solutions u of (WP) and $\{U^n\}_{n=1}^N$ of $(WP)^h$ are such that for all $v^h \in L^\rho(0,T;V^h)$*

$$\begin{aligned}\max_{n=1\to N} \|u(\cdot,n\Delta t) - U^n\|^2_{L^2(\Omega)} &+ \|u - u^h\|^\rho_{L^\rho(0,T;\,|\cdot|_{(p,\alpha)})} \\ &\le C[\,\|u - v^h\|^\rho_{L^\rho(0,T;\,|\cdot|_{(p,\alpha)})} + \|u_0 - U^0\|^2_{L^2(\Omega)}\,] + \\ &\quad + C(\|U^0\|_{W^{1,p}(\Omega)})[\,\|u - v^h\|_{L^2(0,T;L^2(\Omega))} + \Delta t\,],\end{aligned}$$

where $u^h(\cdot,t) \equiv U^n$ for all $t \in (\,(n-1)\Delta t, n\Delta t\,]$, $n = 1 \to N$.

Proof: Adapt the proof of Theorem 3.1 in [26]. □

Let V^h consist of continuous piecewise linears on a regular triangulation of Ω. Let $U^0 \in V^h$ be the interpolant of u_0, assuming $u_0 \in W^{2,p}(\Omega)$ if $p \in (1,2]$. Then combining the relationships (2.3) and (2.4) with the abstract error bound above and noting (4.1) yields for $\sigma = 2 + \alpha(p-2)$ the following error bounds:

If $p \in (1,2]$ and $u \in L^2(0,T;W^{2,2}(\Omega))$ then

$$\max_{n=1\to N} \|u(\cdot,n\Delta t) - U^n\|^2_{L^2(\Omega)} + \|u - u^h\|^2_{L^2(0,T;W^{1,p}(\Omega))} \leq C[h^\sigma + \Delta t]. \qquad (4.2)$$

If $p \in [2,\infty)$ and $u \in L^2(0,T;W^{2,2}(\Omega)) \cap L^\infty(0,T;W^{1,\infty}(\Omega))$ then

$$\max_{n=1\to N} \|u(\cdot,n\Delta t) - U^n\|^2_{L^2(\Omega)} + \|u - u^h\|^\sigma_{L^\sigma(0,T;W^{1,\sigma}(\Omega))} \leq C[h^2 + \Delta t]. \qquad (4.3)$$

The above bounds extend and improve on Wei [31], who proves in the case $k(s) \equiv s^{p-2}$, the p-Laplacian, that if $p \in [2,\infty)$ and $u \in C([0,T],W^{2,p}(\Omega))$ then

$$\max_{n=1\to N} \|u(\cdot,n\Delta t) - U^n\|^2_{L^2(\Omega)} \leq C[h^{\frac{1}{p-1}} + \Delta t].$$

Similar to the elliptic case the bound (4.2) can be improved to $C[h^2+\Delta t]$, if the further regularity $u \in L^\sigma(0,T;C^{2,[2-\sigma]/\sigma}(\overline{\Omega}) \cap W^{3,1}(\Omega))$ is assumed; see [6] for the parabolic p-Laplacian and [26] for general $k(\cdot)$ satisfying (A). In addition, see [6] where these regularity issues are discussed for the parabolic p-Laplacian.

Finally we end this section by noting that the above results and those in section 2 can be adapted to the corresponding variational inequality problems (obstacle problems). Let $K \equiv$ a closed convex subset of V.

The elliptic case, see [25]: Find $u \in K$ such that

$$\int_\Omega k(|\nabla u|)\nabla u.\nabla(v-u) \geq \int_\Omega f\,(v-u) \qquad \forall v \in K.$$

The parabolic case, see [20] for the p-Lapalcian case and [26] for general $k(\cdot)$ satisfying (A): Find $u \in L^\infty(0,T;K) \cap H^1(0,T;L^2(\Omega))$ such that for a.e. $t \in (0,T)$

$$\int_\Omega u_t\,(v-u) + \int_\Omega k(|\nabla u|)\nabla u.\nabla(v-u) \geq \int_\Omega f\,(v-u) \qquad \forall v \in K,$$

$$u(\underline{x},0) = u_0(\underline{x}) \qquad \forall \underline{x} \in \Omega.$$

5 Non-Newtonian Flow Problem

Given $k(\cdot)$ satisfying (A) and $\underline{f} \in [L^2(\Omega)]^2$, we consider the the following problem:

(CP) Find a velocity $\underline{u}$ and a pressure q such that

$$-\sum_{j=1}^{2} \tfrac{\partial}{\partial x_j}[k(|D(\underline{u})|)D_{ij}(\underline{u})] + \tfrac{\partial q}{\partial x_i} = f_i \ \text{ in } \Omega \subset \mathbf{R}^2, \quad i = 1,2,$$

$$\nabla.\underline{u} = 0 \ \text{ in } \Omega, \qquad \underline{u} = \underline{0} \ \text{ on } \partial\Omega, \qquad \int_\Omega q = 0;$$

Here $D(\underline{u})$ is the rate of deformation tensor, i.e. $D_{ij}(\underline{u}) \equiv \frac{1}{2}(\frac{\partial u_i}{\partial x_j} + \frac{\partial u_j}{\partial x_i})$ $i, j = 1, 2$; and $|D(\underline{u})| \equiv [D(\underline{u}).D(\underline{u})]^{1/2}$ its Euclidean norm, where $D(\underline{u}).D(\underline{v}) \equiv \sum_{i,j=1}^{2} D_{ij}(\underline{u})D_{ij}(\underline{v})$.

A Newtonian flow has constant viscosity, i.e. $k(s) \equiv k_0 > 0$ for all $s \geq 0$, in which case

(CP) $\equiv$ Find $\{\underline{u}, q\}$ such that

$$-\tfrac{1}{2}k_0\,\Delta\underline{u} + \nabla q = \underline{f} \quad \text{in } \Omega,$$

$$\nabla.\underline{u} = 0 \ \text{ in } \Omega, \qquad \underline{u} = \underline{0} \ \text{ on } \partial\Omega, \qquad \int_\Omega q = 0.$$

We now study the above non-Newtonian flow problem, where the viscosity $k(\cdot)$ is not constant. Common choices for $k(\cdot)$ with $k_0,\ \lambda > 0$ are:

(a) The Carreau Law $\quad k(s) \equiv k_0(1+\lambda s^2)^{(p-2)/2} \qquad \forall s \geq 0,$

(b) The Power Law $\quad k(s) \equiv k_0 s^{p-2} \quad \forall s \geq 0$

It is easily seen that (a) satisfies Assumptions (A) with $\alpha = 0$ and is therefore non-degenerate, whereas (b) satisfies Assumptions (A) with $\alpha = 1$ and is thus fully degenerate. In either (a) or (b) choosing $p \in (1,2)$, $p = 2$ and $p \in (2,\infty)$ models a pseudo plastic, a Newtonian fluid and a dilatant fluid; respectively.

Let

$$W \equiv [W_0^{1,p}(\Omega)]^2, \quad V \equiv \{\, \underline{v} \in W \ : \ \nabla.\underline{v} = 0 \text{ in } \Omega \,\}$$
$$\text{and} \quad Q \equiv L_0^{p'}(\Omega) \equiv \{\, \theta \in L^{p'}(\Omega) \ : \ \int_\Omega \theta = 0 \,\},$$

where $p' = p/(p-1)$. The weak formulation of (CP) is then

(WP) Find $\{\underline{u}, q\} \in V \times Q$ such that

$$\int_\Omega k(|D(\underline{u})|)D(\underline{u}).D(\underline{w}) \ - \int_\Omega q\nabla.\underline{w} \ = \ \int_\Omega \underline{f}.\underline{w} \qquad \forall \underline{w} \in W.$$

Clearly, a solution $\underline{u} \in V$ of (WP) solves

(WP') Find $\underline{u} \in V$ such that

$$\int_\Omega k(|D(\underline{u})|)D(\underline{u}).D(\underline{v}) \ = \ \int_\Omega \underline{f}.\underline{v} \qquad \forall \underline{v} \in V.$$

It is easily deduced that (WP') is the Euler equation for the minimization problem:

(MP) Find $\underline{u} \in V$ such that

$$J(\underline{u}) \ \leq \ J(\underline{v}) \qquad \forall \underline{v} \in V,$$

where

$$J(\underline{v}) \equiv \int_\Omega [\, \mathcal{K}(|D(\underline{v})|) \ - \ \underline{f}.\underline{v} \,] \qquad \text{and} \qquad \mathcal{K}(s) \equiv \int_0^s k(r)r\,dr \quad \forall s \geq 0.$$

The following lemma extends Lemma 2.1 from vectors to matrices.

Lemma 5.1 *Under Assumptions (A) on $k(\cdot)$ it follows that for all real $d \times d$ matrices $X, Y;\ d \geq 1$,*

$$(i)\ |k(|X|)X - k(|Y|)Y| \leq C\,|X - Y|\,[(|X| + |Y|)^{\alpha}(1 + |X| + |Y|)^{1-\alpha}]^{p-2}$$

and

$$(ii)\ [k(|X|)X - k(|Y|)Y]\,.\,[X - Y] \geq M\,|X - Y|^2\,[(|X| + |Y|)^{\alpha}(1 + |X| + |Y|)^{1-\alpha}]^{p-2}.$$

Proof: See the proof of Lemma 2.1 in [7]. □

The Assumptions (A) on $k(\cdot)$ and Lemma 5.1 imply that $J(\cdot)$ is coercive, strictly convex, continuous and differentiable on V. Hence there exists a unique $\underline{u} \in V$ solving (MP) and (WP'). In addition this is the unique $\underline{u}$ solving (WP). For the well-posedness of q in (WP) we require the Babuska/Brezzi condition

$$\inf_{\theta \in Q} \sup_{\underline{w} \in W} \frac{\int_\Omega \theta\, \nabla . \underline{w}}{\|\theta\|_Q\, \|\underline{w}\|_W} \geq \beta(p) > 0 \tag{5.1}$$

to hold. It follows from Amrouche & Girault [1] that if $\partial\Omega \in C^{0,1}$, then (5.1) holds for all $p \in (1, \infty)$ and hence there exists a unique $\{\underline{u}, q\} \in V \times Q$ solving (WP).

We now consider the discretisation of (WP). Let $W^h \subseteq W$, $Q^h \subseteq Q$ and

$$V^h \equiv \{\underline{v}^h \in W^h : \int_\Omega (\nabla . \underline{v}^h)\, \theta^h = 0 \ \ \forall \theta^h \in Q^h\}.$$

The corresponding discretisation of (WP) is then

$(WP)^h$ Find $\{\underline{u}^h, q^h\} \in V^h \times Q^h$ such that

$$\int_\Omega k(|D(\underline{u}^h)|) D(\underline{u}^h) . D(\underline{w}^h) - \int_\Omega q^h \nabla . \underline{w}^h = \int_\Omega \underline{f} . \underline{w}^h \qquad \forall \underline{w}^h \in W^h.$$

A solution $\underline{u}^h \in V^h$ of $(WP)^h$ solves

$(WP')^h$ Find $\underline{u}^h \in V^h$ such that

$$\int_\Omega k(|D(\underline{u}^h)|) D(\underline{u}^h) . D(\underline{v}^h) = \int_\Omega \underline{f} . \underline{v}^h \quad \forall \underline{v}^h \in V^h.$$

It is easily deduced that $(WP')^h$ is the Euler equation for the minimization problem:

$(MP)^h$ Find $\underline{u}^h \in V^h$ such that

$$J(\underline{u}^h) \leq J(\underline{v}^h) \qquad \forall \underline{v}^h \in V^h.$$

As for (MP), the Assumptions (A) imply that there exists a unique solution $\underline{u}^h \in V^h$ solving $(MP)^h$ and $(WP')^h$. In addition this is the unique $\underline{u}^h$ solving $(WP)^h$. For the well-posedness of q^h in $(WP)^h$ we require the discrete Babuska/Brezzi condition

$$\inf_{\theta^h \in Q^h} \sup_{\underline{w}^h \in W^h} \frac{\int_\Omega \theta^h \, \nabla . \underline{w}^h}{\|\theta^h\|_Q \, \|\underline{w}^h\|_W} \geq \beta^h(p) > 0 \tag{5.2}$$

to hold. We will see below that in deriving error bounds for a given $p \in (1, \infty)$, we require not only $[\beta^h(p)]^{-1}$ but maybe also $[\beta^h(\sigma)]^{-1}$, for some $\sigma \in (1, \infty)$ other than p, to be bounded above independently of h.

Introducing the corresponding quasi-norm on W

$$|\underline{v}|^\rho_{(p,\alpha)} \equiv \int_\Omega [(|D(\underline{u})| + |D(\underline{v})|)^\alpha (1 + |D(\underline{u})| + |D(\underline{v})|)^{1-\alpha}]^{p-2} \, |D(\underline{v})|^2,$$

where $\rho = \max \{p, 2\}$ and $\underline{u}$ is the unique solution of (WP'), we have the following abstract error bounds

Theorem 5.1 *Let $\{ \underline{u}, q \}$ be the unique solution of (WP) and $\underline{u}^h$ be the unique solution of $(WP')^h$. Then for all $\{ \underline{v}^h, \theta^h \} \in V^h \times Q^h$ it follows that*

(i) If $p \in (1, 2]$ and $\sigma \in [p, 2 + \alpha(p-2)]$ then

$$|\underline{u} - \underline{u}^h|_{(p,\alpha)} \leq C(\|\underline{v}^h\|_W)[\, |\underline{u} - \underline{v}^h|_{(p,\alpha)} + \|q - \theta^h\|_Q];$$

and if (5.2) holds then the unique solution $\{ \underline{u}^h, q^h \}$ of $(WP)^h$ is such that

$$\|q - q^h\|_{L^{\sigma'}(\Omega)} \leq C([\beta^h(\sigma)]^{-1})[\, |\underline{u} - \underline{u}^h|^{2/\sigma'}_{(p,\alpha)} + \|q - \theta^h\|_{L^{\sigma'}(\Omega)}\,],$$

where $\sigma' = \sigma/(\sigma - 1)$.

(ii) If $p \in [2, \infty)$ and $\sigma \in [2 + \alpha(p-2), p]$ then

$$|\underline{u} - \underline{u}^h|_{(p,\alpha)} \leq C[\, |\underline{u} - \underline{v}^h|_{(p,\alpha)} + \|q - \theta^h\|^{\sigma'/p}_{L^{\sigma'}(\Omega)}];$$

and if (5.2) holds then the unique solution $\{ \underline{u}^h, q^h \}$ of $(WP)^h$ is such that

$$\|q - q^h\|_Q \leq C([\beta^h(p)]^{-1})[\, |\underline{u} - \underline{u}^h|^{p/2}_{(p,\alpha)} + \|q - \theta^h\|_Q\,],$$

where $\sigma' = \sigma/(\sigma - 1)$.

Proof: See the proof of Theorem 3.1 in [7]. □

We now consider a special choice of W^h and Q^h - the P_1 flux - P_0 finite element, see [8] and also [16]. With $\overline{\Omega} \equiv \bigcup_{\tau \in T^h} \overline{\tau}$, let

$$W^h \equiv \{ \underline{w}^h \in [C(\overline{\Omega})]^2 \; : \; \underline{w}^h \,|_\tau \in \mathcal{P}_1(\tau) \;\; \forall \tau \in T^h, \; \underline{w}^h = \underline{0} \text{ on } \partial\Omega \},$$

$$Q^h \equiv \{ \theta^h \in L^\infty(\Omega) \; : \; \theta^h \,|_\tau = \text{ constant } \forall \tau \in T^h, \; \int_\Omega \theta^h = 0 \};$$

where

$$\mathcal{P}_1(\tau) \equiv \{\ [\text{span}\ \{\ \phi_1,\ \phi_2,\ \phi_3\ \}]^2\ +\ \text{span}\ \{\ \underline{\psi}_1,\ \underline{\psi}_2,\ \underline{\psi}_3\ \}\ \}$$
$$\text{and}\quad \underline{\psi}_1\ \equiv\ \phi_2\,\phi_3\,\underline{n}_1,\ \ \underline{\psi}_2\ \equiv\ \phi_3\,\phi_1\,\underline{n}_2,\ \ \underline{\psi}_3\ \equiv\ \phi_1\,\phi_2\,\underline{n}_3.$$

Here $\underline{n}_i$ is the outward unit normal to the side opposite vertex a_i and ϕ_i is the standard linear basis functions on τ such that $\phi_i(a_j) = \delta_{ij},\ i,\ j = 1 \to 3$. For this element it follows that (5.2) holds and for all $p \in (1, \infty)$

$$\beta^h(p) \geq C\beta(p)\ \ \Rightarrow\ \ [\beta^h(p)]^{-1} \leq C.$$

The abstract error bounds above then yield, see [7], for $\sigma = 2 + \alpha(p-2)$ and $\sigma' = \sigma/(\sigma-1)$ the following error bounds:

If $p \in (1, 2]$ and $\underline{u} \in [W^{2,\sigma}(\Omega)]^2$, $q \in W^{1,p'}(\Omega)$ then

$$\|\underline{u} - \underline{u}^h\|_{W^{1,p}(\Omega)}\ +\ \|q - q^h\|_{L^{\sigma'}(\Omega)}^{\sigma/[2(\sigma-1)]}\ \leq\ Ch^{\sigma/2}. \tag{5.3}$$

If $p \in [2, \infty)$ and $\underline{u} \in [W^{2,2}(\Omega) \cap W^{1,\infty}(\Omega)]^2$, $q \in W^{1,\sigma'}(\Omega)$ then

$$\|\underline{u} - \underline{u}^h\|_{W^{1,\sigma}(\Omega)}\ +\ \|q - q^h\|_{L^{p'}(\Omega)}^{2/\sigma}\ \leq\ Ch^{1/(\sigma-1)}. \tag{5.4}$$

The bound (5.3) can be improved to Ch, if the further regularity $\underline{u} \in [C^{2,[2-\sigma]/\sigma}(\overline{\Omega}) \cap W^{3,1}(\Omega)]^2$ is assumed, see [7]. The above error bounds extend and improve on those given in [3], [4], [13] and [27].

Finally, we note that the above analysis does not exploit the minimization property of $\underline{u}^h$ and therefore can be adapted if the non-linear inertial terms are included and to the corresponding unsteady problem.

References

[1] C. Amrouche and V. Girault, *Propriétés fonctionelles d'opérateurs. Application au problème de Sokes en dimension quelconque*, Publications du Laboratoire d'Analyse Numérique, n. 90025, Universitité Pierre et Marie Curie.

[2] J. Baranger and H. El Amri, *Estimateurs a posteriori d'erreur pour le calcul adaptatif d'écoulements quasi-Newtoniens*, RAIRO M^2AN **25**, (1991), 31-48.

[3] J. Baranger and K. Najib, *Analyse numérique des écoulements quasi-Newtoniens dont la viscositié obéit à la loi puissance ou la loi de Carreau*, Numer. Math. **64**, (1990), 35-49.

[4] J.W. Barrett and W.B. Liu, *Finite element error analysis of a quasi-Newtonian flow obeying the Carreau or power law*, Numer. Math. **64**, (1993), 433-453.

[5] J.W. Barrett and W.B. Liu, *Finite element approximation of the p-Laplacian*, Math. Comp., (1993), (to appear).

[6] J.W. Barrett and W.B. Liu, *Finite element approximation of the parabolic p-Laplacian*, SIAM J. Numer. Anal., (to appear).

[7] J.W. Barrett and W.B. Liu, *Quasi-norm error bounds for the finite element error analysis of a non-Newtonian flow*, (submitted for publication).

[8] C. Bernardi and G. Raugel, *Analysis of some finite element methods for the Stokes problem*, Math. Comp. **44**, (1985), 71-79.

[9] S.S. Chow, *Finite element error estimates for non-linear elliptic equations of monotone type*, Numer. Math. **54**, (1989), 373-393.

[10] P.G. Ciarlet, *The Finite Element Method for Elliptic Problems*, North Holland, Amsterdam 1978.

[11] E. DiBenedetto, *$C^{1+\alpha}$ local regularity of weak solutions of degenerate elliptic equations*, J. Nonlinear Anal. **7**, (1983), 827-850.

[12] M. Dobrowolski and R. Rannacher, *Finite element methods for non-linear elliptic systems of second order*, Math. Nachr. **94**, (1980), 155-172.

[13] Q. Du and M.D. Gunzburger, *Finite-element approximations of a Ladyzhenskaya model for stationary incompressible viscous flows*, SIAM J. Numer. Anal. **27**, (1990), 1-19.

[14] J. Frehse and R. Rannacher, *Asymptotic L^{∞}−error estimates for linear finite element approximations of quasilinear boundary problems*, SIAM J. Numer. Anal. **15**, (1978), 419-431.

[15] D.Gilbarg and N.S. Trudinger, *Elliptic Partial Differential Equtions of Second Order - Second Edition*, Springer-Verlag, Berlin Heidelberg 1983.

[16] V. Girault and P.-A. Raviart *Finite element Methods for Navier-Stokes Equations*, Springer-Verlag, Berlin Heidelberg 1986.

[17] R. Glowinski and A. Marrocco, *Sur l'approximation par éléments finis d'ordre un, et la résolution, par pénalisation-dualitié, d'une classe de problèmes de Dirichlet non linéaires*, R.A.I.R.O. Analyse Numérique **2**, (1975), 41-76.

[18] T. Iwaniec and J.J Manfredi, *Regularity of p-harmonic functions on the plane*, Rev. Math. Iberoamericana **5**, (1989), 1-19.

[19] G.M. Lieberman, *Boundary regularity for solutions of degenerate elliptic equations*, J. Nonlinear Anal. **12**, (1988), 1203-1219.

[20] W.B. Liu and J.W. Barrett, *Error bounds for the finite element approximation of a degenerate quasilinear parabolic variational inequality*, Advances in Comp. Math. **1**, (1993), 223-239.

[21] W.B. Liu and J.W. Barrett, *A remark on the regularity of the solutions of the p-Laplacian and it's applications to their finite element approximation*, J. Math. Anal. Appl., (1993), (to appear).

[22] W.B. Liu and J.W. Barrett, *A further remark on the regularity of the solutions of the p-Laplacian and it's applications to their finite element approximation*, J. Nonlinear Anal., (1993), (to appear).

[23] W.B. Liu and J.W. Barrett, *Higher order regularity for the solutions of some degenerate quasilinear elliptic equations in the plane*, SIAM J. Math. Anal., (1993), (to appear).

[24] W.B. Liu and J.W. Barrett, *Finite element approximation of some degenerate monotone quasilinear elliptic systems*, (submitted for publication).

[25] W.B. Liu and J.W. Barrett, *Quasi-norm error bounds for the finite element approximation of some degenerate quasilinear elliptic equations and variational inequalities*, (submitted for publication).

[26] W.B. Liu and J.W. Barrett, *Quasi-norm error bounds for the finite element approximation of some degenerate quasilinear parabolic equations and variational inequalities*, (submitted for publication).

[27] D. Sandri, *Sur l'approximation numérique des écoulements quasi-Newtoniens dont la viscositié suit la loi puissance ou la loi de Carreau*, RAIRO M^2AN **27**, (1993), 131-155.

[28] G. Simms, Ph.D. Thesis, Imperial College, London, (in preparation).

[29] P. Tolksdorf, *Regularity for a more general class of quasilinear elliptic equations*, J. Diff. Eqns. **51**, (1984), 126-150.

[30] V.B. Tyukhtin, *The rate of convergence of approximation methods of solution of one-sided variational problems*, I. Vestnik Leningrad Univ. Mat. Mekh. Astronom. **13**, (1982), 111-113.

[31] D. Wei, *Existence, uniqueness, and numerical analysis of solutions of a quasilinear parabolic problem*, SIAM J. Numer. Anal. **29**, (1992), 484-497.

Acknowledgement

The financial support of the second author by SERC Research Grant GR/F81255 is gratefully acknowledged

John W. Barrett and W.B. Liu
Department of Mathematics,
Imperial College,
London SW7 2BZ, UK.

j.barrett@ic.ac.uk

R K BEATSON AND M J D POWELL

An iterative method for thin plate spline interpolation that employs approximations to Lagrange functions

Abstract Thin plate spline interpolation to functions of two variables is useful in many applications, because there are few restrictions on the positions of the data points. Further, some smoothness properties are achieved naturally, because the interpolant minimizes a second derivative norm subject to the interpolation conditions. On the other hand, full matrices occur, and the number of data points, n say, may be very large. Therefore we approximate each Lagrange function by a Lagrange function of interpolation to a small subset of the data. Thus each approximation usually has far fewer than n thin plate spline terms, and the approximations provide an initial estimate of the required interpolant which can be improved by iterative refinement. Details are given of an iteration and of the use of updating techniques for the efficient calculation of the coefficients of the approximate Lagrange functions. However, we have not yet developed an automatic way of generating the subsets that have been mentioned. Some numerical results are presented too, which illustrate the number of iterations and the amount of computation of the algorithm. They suggest that interpolation to tens of thousands of scattered data points in two dimensions may soon become a routine calculation.

1 Introduction

The problem of fitting measured values of a real function of two variables occurs in many scientific, engineering, industrial and economic studies. Sometimes it is convenient or suitable to ignore the errors of measurement, or one may wish to consider the properties of certain interpolants to exact data. Therefore in this paper we take the view that we seek a function $s(\cdot)$ from $\mathcal{R}^2$ to $\mathcal{R}$ that satisfies the equations

$$s(\underline{x}_k) = f_k, \quad k=1,2,\ldots,n, \tag{1.1}$$

where $\{\underline{x}_k : k=1,2,\ldots,n\}$ and $\{f_k : k=1,2,\ldots,n\}$ are a given set of points in $\mathcal{R}^2$ and a given set of real numbers, respectively. Of course the points $\{\underline{x}_k : k=1,2,\ldots,n\}$ must all be different. Further, we assume that they are not collinear.

The constraints (1.1) allow many choices of $s(\cdot)$. We prefer to make $s(\cdot)$ as smooth as possible subject to the interpolation conditions, our measure of smoothness being the integral

$$I(s) = \int_{\mathcal{R}^2} \left(\frac{\partial^2 s(\underline{x})}{\partial x^2}\right)^2 + 2\left(\frac{\partial^2 s(\underline{x})}{\partial x \partial y}\right)^2 + \left(\frac{\partial^2 s(\underline{x})}{\partial y^2}\right)^2 d\underline{x}, \tag{1.2}$$

where x and y are the components of $\underline{x}$. There are three main reasons for this choice. Firstly, because it is proved by Duchon [3] that $s(\cdot)$ has the form

$$s(\underline{x}) = \sum_{k=1}^{n} \lambda_k \, \|\underline{x}-\underline{x}_k\|_2^2 \log \|\underline{x}-\underline{x}_k\|_2 + p(\underline{x}), \quad \underline{x} \in \mathcal{R}^2, \tag{1.3}$$

where each λ_k is a real coefficient and where $p(\cdot)$ is a linear polynomial, it is straightforward to represent $s(\cdot)$ on a computer so that $s(\underline{x})$ can be calculated for any $\underline{x}$ in $\mathcal{R}^2$. Secondly, it is also shown in [3] that, for every set of points $\{\underline{x}_k : k = 1,2,\ldots,n\}$ satisfying the assumptions at the end of the previous paragraph, the minimization of the integral (1.2) subject to the equations (1.1) defines $s(\cdot)$ uniquely. Thirdly, this way of taking up the freedom in $s(\cdot)$ has been found to provide interpolating functions that are very useful in many applications.

We see that we are minimizing a homogeneous quadratic functional of $s(\cdot)$ subject to linear constraints. Therefore the coefficients of expression (1.3) are defined by a linear system of equations. In order to display this system explicitly, we let A be the nxn matrix that has the elements

$$A_{ij} = \|\underline{x}_i-\underline{x}_j\|_2^2 \log \|\underline{x}_i-\underline{x}_j\|_2, \quad i,j=1,2,\ldots,n, \tag{1.4}$$

and we let P be the $n \times 3$ matrix whose k-th row is the vector $(1 \; x_k \; y_k)$, where x_k and y_k are the components of $\underline{x}_k$. Further, we note that the integral (1.2) is finite if and only if the coefficients of $s(\cdot)$ have the property

$$\sum_{k=1}^{n} \lambda_k = \sum_{k=1}^{n} \lambda_k \, x_k = \sum_{k=1}^{n} \lambda_k \, y_k = 0. \tag{1.5}$$

Further, we let the components of $\underline{c} \in \mathcal{R}^3$ be the multipliers of the constant, x and y terms of $p(\cdot)$. Thus the interpolation conditions (1.1) and the constraints (1.5) provide the linear system of equations

$$\left(\begin{array}{c|c} A & P \\ \hline P^T & 0 \end{array}\right) \left(\begin{array}{c} \underline{\lambda} \\ \hline \underline{c} \end{array}\right) = \left(\begin{array}{c} \underline{f} \\ \hline 0 \end{array}\right), \tag{1.6}$$

where $\underline{\lambda}$ and $\underline{f}$ are the vectors in $\mathcal{R}^n$ with the components $\{\lambda_k : k = 1,2,\ldots,n\}$ and $\{f_k : k=1,2,\ldots,n\}$, respectively.

The matrix A has the property that, if $\underline{v}$ is any nonzero vector in $\mathcal{R}^n$ such that $P^T\underline{v}=0$, then $\underline{v}^TA\underline{v}$ is positive (see [6], for instance). It follows that, if Q is any $n \times (n-3)$ matrix whose columns span the null space of P^T, then the symmetric matrix Q^TAQ is positive definite. Further, both expression (1.5) and the last part of expression (1.6) show that $\underline{\lambda}$ is in this null space. Therefore, having chosen Q, we can eliminate the constraints (1.5) by writing $\underline{\lambda}$ in the form $\underline{\lambda}=Q\underline{\mu}$, where $\underline{\mu}$ is a vector in $\mathcal{R}^{n-3}$ that has to be calculated. Further, by pre-multiplying the first part of expression (1.6), namely $A\underline{\lambda}+P\underline{c}=\underline{f}$, by Q^T, we deduce the positive definite linear system

$$Q^TAQ\underline{\mu} = Q^T\underline{f}. \tag{1.7}$$

Thus many algorithms are available for the computation of $\underline{\mu}$. Then, after forming $\underline{\lambda}=Q\underline{\mu}$, it is convenient to deduce the coefficients of the linear polynomial part of the function (1.3) from three of the interpolation equations (1.1) whose points $\{\underline{x}_k\}$ are not collinear.

The use of the positive definite system (1.7) for calculating the interpolant (1.3) to the data is recommended by Sibson and Stone [9]. Further, Powell [8] studies the solution of this system by the method of conjugate gradients, when one tries to improve the conditioning of Q^TAQ by taking advantage of the freedom in Q, and he identifies some successful ways of choosing Q for special positions of the data points $\{\underline{x}_k : k=1,2,\ldots,n\}$. His pre-conditioners are closely related to the ones that are proposed by Dyn, Levin and Rippa [4], because they are derived from the fact that the function (1.3) satisfies the biharmonic equation $\nabla^4 s(\underline{x}) = 0$ when $\underline{x}$ is not in the set $\{\underline{x}_k : k = 1,2,\ldots,n\}$. The technique that we are going to consider, however, does not depend on such details, although we are also trying to achieve the calculation of the coefficients of $s(\cdot)$ in far fewer than $\mathcal{O}(n^3)$ operations when n is large.

We regard the data points as fixed, but we take the view that the right hand sides $\{f_k : k=1,2,\ldots,n\}$ are parameters. It follows from the system (1.6) that $s(\cdot)$ is a linear function of these parameters. Therefore there is a Lagrange form of the solution to the interpolation problem. Specifically, for $k=1,2,\ldots,n$, we let $\ell_k(\cdot\,;\mathcal{L}_*)$ be the interpolant to the parameter values $f_i = \delta_{ik}$, $i = 1,2,\ldots,n$, the purpose of the argument $\mathcal{L}_*$ being explained below. Then the interpolant to general right hand sides is the expression

$$s(\cdot) = \sum_{k=1}^{n} f_k\, \ell_k(\cdot\,;\mathcal{L}_*). \tag{1.8}$$

We are going to save work by making the approximations

$$\ell_k(\cdot\,;\mathcal{L}_k) \approx \ell_k(\cdot\,;\mathcal{L}_*), \quad k=1,2,\ldots,n, \tag{1.9}$$

to the Lagrange functions, where our notation is as follows. Let $\mathcal{L}_*$ be the set $\{1,2,\ldots,n\}$ and let each $\mathcal{L}_k$ be a subset of $\mathcal{L}_*$, where $\mathcal{L}_k$ includes k and has the property that the points $\{\underline{x}_i : i\in\mathcal{L}_k\}$ are not collinear. Then the "thin plate spline" interpolation method defines the function $s(\cdot)$ that minimizes the integral (1.2) subject to the conditions

$$s(\underline{x}_i) = \delta_{ik}, \quad i\in\mathcal{L}_k. \tag{1.10}$$

We let $\ell_k(\cdot\,;\mathcal{L}_k)$ be this function, and we note that its coefficients can be found in $\mathcal{O}(|\mathcal{L}_k|^3)$ operations, where $|\mathcal{L}_k|$ is the number of elements in the set $\mathcal{L}_k$. Therefore the efficiency of our procedure is due to the remark that, when n is large, the estimates (1.9) can provide adequate accuracy for values of $|\mathcal{L}_k|$ that are much less than n.

This last remark is a consequence of the variational calculation that provides the thin plate spline interpolant. Specifically, it follows that $\ell_k(\cdot\,;\mathcal{L}_k)$ does not include any oscillations that increase the integral $I(\ell_k)$ unnecessarily. Therefore, if $\underline{x}_j$ is an interpolation point that is not in the set $\{\underline{x}_i : i\in\mathcal{L}_k\}$, and if the conditions

$$\ell_k(\underline{x}_i;\mathcal{L}_k) = \delta_{ik}, \quad i\in\mathcal{L}_k, \tag{1.11}$$

force $\ell_k(\cdot\,;\mathcal{L}_k)$ to be zero at several points that are between $\underline{x}_j$ and $\underline{x}_k$, then we expect $|\ell_k(\underline{x}_j;\mathcal{L}_k)|$ to be small. This suggestion is investigated in Section 2.

Having chosen the sets $\{\mathcal{L}_k : k=1,2,\ldots,n\}$, the approximations (1.9) are used in an iterative refinement procedure. Specifically, we let the function

$$s^{(1)}(\cdot) = \sum_{k=1}^{n} f_k \ell_k(\cdot\,;\mathcal{L}_k) \tag{1.12}$$

be the initial estimate of the interpolant (1.8). Then we apply the correction formula

$$s^{(j+1)}(\cdot) = s^{(j)}(\cdot) + \sum_{k=1}^{n} [f_k - s^{(j)}(\underline{x}_k)]\, \ell_k(\cdot\,;\mathcal{L}_k), \tag{1.13}$$

for each positive integer j, until all of the residuals $\{f_k - s^{(j)}(\underline{x}_k) : k = 1,2,\ldots,n\}$ are sufficiently close to zero. Section 3 presents some numerical examples that demonstrate the number of iterations that are required by this procedure.

The details of the calculations of Section 3 are described in Sections 4–6. We will find that it is advantageous to let the sets $\{\mathcal{L}_k : k=1,2,\ldots,n\}$ have the property that their intersection, $\mathcal{L}_0$ say, is nonempty. The reason is that we need to store only the coefficients $\{\nu_{ki} : i \in \bar{\mathcal{L}}_k\}$ that occur in the thin plate spline expression

$$\ell_k(\cdot\,;\mathcal{L}_k) = \sum_{i\in\mathcal{L}_k} \nu_{ki}\, \|\cdot - \underline{x}_i\|_2^2 \log \|\cdot - \underline{x}_i\|_2 + \text{linear polynomial}, \tag{1.14}$$

where $\bar{\mathcal{L}}_k$ is the set $\mathcal{L}_k \backslash \mathcal{L}_0$. This technique is explained in Section 4. Then the calculation of the coefficients $\{\nu_{ki} : i \in \bar{\mathcal{L}}_k\}$ for each k, which is completed before the start of the iterative procedure, is the subject of the following two sections, the basic method being described in Section 5 and some useful refinements being given in Section 6. Finally, there is a discussion of our work in Section 7, some of the conclusions being drawn from timings of numerical experiments.

One property of our method is so fundamental that we mention it now, in order to argue that we are doing more than proposing yet another pre-conditioned conjugate gradient procedure. It is that the iterative method (1.12)–(1.13) depends on the approximations (1.9) to the Lagrange functions, which are determined by the sets $\{\mathcal{L}_k : k=1,2,\ldots,n\}$. On the other hand, the purpose of a pre-conditioner is to make a linear transformation to the required vector of parameters that improves the matrix that occurs in the linear system of interpolation equations. Indeed, the description of our method does not demand a particular basis of the n-dimensional linear space of functions of the form (1.3) whose parameters satisfy the conditions (1.5), but the choice of basis is vital to the success of a pre-conditioner. An objection to this argument is that the functions $\{\ell_k(\cdot\,;\mathcal{L}_k) : k = 1,2,\ldots,n\}$ are a basis of the n-dimensional linear space, so perhaps we should regard the iteration (1.13) as a variation on the conjugate gradient algorithm that is numbered (10.3-3) in the book by Golub and Van Loan [5]. Then the parameters ν_{ki} of our expression (1.14) would correspond to elements of the inverse of their M-matrix. The analogy breaks down, however, because M is symmetric, but our procedure makes use of the remark that $i \in \mathcal{L}_k$ need not imply $k \in \mathcal{L}_i$.

2 On the choice of the sets $\mathcal{L}_k$

It is necessary for the approximations (1.9) to be so accurate that the iteration (1.13) converges. Therefore we note that the iteration implies the identity

$$\begin{aligned} f_i - s^{(j+1)}(\underline{x}_i) &= f_i - s^{(j)}(\underline{x}_i) - \sum_{k=1}^{n} [f_k - s^{(j)}(\underline{x}_k)]\, \ell_k(\underline{x}_i; \mathcal{L}_k) \\ &= \sum_{k=1}^{n} [f_k - s^{(j)}(\underline{x}_k)]\, [\delta_{ik} - \ell_k(\underline{x}_i; \mathcal{L}_k)], \quad i = 1, 2, \ldots, n, \end{aligned} \tag{2.1}$$

which we write in the form

$$\underline{r}^{(j+1)} = R\,\underline{r}^{(j)}, \tag{2.2}$$

where $\underline{r}^{(j)}$ is the vector of residuals $\{f_k - s^{(j)}(\underline{x}_k) : k = 1, 2, \ldots, n\}$ and where R is the $n \times n$ matrix that has the elements

$$R_{ik} = \delta_{ik} - \ell_k(\underline{x}_i; \mathcal{L}_k), \quad i, k = 1, 2, \ldots, n. \tag{2.3}$$

It follows that the conditions

$$\sum_{i=1}^{n} |R_{ik}| < 1, \quad k = 1, 2, \ldots, n, \tag{2.4}$$

are sufficient for the convergence of the algorithm. In view of expressions (1.11) and (2.3), they are equivalent to the inequalities

$$\sum_{i \notin \mathcal{L}_k} |\ell_k(\underline{x}_i; \mathcal{L}_k)| < 1, \quad k = 1, 2, \ldots, n. \tag{2.5}$$

Thus we have some guidance on the suitability of trial choices of the sets $\mathcal{L}_k$, $k = 1, 2, \ldots, n$.

For example, we study the magnitude of the sum (2.5) when the interpolation points $\{\underline{x}_i : i = 1, 2, \ldots, n\}$ form a regular square grid in $\mathcal{R}^2$, in the case when $\underline{x}_k$ is at the centre of the grid. It is shown by Buhmann [2] that, for $n = \infty$, the Lagrange function $\ell_k(\cdot\,; \mathcal{L}_*)$ has an asymptotic decay rate that is exponential. Therefore we hope to achieve adequate accuracy with $|\mathcal{L}_k| \sim \log n$ when n is large and finite, by allowing the distance from $\underline{x}_i$, $i \in \mathcal{L}_k$, to its nearest neighbour in $\{\underline{x}_j : j \in \mathcal{L}_k\}$ to grow exponentially as $\|\underline{x}_i - \underline{x}_k\|$ increases.

Fortunately, the following numerical experiment shows that this hope is realised. The central position of $\underline{x}_k$ in the grid implies $n = (2m+1)^2$ for some positive integer m. Let $\{\underline{h}_j : j = 1, 2, \ldots, 8\}$ be the displacements from $\underline{x}_k$ to the other corners of the four grid squares that meet at $\underline{x}_k$, and let $\{\underline{x}_i : i \in \mathcal{L}_k\}$ include only $\underline{x}_k$ and the points $\{\underline{x}_k + m_t\,\underline{h}_j : j = 1, 2, \ldots, 8,\ t = 1, 2, \ldots, \hat{t}\}$, where $\{m_t : t = 1, 2, \ldots, \hat{t}\}$ is the following increasing sequence of positive integers. We set $m_1 = 1$, and for $t \geq 1$ we apply the formula

$$m_{t+1} = \min[m,\ \max\{m_t + 1,\ \mathrm{int}\,(1.7 m_t)\}], \tag{2.6}$$

where $\mathrm{int}\,(1.7 m_t)$ denotes the greatest integer that is at most $1.7 m_t$. Further, we let $\hat{t}$ be the least integer such that $m_{\hat{t}} = m$. Thus, when $m = 15$ for example, we find the sequence $\{m_t\} = \{1,\ 2,\ 3,\ 5,\ 8,\ 13,\ 15\}$.

m	n	$\lvert\mathcal{L}_k\rvert$	$\sum_{i\notin\mathcal{L}_k} \lvert\ell_k(\underline{x}_i;\mathcal{L}_k)\rvert$
15	961	57	0.2809
30	3721	65	0.2938
60	14641	73	0.3026
120	58081	89	0.3076
240	231361	97	0.3112

Table 2.1: $|\mathcal{L}_k|$ when $\underline{x}_k$ is at the centre of a square grid

We calculated the approximate Lagrange function (1.14) in this case for several values of m, and then we formed the sum (2.5). The results are presented in Table 2.1. We see that the convergence condition (2.5) allows $|\mathcal{L}_k| \sim \log n$ for very large values of n, when k is the index at the centre of a square grid of interpolation points.

The choice of the set $\mathcal{L}_k$ in Table 2.1 was guided by the known properties of the true Lagrange function $\ell_k(\cdot\,;\mathcal{L}_*)$ for large n. In order to take this work further, we continue to assume that the interpolation points $\{\underline{x}_k : k \in \mathcal{L}_*\}$ form a regular square grid, and we consider the asymptotic properties of $\ell_1(\cdot\,;\mathcal{L}_*)$, where $\underline{x}_1$ is at the bottom left hand corner of the grid. We do not know of any published work on this subject, except that the one-dimensional analogue has been analysed thoroughly. Indeed, in that case the integral (1.2) is replaced by the functional $\int_{\mathcal{R}}[s''(x)]^2dx$, and it is well-known that the resultant interpolant is the "natural cubic spline" (see [7], for instance). Hence, if the points $\{x_k : k=1,2,\ldots,n\}$ are equally spaced, then, for each value of k, the Lagrange function $\ell_k(x;\mathcal{L}_*)$, $x \in \mathcal{R}$, enjoys exponential decay as $|x-x_k|$ increases, provided that x remains within the range of the interpolation points.

In two dimensions, however, numerical experiments show clearly that the function $\ell_1(\cdot\,;\mathcal{L}_*)$ does not decay exponentially. Indeed, we let the grid cover the unit square, and we calculate the gradient $\underline{\nabla}\ell_1(\underline{x}_n;\mathcal{L}_*)$, where $\underline{x}_n$ is at the top right hand corner of the grid. The components of this vector are given in the middle column of Table 2.2 for three values of n. It seems that these components remain bounded away from zero as $n\to\infty$. Therefore we conjecture that the maximum value of $|\ell_1(\underline{x};\mathcal{L}_*)|$, as $\underline{x}$ ranges over the top right grid square, is at least $\mathcal{O}(n^{-1/2})$. On the other hand, the analogous magnitude in the one-dimensional case is the tiny number $(2-\sqrt{3})^n$.

The reason for the exponential decay in one dimension is that the path from x_1 to x_n passes through all the grid points. In two dimensions, however, there is room for all the interior points of the path to be outside the unit square, so there is a sense in which $\underline{x}_n$ is a neighbour of $\underline{x}_1$. Further, because $\ell_1(\cdot\,;\mathcal{L}_*)$ has the form (1.3), where the coefficients $\{\lambda_k : k=1,2,\ldots,n\}$ satisfy the equations (1.5), the gradient of $\ell_1(\cdot\,;\mathcal{L}_*)$ at infinity is the constant vector $\underline{\nabla}p(\cdot)$. The components of this vector are given in the last column of Table 2.2. They suggest that $\|\underline{\nabla}\ell_1(\underline{\infty};\mathcal{L}_*)\|$ becomes unbounded as $n\to\infty$, which is not unexpected, as an application of the mean value theorem proves that $\|\underline{\nabla}\ell_1(\cdot\,;\mathcal{L}_*)\|$ is at least $\mathcal{O}(n^{1/2})$ in the bottom left square of the grid. We believe that the size of $\|\underline{\nabla}\ell_1(\underline{\infty};\mathcal{L}_*)\|$ does not allow $\|\underline{\nabla}\ell_1(\underline{x}_n;\mathcal{L}_*)\|$ to become small as $n\to\infty$, because there are

n	$\underline{\nabla}\ell_1(\underline{x}_n;\mathcal{L}_*)^T$	$\underline{\nabla}\ell_1(\underline{\infty};\mathcal{L}_*)^T$
25	$(-0.1210\ -0.1210)$	$(-0.757\ -0.757)$
81	$(-0.1016\ -0.1016)$	$(-1.016\ -1.016)$
289	$(-0.0997\ -0.0997)$	$(-1.378\ -1.378)$

Table 2.2: $\underline{\nabla}\ell_1(\cdot\,;\mathcal{L}_*)$ when $\underline{x}_1$ is at the corner of a square grid

no interpolation points between $\underline{x}_n$ and $(+\infty, +\infty)$.

We infer from the magnitudes of the gradients in Table 2.2 that the approximation (1.9), for $k=1$, may be sufficiently accurate only if $\mathcal{L}_1$ includes several indices of interpolation points that are close to $\underline{x}_n$. Further, edge effects may be troublesome all round the boundary of the square grid. Perhaps, therefore, the good values of $|\mathcal{L}_k|$ in Table 2.1 are attainable only when $\underline{x}_k$ is well inside the convex hull of the interpolation points. Fortunately, however, the numerical results in the next section indicate that this view is too pessimistic.

In any case, we recall from the penultimate paragraph of Section 1 that there are advantages in letting each $\mathcal{L}_k$ have the form

$$\mathcal{L}_k = \mathcal{L}_0 \cup \bar{\mathcal{L}}_k, \quad k=1,2,\ldots,n, \tag{2.7}$$

where $\mathcal{L}_0$ and $\bar{\mathcal{L}}_k$ are disjoint. We recommend choosing an $\mathcal{L}_0$ that allows all of the points $\{\underline{x}_i : i \in \bar{\mathcal{L}}_k\}$ to be relatively close to $\underline{x}_k$, so we would compensate for edge effects, if necessary, by adding points to $\mathcal{L}_0$, instead of introducing some relatively large values of $|\bar{\mathcal{L}}_k|$. Therefore the remainder of our work addresses the efficiency and implementation of the iterative procedure (1.12)–(1.13) when the sets $\{\mathcal{L}_k : k = 1,2,\ldots,n\}$ have the structure (2.7). We require $\mathcal{L}_0$ to have the property that the points $\{\underline{x}_i : i \in \mathcal{L}_0\}$ are not collinear, which implies $|\mathcal{L}_0| \geq 3$.

3 On the numbers of iterations

Some results of numerical calculations are presented in this section. We find that the numbers of iterations of our procedure increase very slowly with n in most of the cases that are studied. In every experiment the right hand sides $\{f_k : k=1,2,\ldots,n\}$ are random numbers that are taken from the uniform distribution on $[0,1]$. Further, we terminate the applications of the correction formula (1.13) as soon as j satisfies the condition

$$|f_k - s^{(j)}(\underline{x}_k)| \leq 10^{-12}, \quad k=1,2,\ldots,n. \tag{3.1}$$

In the first experiments the data points have the positions

$$\underline{x}_k = \begin{pmatrix} \cos(2\pi k/n) \\ \sin(2\pi k/n) \end{pmatrix}, \quad k=1,2,\ldots,n, \tag{3.2}$$

$\lvert\mathcal{L}_0\rvert$	Δ	$n=128$	$n=256$	$n=512$	$n=1024$	$n=2048$
8	1.5	29 (5–8)	29 (6–10)	29 (7–12)	29 (8–14)	29 (9–16)
16	1.5	29 (4–6)	29 (5–8)	29 (6–10)	29 (7–12)	29 (8–14)
32	1.5	28 (3–4)	29 (4–6)	29 (5–8)	29 (6–10)	29 (7–12)
8	2.0	8 (8–9)	8 (10–11)	10 (12–13)	11 (14–15)	11 (16–17)
16	2.0	8 (6–7)	8 (8–9)	9 (10–11)	9 (12–13)	10 (14–15)
32	2.0	8 (4–5)	8 (6–7)	8 (8–9)	9 (10–11)	9 (12–13)
8	4.0	6 (16–17)	6 (20–21)	6 (24–25)	7 (28–29)	7 (32–33)
16	4.0	5 (12–13)	6 (16–17)	6 (20–21)	6 (24–25)	6 (28–29)
32	4.0	5 (8–9)	5 (12–13)	6 (16–17)	6 (20–21)	6 (24–25)

Table 3.1: Iterations (Range of $\lvert\bar{\mathcal{L}}_k\rvert$) for points on the circumference of a circle

where n is a power of 2. Thus they are equally spaced on the unit circle, so there is no need to give special attention to edge effects. Further, we let $\lvert\mathcal{L}_0\rvert$ be a smaller power of 2, and we let $\mathcal{L}_0$ be the set of integers $\{jn/\lvert\mathcal{L}_0\rvert : j=1,2,\ldots,\lvert\mathcal{L}_0\rvert\}$, in order that the points $\{\underline{x}_i : i\in\mathcal{L}_0\}$ are also equally spaced on the unit circle. Further, the sets $\{\bar{\mathcal{L}}_k : k=1,2,\ldots,n\}$ of expression (2.7) are constructed in the following way, which depends on a parameter Δ that satisfies $\Delta\geq 1$. For each integer i in $[1,n]$, we define $\hat{\imath}$ to be the greatest power of 2 such that $i/\hat{\imath}$ is an integer. Then $i\notin\mathcal{L}_0$ is included in $\bar{\mathcal{L}}_k$ if and only if one of the conditions

$$|i-k|\leq\Delta\,\hat{\imath}\quad\text{or}\quad|i-k|\geq n-\Delta\,\hat{\imath}\tag{3.3}$$

is satisfied. In other words, i is in $\bar{\mathcal{L}}_k$ if the number of interpolation points between $\underline{x}_i$ and $\underline{x}_k$ is less than $\Delta\,\hat{\imath}$. Thus, when $n=16$, $\lvert\mathcal{L}_0\rvert=4$ and $\Delta=1.5$, for example, the sets $\mathcal{L}_0$, $\bar{\mathcal{L}}_1$ and $\bar{\mathcal{L}}_4$ have the elements $\{4, 8, 12, 16\}$, $\{14, 1, 2\}$ and $\{2, 3, 5, 6\}$, respectively.

The numerical results in the case (3.2) for several values of $\lvert\mathcal{L}_0\rvert$, Δ and n are shown in Table 3.1. The entries in the main part of the table are in the form $a\,(b-c)$, where a, b and c are the number of iterations, $\min\{\lvert\bar{\mathcal{L}}_k\rvert : k=1,2,\ldots,n\}$ and $\max\{\lvert\bar{\mathcal{L}}_k\rvert : k=1,2,\ldots,n\}$, respectively. We see that our procedure requires far fewer than n iterations to achieve the termination condition (3.1). Further, for $\Delta\geq 2$ the iteration counts are much less than the best results that are given by in [8] for pre-conditioned conjugate gradient algorithms. Further, it seems that the spectral radius of the iteration (1.13) is insensitive to $\lvert\mathcal{L}_0\rvert$ but depends strongly on the choice of Δ. On the other hand, we expect $\mathcal{L}_0$ to be useful when edge effects are substantial.

The second batch of experiments introduces edge effects in a mild way. Specifically, we let the points $\{\underline{x}_k : k=1,2,\ldots,n\}$ and $\{\underline{x}_i : i\in\mathcal{L}_0\}$ be equally spaced around the perimeter of a square, where, as before, both n and $\lvert\mathcal{L}_0\rvert$ are powers of 2. We also impose the condition that $\{\underline{x}_i : i\in\mathcal{L}_0\}$ includes the corners of the square. Then the sets $\{\bar{\mathcal{L}}_k : k=1,2,\ldots,n\}$ are generated by the method that is described in the second paragraph of this section. Thus the values of $\lvert\bar{\mathcal{L}}_k\rvert$ are the same as before, but different

$\lvert\mathcal{L}_0\rvert$	Δ	$n=128$	$n=256$	$n=512$	$n=1024$	$n=2048$
8	1.5	28 (5–8)	28 (6–10)	29 (7–12)	29 (8–14)	29 (9–16)
16	1.5	28 (4–6)	28 (5–8)	29 (6–10)	29 (7–12)	29 (8–14)
32	1.5	28 (3–4)	28 (4–6)	29 (5–8)	29 (6–10)	29 (7–12)
8	2.0	13 (8–9)	14 (10–11)	15 (12–13)	15 (14–15)	16 (16–17)
16	2.0	10 (6–7)	11 (8–9)	12 (10–11)	12 (12–13)	13 (14–15)
32	2.0	8 (4–5)	9 (6–7)	9 (8–9)	9 (10–11)	11 (12–13)
8	4.0	12 (16–17)	12 (20–21)	13 (24–25)	13 (28–29)	13 (32–33)
16	4.0	9 (12–13)	9 (16–17)	9 (20–21)	9 (24–25)	9 (28–29)
32	4.0	7 (8–9)	7 (12–13)	8 (16–17)	8 (20–21)	8 (24–25)

Table 3.2: Iterations (Range of $\lvert\bar{\mathcal{L}}_k\rvert$) for points on the circumference of a square

numbers of iterations are performed by our algorithm, as shown in Table 3.2. Indeed, the edge effects impair the reductions in the numbers of iterations that occur when Δ is increased. Thus the iteration counts become sensitive to $\lvert\mathcal{L}_0\rvert$ for the larger values of Δ.

The final experiments of this section are in the case when the points $\{\underline{x}_k : k = 1,2,\ldots,n\}$ form a regular square grid, but they are crude, because we have not yet developed a satisfactory way of compensating for edge effects by the choice of $\mathcal{L}_0$. Indeed, in the experiments we let $\{\underline{x}_i : i \in \mathcal{L}_0\}$ be a regular square subgrid that includes the four corners. Further, n and $\lvert\mathcal{L}_0\rvert$ have the values $(m+1)^2$ and $(m_0+1)^2$, respectively, where m and m_0 are powers of 2. Further, the sets $\{\bar{\mathcal{L}}_k : k=1,2,\ldots,n\}$ are generated by an extension to two dimensions of the construction in the second paragraph of this section, so we continue to employ a parameter Δ that satisfies $\Delta \geq 1$. In order to specify this extension, we distinguish the grid points by integer coordinates from the interval $[0,m]$. Specifically, for $i=1,2,\ldots,n$, we let the coordinates of $\underline{x}_i$ be α_i and β_i, so $\underline{x}_i$ is the point of intersection of the (α_i+1)-th grid line in one direction with the (β_i+1)-th grid line in the orthogonal direction. Further, let $\hat{\imath}$ be the greatest power of 2 such that both α_i and β_i are integer multiples of $\hat{\imath}$. Thus i is in $\mathcal{L}_0$ if and only if $\hat{\imath}$ is at least m/m_0. Otherwise, we include i in the set $\bar{\mathcal{L}}_k$ if and only if both of the inequalities

$$\lvert\alpha_i - \alpha_k\rvert \leq \Delta\,\hat{\imath} \quad \text{and} \quad \lvert\beta_i - \beta_k\rvert \leq \Delta\,\hat{\imath} \tag{3.4}$$

hold. In other words, assuming that the grid lines are parallel to the coordinate axes, the index $i \notin \mathcal{L}_0$ is in the set $\bar{\mathcal{L}}_k$ for all of the integers k that satisfy $\|\underline{x}_k - \underline{x}_i\|_\infty \leq \Delta\,\hat{\imath}h$, where h is the mesh size of the grid.

The results of the square grid experiments are given in Table 3.3. The iteration counts are given as "$\star$" when the calculated sequence $\{s^{(j)}(\cdot) : j=1,2,3,\ldots\}$ diverges, but we had expected more failures, because our choice of the sets $\{\bar{\mathcal{L}}_k : k=1,2,\ldots,n\}$ ignores the edge effects that seem to be important in Section 2. Therefore, for $\Delta \geq 2$, the entries in the table provide much encouragement for further developments of our algorithm.

$\lvert\mathcal{L}_0\rvert$	Δ	$n=81$	$n=289$	$n=1089$
9	1.5	63 (6 − 17)	⋆ (9 − 25)	⋆ (12 − 34)
25	1.5	95 (3 − 8)	165 (6 − 17)	⋆ (9 − 25)
81	1.5	1 (0 − 0)	149 (3 − 8)	353 (6 − 17)
9	2.0	14 (10 − 33)	18 (15 − 49)	29 (20 − 65)
25	2.0	12 (5 − 21)	16 (10 − 37)	25 (15 − 53)
81	2.0	1 (0 − 0)	14 (5 − 21)	18 (10 − 37)
9	4.0	8 (32 − 72)	9 (48 − 129)	12 (64 − 185)
25	4.0	8 (16 − 56)	9 (32 − 113)	12 (48 − 169)
81	4.0	1 (0 − 0)	8 (16 − 65)	11 (32 − 121)

Table 3.3: Iterations (Range of $\lvert\bar{\mathcal{L}}_k\rvert$) for points on a square grid

4 Details of an iteration

Equations (1.12)–(1.14) show that each approximation $s^{(j)}(\cdot)$ to the required interpolating function has the form

$$s^{(j)}(\underline{x}) = \sum_{i=1}^{n} \lambda_i^{(j)} \|\underline{x}-\underline{x}_i\|_2^2 \log \|\underline{x}-\underline{x}_i\|_2 + p^{(j)}(\underline{x}), \quad \underline{x}\in\mathcal{R}^2, \tag{4.1}$$

where $p^{(j)}(\cdot)$ is a linear polynomial. Further, our definitions of the approximate Lagrange functions (1.9) give the properties

$$\sum_{i=1}^{n} \lambda_i^{(j)} = \sum_{i=1}^{n} \lambda_i^{(j)} x_i = \sum_{i=1}^{n} \lambda_i^{(j)} y_i = 0. \tag{4.2}$$

We let $\underline{\lambda}^{(j)} \in \mathcal{R}^n$ be the vector whose components are $\{\lambda_i^{(j)} : i = 1,2,\ldots,n\}$ and we let $\underline{c}^{(j)} \in \mathcal{R}^3$ be the vector of coefficients of $p^{(j)}(\cdot)$. This section describes our way of calculating $\underline{\lambda}^{(j)}$ and $\underline{c}^{(j)}$ for $j=1,2,3,\ldots$, which takes advantage of the structure (2.7).

We extend the notation of equation (1.14) by defining $\nu_{ki}=0$ for $i\notin\mathcal{L}_k$. Therefore expressions (1.12)–(1.14) and (4.1) provide the relations

$$\left.\begin{array}{rl} \lambda_i^{(1)} &= \sum_{k=1}^{n} f_k\, \nu_{ki} \\ \lambda_i^{(j+1)} &= \lambda_i^{(j)} + \sum_{k=1}^{n} [f_k - s^{(j)}(\underline{x}_k)]\, \nu_{ki} \end{array}\right\}, \quad i=1,2,\ldots,n, \tag{4.3}$$

but we recall from the penultimate paragraph of Section 1 that we store only the coefficients $\{\nu_{ki} : i\in\bar{\mathcal{L}}_k\}$ for each k, their values being generated by the procedure in Sections 5 and 6. It follows that the numbers $\{\nu_{ki} : k=1,2,\ldots,n\}$ are available for every index i that is not in $\mathcal{L}_0$, so we calculate the parameters $\{\lambda_i^{(1)} : i\notin\mathcal{L}_0\}$ using the first line of formula (4.3). Then the remaining components of $\underline{\lambda}^{(1)}$ are derived by solving the interpolation problem that is defined below.

The structure (2.7) states that $i \in \mathcal{L}_0$ implies $i \in \mathcal{L}_k$ for every k. It follows from the conditions (1.11) that such an i enjoys the properties

$$\ell_k(\underline{x}_i; \mathcal{L}_k) = \delta_{ik}, \quad k = 1, 2, \ldots, n. \tag{4.4}$$

Therefore formula (1.12) gives the equations

$$s^{(1)}(\underline{x}_k) = f_k, \quad k \in \mathcal{L}_0, \tag{4.5}$$

which, in view of expressions (4.1) and (1.4) can be written in the form

$$\sum_{i \in \mathcal{L}_0} A_{ki} \, \lambda_i^{(1)} + p^{(1)}(\underline{x}_k) = f_k - \sum_{i \notin \mathcal{L}_0} A_{ki} \, \lambda_i^{(1)}, \quad k \in \mathcal{L}_0, \tag{4.6}$$

where we have taken to the right hand side the part of the sum of expression (4.1) that includes the known coefficients $\{\lambda_i^{(1)} : i \notin \mathcal{L}_0\}$. Remembering the conditions (4.2), we now have $|\mathcal{L}_0|+3$ equations in $|\mathcal{L}_0|+3$ unknowns, namely $\{\lambda_i^{(1)} : i \in \mathcal{L}_0\}$ and the three components of $\underline{c}^{(1)}$.

This system differs slightly from the usual thin plate spline interpolation equations, because expression (4.2) allows the sums $\sum_{i \in \mathcal{L}_0} \lambda_i^{(1)}$, $\sum_{i \in \mathcal{L}_0} \lambda_i^{(1)} x_i$ and $\sum_{i \in \mathcal{L}_0} \lambda_i^{(1)} y_i$ to be nonzero. Therefore we pick coefficients $\{\hat{\lambda}_i^{(1)} : i \in \mathcal{L}_0\}$ that provide the properties

$$\sum_{i \in \mathcal{L}_0} \hat{\lambda}_i^{(1)} + \sum_{i \notin \mathcal{L}_0} \lambda_i^{(1)} = \sum_{i \in \mathcal{L}_0} \hat{\lambda}_i^{(1)} x_i + \sum_{i \notin \mathcal{L}_0} \lambda_i^{(1)} x_i = \sum_{i \in \mathcal{L}_0} \hat{\lambda}_i^{(1)} y_i + \sum_{i \notin \mathcal{L}_0} \lambda_i^{(1)} y_i = 0, \tag{4.7}$$

taking advantage of the fact that at most three of these coefficients need be nonzero. Thus the system (4.6) and (4.2) is equivalent to the equations

$$\sum_{i \in \mathcal{L}_0} A_{ki} \, (\lambda_i^{(1)} - \hat{\lambda}_i^{(1)}) + p^{(1)}(\underline{x}_k) = f_k - \sum_{i \in \mathcal{L}_0} A_{ki} \, \hat{\lambda}_i^{(1)} - \sum_{i \notin \mathcal{L}_0} A_{ki} \, \lambda_i^{(1)}, \quad k \in \mathcal{L}_0, \tag{4.8}$$

and

$$\sum_{i \in \mathcal{L}_0} (\lambda_i^{(1)} - \hat{\lambda}_i^{(1)}) = \sum_{i \in \mathcal{L}_0} (\lambda_i^{(1)} - \hat{\lambda}_i^{(1)}) \, x_i = \sum_{i \in \mathcal{L}_0} (\lambda_i^{(1)} - \hat{\lambda}_i^{(1)}) \, y_i = 0. \tag{4.9}$$

It follows that the differences $\{\lambda_i^{(1)} - \hat{\lambda}_i^{(1)} : i \in \mathcal{L}_0\}$ and the coefficients of $p^{(1)}(\cdot)$ are defined by thin plate spline interpolation to the known right hand sides of expression (4.8) at the points $\{\underline{x}_k : k \in \mathcal{L}_0\}$.

Therefore the preliminary calculations of the next section include the construction of a $|\mathcal{L}_0| \times (|\mathcal{L}_0|-3)$ matrix, Q_0 say, whose columns are linearly independent and orthogonal to values of linear polynomials on the point set $\{\underline{x}_k : k \in \mathcal{L}_0\}$. Further, we let A_0 be the $|\mathcal{L}_0| \times |\mathcal{L}_0|$ matrix whose elements are the numbers $\{A_{ij} : i, j \in \mathcal{L}_0\}$, and we let $\underline{f}_0^{(1)}$ be the vector whose components are the right hand sides of equation (4.8). Thus, corresponding to expression (1.7), the required differences $\{\lambda_i^{(1)} - \hat{\lambda}_i^{(1)} : i \in \mathcal{L}_0\}$ are the components of $Q_0 \, \underline{\mu}^{(1)}$, where $\underline{\mu}^{(1)}$ is the solution of the positive definite linear system

$$Q_0^T A_0 Q_0 \, \underline{\mu}^{(1)} = Q_0^T \underline{f}_0^{(1)}. \tag{4.10}$$

We also generate the Cholesky factorization of $Q_0^T A_0 Q_0$ before beginning the iterative part of our procedure. Thus, after calculating $\underline{f}_0^{(1)}$, the system (4.10) provides $\underline{\mu}^{(1)}$ in $\mathcal{O}(|\mathcal{L}_0|^2)$ operations. Then the determination of $\{\lambda_i^{(1)} : i \in \mathcal{L}_0\}$ is straightforward. Then, as in Section 1, we obtain the coefficients of $p^{(1)}(\cdot)$ from three of the equations (4.8). The description of our calculation of the parameters of the function (4.1) when $j=1$ is now complete.

Let j be any positive integer such that $s^{(j)}(\cdot)$ is available. Our algorithm generates the residuals

$$f_k - s^{(j)}(\underline{x}_k), \quad k=1,2,\ldots,n, \tag{4.11}$$

in order to decide whether $s^{(j)}(\cdot)$ is a sufficiently accurate approximation to the interpolating function. If it is not, then we seek the coefficients of $s^{(j+1)}(\cdot)$. The second line of equation (4.3) provides the parameters $\{\lambda_i^{(j+1)} : i \notin \mathcal{L}_0\}$, because the relevant values of $\{\nu_{ki}\}$ are available. The remaining parameters of $s^{(j+1)}(\cdot)$, however, are obtained by solving an interpolation problem that is similar to the one we have just studied.

Indeed, because the properties (4.4) are valid for every i in $\mathcal{L}_0$, formula (1.13) implies the identities

$$s^{(j+1)}(\underline{x}_k) = f_k, \quad k \in \mathcal{L}_0, \tag{4.12}$$

which are analogous to the equations (4.5). Therefore, expression (4.6) remains true if each superscript "(1)" is replaced by "$(j+1)$". Further, this replacement can be made throughout the three paragraphs that include equations (4.4)–(4.10), which provides a complete description of our calculation of $\{\lambda_i^{(j+1)} : i \in \mathcal{L}_0\}$ and $\underline{c}^{(j+1)}$.

We note that our procedure employs the relations (4.3) only for values of i that are not in $\mathcal{L}_0$, which is why there is no need to store the coefficients $\{\nu_{ki} : i \in \mathcal{L}_0,\ k=1,2,\ldots,n\}$. Then the remaining coefficients of the next approximate interpolating function are derived from the properties (4.2) and the equations (4.5) or (4.12). Because this method requires the solution of the positive definite system (4.10), some work is saved by the generation of the Cholesky factorization of $Q_0^T A_0 Q_0$ in advance, this matrix being independent of j.

In the numerical experiments of Section 3, the most expensive part of each iteration is the calculation of the residuals (4.11). Indeed, the numbers $\{s^{(j)}(\underline{x}_k) : k=1,2,\ldots,n\}$ are obtained directly from expression (4.1), so this task requires $\mathcal{O}(n^2)$ operations. On the other hand, our use of formula (4.3) takes only $\mathcal{O}(\sum_{k=1}^n |\bar{\mathcal{L}}_k|)$ operations. Further, the right hand sides (4.8) are generated in $\mathcal{O}(n\,|\mathcal{L}_0|)$ operations, and finally the work of solving the thin plate spline subproblem (4.8)–(4.9) when the right hand sides are available is only $\mathcal{O}(|\mathcal{L}_0|^2)$. Thus the calculation of the residuals is dominant in the usual case when $|\mathcal{L}_0|$ is much less than n. This task is also the most expensive part of each iteration of the conjugate gradient method. Therefore it is suitable to compare the iteration counts in the experiments of Section 3 with those of a pre-conditioned conjugate gradient algorithm. Fortunately, however, Laurent expansions can good provide approximations to all of the residuals (4.11) in far fewer than $\mathcal{O}(n^2)$ operations when n is large [1].

5 The basic method of the preliminary calculations

Let P_0 be the $|\mathcal{L}_0|\times 3$ matrix whose i-th row is the vector $(1\ x_{t(i)}\ y_{(t(i))})$, where $t(i)$ is the i-th integer in the set $\mathcal{L}_0$ and where $x_{t(i)}$ and $y_{t(i)}$ are the components of $\underline{x}_{t(i)}$. We seek an $|\mathcal{L}_0|\times|\mathcal{L}_0|$ orthogonal matrix Ω_0 such that the product $\Omega_0^T P_0$ is upper triangular, because then the last $(|\mathcal{L}_0|-3)$ columns of Ω_0 are orthogonal to values of linear polynomials on the point set $\{\underline{x}_t : t\in\mathcal{L}_0\}$. Therefore, having chosen Ω_0, the matrix Q_0 of equation (4.10) is generated by deleting the first three columns of Ω_0.

The preliminary calculations of our algorithm pick an Ω_0 that is a product of $(3|\mathcal{L}_0|-6)$ Givens rotations. Specifically, Ω_0^T is the expression

$$\Omega_0^T = \Omega_0^{(3,|\mathcal{L}_0|)}\Omega_0^{(3,|\mathcal{L}_0|-1)}\cdots\Omega_0^{(3,4)}\Omega_0^{(2,|\mathcal{L}_0|)}\Omega_0^{(2,|\mathcal{L}_0|-1)}\cdots\Omega_0^{(2,3)} \times\Omega_0^{(1,|\mathcal{L}_0|)}\Omega_0^{(1,|\mathcal{L}_0|-1)}\cdots\Omega_0^{(1,2)}, \tag{5.1}$$

where the notation $\Omega_0^{(i,j)}$ denotes an $|\mathcal{L}_0|\times|\mathcal{L}_0|$ orthogonal matrix that differs from the $|\mathcal{L}_0|\times|\mathcal{L}_0|$ unit matrix only in its (i,i)-th, (i,j)-th (j,i)-th and (j,j)-th elements. It is well-known that we can make $\Omega_0^T P_0$ upper triangular. Indeed, the usual way begins by setting an $|\mathcal{L}_0|\times 3$ working space array, $\bar{P}$ say, to P_0. Then the Givens rotations of expression (5.1) are generated in sequence from right to left. The freedom in each $\Omega_0^{(i,j)}$ is used to make the (j,i)-th element of $\Omega_0^{(i,j)}\bar{P}$ zero, and then $\bar{P}$ is overwritten by $\Omega_0^{(i,j)}\bar{P}$ before the next Givens rotation is calculated. Thus we specify Q_0 in $\mathcal{O}(|\mathcal{L}_0|)$ operations, there being no need to calculate its elements explicitly. On the other hand, Powell [8] generates Q_0 in a similar way that employs Householder instead of Givens rotations. We prefer the form (5.1), because we will find that some of the commutivity properties of the Givens matrices are very useful.

The relation between Q_0 and Ω_0 provides an identity of the form

$$\Omega_0^T A_0 \Omega_0 = \left(\begin{array}{c|c} X_0 & Y_0^T \\ \hline Y_0 & Q_0^T A_0 Q_0 \end{array}\right), \tag{5.2}$$

where X_0 and Y_0 are 3×3 and $(|\mathcal{L}_0|-3)\times 3$, respectively, and where A_0 is introduced before equation (4.10). Using the structure (5.1), we determine all the elements of the matrix (5.2) in $\mathcal{O}(|\mathcal{L}_0|^2)$ operations. The submatrices X_0 and Y_0 are stored because they are required later. Instead of saving $Q_0^T A_0 Q_0$, however, we generate and store the lower triangular matrix L_0 of the Cholesky factorization

$$L_0 L_0^T = Q_0^T A_0 Q_0, \tag{5.3}$$

the work of calculating the factorization being $\mathcal{O}(|\mathcal{L}_0|^3)$. Our only safeguard against loss of positive definiteness is to stop this calculation if rounding errors cause any pivot to be nonpositive, but no premature termination occurred in the experiments of Section 3.

The description of the preliminary calculations for the interpolation part of our iterative procedure is now complete. Therefore we note that the right hand side of the system (4.10) can be generated in $\mathcal{O}(|\mathcal{L}_0|)$ operations by forming the vector $\Omega_0^T \underline{f}_0^{(1)}$ and then deleting its first three components. Moreover, we obtain the required product $Q_0\underline{\mu}^{(1)}$

by adding three zero components to the beginning of $\underline{\mu}^{(1)}$ and then pre-multiplying the resultant vector by Ω_0, so again the amount of work is $\mathcal{O}(|\mathcal{L}_0|)$. Therefore we turn to the calculation of the coefficients $\{\nu_{ki} : i \in \bar{\mathcal{L}}_k\}$ for $k=1,2,\ldots,n$.

We extend the notation P_0, Ω_0, Q_0, A_0, X_0 and Y_0 in the following way. We let P_k be the $|\mathcal{L}_k|\times 3$ matrix whose i-th row is the vector $(1 \; x_{t(i)} \; y_{(t(i))})$, where $\underline{x}_{t(i)}$ is now the i-th element of the set $\{\underline{x}_t : t \in \mathcal{L}_k\}$. We let Ω_k^T be the product

$$\Omega_k^T = \Omega_k^{(3,|\mathcal{L}_k|)}\Omega_k^{(3,|\mathcal{L}_k|-1)}\cdots\Omega_k^{(3,4)}\Omega_k^{(2,|\mathcal{L}_k|)}\Omega_k^{(2,|\mathcal{L}_k|-1)}\cdots\Omega_k^{(2,3)} \times \Omega_k^{(1,|\mathcal{L}_k|)}\Omega_k^{(1,|\mathcal{L}_k|-1)}\cdots\Omega_k^{(1,2)}, \tag{5.4}$$

where each $\Omega_k^{(i,j)}$ is a $|\mathcal{L}_k|\times|\mathcal{L}_k|$ Givens rotation and where $\Omega_k^T P_k$ is upper triangular. We form Q_k by deleting the first three columns of Ω_k. We let A_k be the $|\mathcal{L}_k|\times|\mathcal{L}_k|$ matrix whose elements are the numbers $\{A_{ij} : i,j \in \mathcal{L}_k\}$, where A_{ij} has the value (1.4). And, corresponding to expression (5.2), X_k and Y_k are defined by the equation

$$\Omega_k^T A_k \Omega_k = \left(\begin{array}{c|c} X_k & Y_k^T \\ \hline Y_k & Q_k^T A_k Q_k \end{array}\right). \tag{5.5}$$

For $k>0$, the structure (2.7) allows the elements of $\mathcal{L}_0$ to be placed at the beginning of $\mathcal{L}_k$. Thus P_0 is the leading $|\mathcal{L}_0|\times 3$ submatrix of P_k. Moreover, when constructing the product (5.4), the required zeros below the diagonal of $\Omega_k^T P_k$ can be introduced row by row, so the product takes the form

$$\Omega_k^T = (\Omega_k^{(3,|\mathcal{L}_k|)}\Omega_k^{(2,|\mathcal{L}_k|)}\Omega_k^{(1,|\mathcal{L}_k|)})\cdots(\Omega_k^{(3,4)}\Omega_k^{(2,4)}\Omega_k^{(1,4)})\Omega_k^{(2,3)}\Omega_k^{(1,3)}\Omega_k^{(1,2)}, \tag{5.6}$$

where the same Givens rotations occur in expressions (5.4) and (5.6), because $\Omega_k^{(\alpha,\beta)}$ commutes with $\Omega_k^{(i,j)}$ if the integers $\{\alpha,\beta,i,j\}$ are all different. Therefore $\Omega_k^{(i,j)}$ is independent of the last $(|\mathcal{L}_k|-j)$ rows of P_k. Hence every rotation $\Omega_k^{(i,j)}$ with $j \le |\mathcal{L}_0|$ has $\Omega_0^{(i,j)}$ as its leading $|\mathcal{L}_0|\times|\mathcal{L}_0|$ submatrix. It follows from equation (5.6) that, if j is any integer from $[4,|\mathcal{L}_0|]$ and if $\hat{\underline{e}}_j$ and $\underline{e}_j$ are the j-th coordinate vectors with $|\mathcal{L}_0|$ and $|\mathcal{L}_k|$ components, respectively, then the elements of $\underline{e}_j^T\Omega_k^T$ are the $|\mathcal{L}_0|$ components of $\hat{\underline{e}}_j^T\Omega_0^T$ followed by $|\bar{\mathcal{L}}_k|$ zeros. In other words, because $\Omega_0\hat{\underline{e}}_j$ and $\Omega_k\underline{e}_j$ are the j-th columns of Ω_0 and Ω_k, we have found that Q_0 is the leading $(|\mathcal{L}_0|-3)\times(|\mathcal{L}_0|-3)$ submatrix of Q_k, and that the remaining elements in the first $(|\mathcal{L}_0|-3)$ columns of Q_k are zero. Therefore $Q_0^T A_0 Q_0$ is the leading $(|\mathcal{L}_0|-3)\times(|\mathcal{L}_0|-3)$ submatrix of $Q_k^T A_k Q_k$.

Our calculation of the coefficients $\{\nu_{ki} : i \in \bar{\mathcal{L}}_k\}$ requires the Cholesky factorization of the matrix $Q_k^T A_k Q_k$. Fortunately, the property of $Q_k^T A_k Q_k$ that has just been mentioned allows this factorization to be generated in the following way. We write A_k in the form

$$A_k = \left(\begin{array}{c|c} A_0 & \dot{A}_k^T \\ \hline \dot{A}_k & \bar{A}_k \end{array}\right), \tag{5.7}$$

where the dimensions of $\dot{A}_k$ and $\bar{A}_k$ are $|\bar{\mathcal{L}}_k|\times|\mathcal{L}_0|$ and $|\bar{\mathcal{L}}_k|\times|\bar{\mathcal{L}}_k|$, respectively, and we let $\bar{\Omega}_k^T$ be the orthogonal matrix

$$\bar{\Omega}_k^T = (\Omega_k^{(3,|\mathcal{L}_k|)}\Omega_k^{(2,|\mathcal{L}_k|)}\Omega_k^{(1,|\mathcal{L}_k|)})\cdots(\Omega_k^{(3,|\mathcal{L}_0|+1)}\Omega_k^{(2,|\mathcal{L}_0|+1)}\Omega_k^{(1,|\mathcal{L}_0|+1)}), \tag{5.8}$$

so it is the product of the $3\,|\bar{\mathcal{L}}_k|$ Givens rotations of expression (5.6) that do not occur in Ω_0^T. Then the definition (5.1) and equations (5.6)–(5.8) imply the identity

$$\Omega_k^T A_k \Omega_k = \bar{\Omega}_k^T \left(\begin{array}{c|c} \Omega_0^T A_0 \Omega_0 & \Omega_0^T \dot{A}_k^T \\ \hline \dot{A}_k \Omega_0 & \bar{A}_k \end{array} \right) \bar{\Omega}_k. \tag{5.9}$$

The structure of Ω_0 allows us to calculate the product $\dot{A}_k\Omega_0$ in $\mathcal{O}(|\bar{\mathcal{L}}_k|\,|\mathcal{L}_0|)$ operations. Then the last $|\bar{\mathcal{L}}_k|$ rows of expression (5.9) are generated, using equation (5.8) and the submatrices X_0 and Y_0 that were preserved from the partition (5.2), which requires $\mathcal{O}(|\bar{\mathcal{L}}_k|\,|\mathcal{L}_k|)$ operations. Thus, because we can take the view that this calculation gives the last $|\bar{\mathcal{L}}_k|$ rows of the matrix (5.5), we obtain the elements of the submatrices $\dot{B}_k$ and $\bar{B}_k$ of the equation

$$Q_k^T A_k Q_k = \left(\begin{array}{c|c} Q_0^T A_0 Q_0 & \dot{B}_k^T \\ \hline \dot{B}_k & \bar{B}_k \end{array} \right), \tag{5.10}$$

the presence of $Q_0^T A_0 Q_0$ being justified by the last remark of the previous paragraph. Now we are seeking the Cholesky factorization

$$Q_k^T A_k Q_k = L_k L_k^T \tag{5.11}$$

of the matrix (5.10) and the factorization (5.3) is available. It follows from expression (5.10) that L_0 is the leading $(|\mathcal{L}_0|-3)\times(|\mathcal{L}_0|-3)$ submatrix of L_k. Therefore it is sufficient to calculate only the last $|\bar{\mathcal{L}}_k|$ rows of L_k, which requires $\mathcal{O}(|\bar{\mathcal{L}}_k|\,|\mathcal{L}_k|^2)$ operations.

It is now straightforward to generate the coefficients $\{\nu_{ki} : i \in \bar{\mathcal{L}}_k\}$ that are used in formula (4.3). Specifically, we recall from Section 1 that the function (1.14) is defined by thin plate spline interpolation to the right hand sides (1.10). Hence, as in equation (1.7), the coefficients $\{\nu_{ki} : i \in \mathcal{L}_k\}$ are the components of the vector $Q_k \underline{\mu}_k = \underline{\nu}_k$, say, where $\underline{\mu}_k$ is the solution of the linear system

$$Q_k^T A_k Q_k \underline{\mu}_k = L_k L_k^T \underline{\mu}_k = Q_k^T \underline{e}_\kappa, \tag{5.12}$$

and where $\underline{e}_\kappa$ is the coordinate vector in $\mathcal{R}^{|\mathcal{L}_k|}$ whose nonzero component corresponds to the position of k in the set $\mathcal{L}_k$. We derive the right hand side $Q_k^T \underline{e}_\kappa$ by deleting the first three components of $\Omega_k^T \underline{e}_\kappa$, and, when generating $\Omega_k^T \underline{e}_\kappa$ with the help of expression (5.6), we ignore any early Givens rotations that act on the zero part of $\underline{e}_\kappa$. Further, when calculating the vector

$$L_k^T \underline{\mu}_k = L_k^{-1} Q_k^T \underline{e}_\kappa \tag{5.13}$$

by forward substitution, we take advantage of any leading zeros of $Q_k^T \underline{e}_\kappa$ that are induced by the fact that $\underline{e}_\kappa$ is a coordinate vector. Further, there is no need to complete the back substitution that determines $\underline{\mu}_k$, because we require only the last $|\bar{\mathcal{L}}_k|$ components of the vector

$$\underline{\nu}_k = Q_k \underline{\mu}_k = \Omega_k \underline{\hat{\mu}}_k, \tag{5.14}$$

where $\underline{\hat{\mu}}_k$ is formed by adding three zeros to the beginning of $\underline{\mu}_k$. Indeed, it follows from the structure (5.6) that the last $|\bar{\mathcal{L}}_k|$ elements of $\Omega_k \underline{\hat{\mu}}_k$ are independent of the first $(|\mathcal{L}_0|-3)$ elements of $\underline{\mu}_k$, so the back substitution is stopped when the last $|\bar{\mathcal{L}}_k|$ components of $\underline{\mu}_k$

are found. Then it is easy to obtain the required parameters $\{\nu_{ki} : i \in \bar{\mathcal{L}}_k\}$ from equation (5.14). Thus the work of the procedure of this paragraph is only $\mathcal{O}(|\bar{\mathcal{L}}_k|^2)$ in the usual case when $k \notin \mathcal{L}_0$, and otherwise it is bounded above by a multiple of $|\mathcal{L}_k|^2$.

6 Some refinements of the preliminary calculations

The most expensive task that is mentioned in the previous section is the calculation of the Cholesky factorization (5.11) from the factorization (5.3). Indeed, we have noted that the work for each k is $\mathcal{O}(|\bar{\mathcal{L}}_k|\,|\mathcal{L}_k|^2)$. Thus, if $n = 10^4$, $|\mathcal{L}_k| \approx 200$ and $|\bar{\mathcal{L}}_k| \approx 25$, for example, then about 10^{10} computer operations occur. Therefore it is highly valuable that updating techniques usually allow the factor $|\bar{\mathcal{L}}_k|$ to be replaced by a smaller number. In order to explain this idea, we let $\mathcal{L}_{k(1)}$ and $\mathcal{L}_{k(2)}$ be two of the sets $\{\mathcal{L}_k : k = 1, 2, \ldots, n\}$ such that their intersection, $\mathcal{L}_{k(0)}$ say, is larger than $\mathcal{L}_0$. Further, we suppose that the indices in the sets $\mathcal{L}_{k(1)}$ and $\mathcal{L}_{k(2)}$ are ordered so that the elements of $\mathcal{L}_{k(0)}$ come first. Then, corresponding to expression (5.10), the matrix $Q_{k(0)}^T A_{k(0)} Q_{k(0)}$ is the leading $(|\mathcal{L}_{k(0)}|-3) \times (|\mathcal{L}_{k(0)}|-3)$ submatrix of both $Q_{k(1)}^T A_{k(1)} Q_{k(1)}$ and $Q_{k(2)}^T A_{k(2)} Q_{k(2)}$, where we are using an obvious notation. Therefore the Cholesky factorization of $Q_{k(0)}^T A_{k(0)} Q_{k(0)}$ is generated automatically during the calculation of the factorization of $Q_{k(1)}^T A_{k(1)} Q_{k(1)}$ from the factorization of $Q_0^T A_0 Q_0$. Further, if this intermediate factorization is stored, then it can replace expression (5.3) when the matrix $Q_{k(2)}^T A_{k(2)} Q_{k(2)}$ is factored. Thus, because we know in advance the first $(|\mathcal{L}_{k(0)}|-3)$ rows of the required Cholesky factor of $Q_{k(2)}^T A_{k(2)} Q_{(k(2)}$ instead of only the first $(|\mathcal{L}_0|-3)$ rows, the work of the factorization is reduced from $\mathcal{O}(|\bar{\mathcal{L}}_{k(2)}|\,|\mathcal{L}_{k(2)}|^2)$ to $\mathcal{O}(\{|\mathcal{L}_{k(2)}| - |\mathcal{L}_{k(0)}|\}|\mathcal{L}_{k(2)}|^2)$ operations. The choices of the sets $\{\mathcal{L}_k : k = 1, 2, \ldots, n\}$ in the experiments of Section 3 allow this technique to be applied often.

More generally, having calculated the factorization (5.11), we might wish to save the factorization of $Q_{k(0)}^T A_{k(0)} Q_{k(0)}$, where $\mathcal{L}_{k(0)}$ is a subset of $\mathcal{L}_k$, but the elements of $\mathcal{L}_{k(0)}$ are not necessarily the leading elements of $\mathcal{L}_k$. If they were the leading elements, then it would be suitable to store the lower triangular part of the first $(|\mathcal{L}_{k(0)}| - 3)$ rows of L_k. Otherwise, we would reorder the elements of $\mathcal{L}_k$, which would induce some modifications to Ω_k and L_k. Because any reordering can be expressed as a sequence of exchanges of adjacent integers of $\mathcal{L}_k$, we consider the updating of Ω_k and L_k when we swap the t-th and $(t+1)$-th elements of $\mathcal{L}_k$, where we assume $t > 3$, smaller values of t being irrelevant due to the condition $|\mathcal{L}_0| \geq 3$ that is given at the end of Section 2. It is remarkable that our updating method is going to depend only on L_k and on the Givens rotations in the product (5.4).

Let Π_t be the $|\mathcal{L}_k| \times |\mathcal{L}_k|$ permutation matrix that corresponds to the exchange, so the $(t, t+1)$-th and $(t+1, t)$-th elements of Π_t are one, but all other elements of one are on the diagonal. The notation for matrices that has been used already will denote matrices that occur before the exchange, while updated quantities will be distinguished by "hats". Therefore, for example, we have the identity $\hat{P}_k = \Pi_t P_k$, where P_k is introduced just before equation (5.4). Further, because the Givens rotation $\Omega_k^{(i,j)}$ is derived from the first j rows of P_k, we also have the identities

$$\hat{\Omega}_k^{(i,j)} = \Omega_k^{(i,j)}, \quad i = 1, 2, 3, \quad j = i+1, i+2, \ldots, t-1. \tag{6.1}$$

Now $\Omega_k^{(i,j)}$ depends just on the i-th and j-th rows of the current working space array $\bar{P}$ that is mentioned in the second paragraph of Section 5. Further, for $j>4$, the previous Givens rotations make the first $j-1$ rows of $\bar{P}$ upper triangular, with positive diagonal elements. Thus the first three rows of $\bar{P}$ coincide with the upper triangular Cholesky factor of the product $P_k^{(j-1)T} P_k^{(j-1)}$, where $P_k^{(j-1)}$ is the $(j-1)\times 3$ matrix whose rows are the first $(j-1)$ rows of P_k. Hence the rotations $\{\Omega_k^{(i,j)} : i=1,2,3\}$ are defined by the product $P_k^{(j-1)T} P_k^{(j-1)}$ and by the j-th row of P_k. It follows from the identity

$$\hat{P}_k^{(j-1)T} \hat{P}_k^{(j-1)} = P_k^{(j-1)T} P_k^{(j-1)}, \quad j>t+1, \tag{6.2}$$

that we also have the coincidences

$$\hat{\Omega}_k^{(i,j)} = \Omega_k^{(i,j)}, \quad i=1,2,3, \quad j=t+2,t+3,\ldots,|\mathcal{L}_k|. \tag{6.3}$$

Therefore, when we update the product (5.4), we alter only the six terms

$$\Omega_k^{(3,t+1)}\Omega_k^{(3,t)}\Omega_k^{(2,t+1)}\Omega_k^{(2,t)}\Omega_k^{(1,t+1)}\Omega_k^{(1,t)}, \tag{6.4}$$

which can be treated as adjacent because of the commutivity properties of the Givens rotations.

We continue to let $\underline{e}_j$ denote the j-th coordinate vector in $\mathcal{R}^{|\mathcal{L}_k|}$. Therefore the numbers that determine $\Omega_k^{(1,t+1)}$ and $\Omega_k^{(1,t)}$ are the ratios of the first, t-th and $(t+1)$-th components of the vector $\Omega_k^{(1,t)T}\Omega_k^{(1,t+1)T}\underline{e}_1$, so these ratios are available in the first column of the current $\bar{P}$ immediately before $\Omega_k^{(1,t)}$ and $\Omega_k^{(1,t+1)}$ are calculated. Further, the analogous vector after the switch is $\Pi_t\Omega_k^{(1,t)T}\Omega_k^{(1,t+1)T}\underline{e}_1$. It follows that the new rotations satisfy the condition

$$\hat{\Omega}_k^{(1,t+1)}\hat{\Omega}_k^{(1,t)}\Pi_t\Omega_k^{(1,t)T}\Omega_k^{(1,t+1)T}\underline{e}_1 = \underline{e}_1, \tag{6.5}$$

which provides a highly convenient way of generating $\hat{\Omega}_k^{(1,t)}$ and $\hat{\Omega}_k^{(1,t+1)}$ from $\Omega_k^{(1,t)}$ and $\Omega_k^{(1,t+1)}$. Equation (6.5), with the structures of Π_t and the Givens rotations, imply that the product

$$\hat{\Omega}_k^{(1,t+1)}\hat{\Omega}_k^{(1,t)}\Pi_t\Omega_k^{(1,t)T}\Omega_k^{(1,t+1)T} = \check{\Omega}_k^{(1,t)}, \tag{6.6}$$

say, is an orthogonal matrix that differs from the $|\mathcal{L}_k|\times|\mathcal{L}_k|$ unit matrix only in its (t,t)-th, $(t,t+1)$-th, $(t+1,t)$-th and $(t+1,t+1)$-th elements. We calculate the values of these elements, because, by extending the argument that yields equation (6.5), one can deduce the identity

$$\hat{\Omega}_k^{(2,t+1)}\hat{\Omega}_k^{(2,t)}\check{\Omega}_k^{(1,t)}\Omega_k^{(2,t)T}\Omega_k^{(2,t+1)T}\underline{e}_2 = \underline{e}_2, \tag{6.7}$$

which we use to generate $\hat{\Omega}_k^{(2,t)}$ and $\hat{\Omega}_k^{(2,t+1)}$. Further, it follows that the product

$$\hat{\Omega}_k^{(2,t+1)}\hat{\Omega}_k^{(2,t)}\check{\Omega}_k^{(1,t)}\Omega_k^{(2,t)T}\Omega_k^{(2,t+1)T} = \check{\Omega}_k^{(2,t)}, \tag{6.8}$$

say, is an orthogonal matrix that has the same structure as $\check{\Omega}_k^{(1,t)}$. We complete the updating of the Givens rotations by calculating $\hat{\Omega}_k^{(3,t)}$ and $\hat{\Omega}_k^{(3,t+1)}$ from the condition

$$\hat{\Omega}_k^{(3,t+1)}\hat{\Omega}_k^{(3,t)}\check{\Omega}_k^{(2,t)}\Omega_k^{(3,t)T}\Omega_k^{(3,t+1)T}\underline{e}_3 = \underline{e}_3. \tag{6.9}$$

Thus the amount of work of these updates is bounded above by a constant.

The details of the last paragraph are also important to the updating of the factorization (5.11) when the t-th and $(t+1)$-th elements of $\mathcal{L}_k$ are switched. Specifically, corresponding to the definitions (6.6) and (6.8), we introduce the matrix

$$\check{\Omega}_k^{(3,t)} = \hat{\Omega}_k^{(3,t+1)}\hat{\Omega}_k^{(3,t)}\check{\Omega}_k^{(2,t)}\Omega_k^{(3,t)T}\Omega_k^{(3,t+1)T}, \tag{6.10}$$

which has the same structure as $\check{\Omega}_k^{(1,t)}$ and $\check{\Omega}_k^{(2,t)}$. Equations (6.10), (6.8) and (6.6) imply the identity

$$\begin{aligned}\check{\Omega}_k^{(3,t)}&\Omega_k^{(3,t+1)}\Omega_k^{(3,t)}\Omega_k^{(2,t+1)}\Omega_k^{(2,t)}\Omega_k^{(1,t+1)}\Omega_k^{(1,t)}\\ &= \hat{\Omega}_k^{(3,t+1)}\hat{\Omega}_k^{(3,t)}\check{\Omega}_k^{(2,t)}\Omega_k^{(2,t+1)}\Omega_k^{(2,t)}\Omega_k^{(1,t+1)}\Omega_k^{(1,t)}\\ &= \hat{\Omega}_k^{(3,t+1)}\hat{\Omega}_k^{(3,t)}\hat{\Omega}_k^{(2,t+1)}\hat{\Omega}_k^{(2,t)}\check{\Omega}_k^{(1,t)}\Omega_k^{(1,t+1)}\Omega_k^{(1,t)}\\ &= \hat{\Omega}_k^{(3,t+1)}\hat{\Omega}_k^{(3,t)}\hat{\Omega}_k^{(2,t+1)}\hat{\Omega}_k^{(2,t)}\hat{\Omega}_k^{(1,t+1)}\hat{\Omega}_k^{(1,t)}\Pi_t,\end{aligned} \tag{6.11}$$

which we write in the form

$$\check{\Omega}_k^{(3,t)}\Omega_k^{(*,t+1)}\Omega_k^{(*,t)} = \hat{\Omega}_k^{(*,t+1)}\hat{\Omega}_k^{(*,t)}\Pi_t, \tag{6.12}$$

where we are using the notation

$$\left.\begin{aligned}\Omega_k^{(*,j)} &= \Omega_k^{(3,j)}\Omega_k^{(2,j)}\Omega_k^{(1,j)}\\ \hat{\Omega}_k^{(*,j)} &= \hat{\Omega}_k^{(3,j)}\hat{\Omega}_k^{(2,j)}\hat{\Omega}_k^{(1,j)}\end{aligned}\right\}, \quad j=4,5,\ldots,|\mathcal{L}_k|. \tag{6.13}$$

It follows from expressions (5.6), (6.1), (6.3), (6.12) and (6.13) and from some commutivity properties that we have the relation

$$\begin{aligned}\check{\Omega}_k^{(3,t)}\Omega_k^T &= \check{\Omega}_k^{(3,t)}\Omega_k^{(*,|\mathcal{L}_k|)}\Omega_k^{(*,|\mathcal{L}_k|-1)}\cdots\Omega_k^{(*,4)}\Omega_k^{(2,3)}\Omega_k^{(1,3)}\Omega_k^{(1,2)}\\ &= \Omega_k^{(*,|\mathcal{L}_k|)}\cdots\Omega_k^{(*,t+2)}\check{\Omega}_k^{(3,t)}\Omega_k^{(*,t+1)}\cdots\Omega_k^{(*,4)}\Omega_k^{(2,3)}\Omega_k^{(1,3)}\Omega_k^{(1,2)}\\ &= \hat{\Omega}_k^{(*,|\mathcal{L}_k|)}\cdots\hat{\Omega}_k^{(*,t)}\Pi_t\hat{\Omega}_k^{(*,t-1)}\cdots\hat{\Omega}_k^{(*,4)}\hat{\Omega}_k^{(2,3)}\hat{\Omega}_k^{(1,3)}\hat{\Omega}_k^{(1,2)}\\ &= \hat{\Omega}_k^{(*,|\mathcal{L}_k|)}\cdots\hat{\Omega}_k^{(*,4)}\hat{\Omega}_k^{(2,3)}\hat{\Omega}_k^{(1,3)}\hat{\Omega}_k^{(1,2)}\Pi_t = \hat{\Omega}_k^T\Pi_t.\end{aligned} \tag{6.14}$$

Further, because Q_k is generated by deleting the first three columns of Ω_k, the structure of $\check{\Omega}_k^{(3,t)}$ and equation (6.14) provide the formula

$$\dot{\Omega}_k^{(t-3,t-2)}Q_k^T = \hat{Q}_k^T\Pi_t, \tag{6.15}$$

where $\dot{\Omega}_k^{(t-3,t-2)}$ is the bottom right $(|\mathcal{L}_k|-3)\times(|\mathcal{L}_k|-3)$ submatrix of $\check{\Omega}_k^{(3,t)}$. Therefore the only elements of $\dot{\Omega}_k^{(t-3,t-2)}$ that differ from those of the unit matrix are in the 2×2 block that includes the $(t-3)$-th and $(t-2)$-th diagonal elements. Further, we obtain the values of these elements from the definitions (6.10), (6.8) and (6.6), which is a straightforward calculation.

Now, corresponding to equation (5.11), we require the Cholesky factorization of the matrix $\hat{Q}_k^T\hat{A}_k\hat{Q}_k$, where $\hat{A}_k = \Pi_t A_k \Pi_t$. It is therefore highly useful that expressions (6.15) and (5.11) imply the identity

$$\begin{aligned}\hat{Q}_k^T\hat{A}_k\hat{Q}_k = \hat{Q}_k^T\Pi_t A_k \Pi_t \hat{Q}_k &= \dot{\Omega}_k^{(t-3,t-2)} Q_k^T A_k Q_k \dot{\Omega}_k^{(t-3,t-2)T} \\ &= (\dot{\Omega}_k^{(t-3,t-2)} L_k)(\dot{\Omega}_k^{(t-3,t-2)} L_k)^T. \end{aligned} \tag{6.16}$$

Indeed, we form the matrix $\dot{\Omega}_k^{(t-3,t-2)} L_k$, which alters only two rows of L_k and which introduces only one departure from lower triangularity, namely the introduction of a $(t-3,t-2)$-th nonzero element. Therefore the new Cholesky factor has the form

$$\hat{L}_k = \dot{\Omega}_k^{(t-3,t-2)} L_k \bar{\Omega}_k^{(t-3,t-2)}, \tag{6.17}$$

where $\bar{\Omega}_k^{(t-3,t-2)}$ is the Givens rotation that restores the required triangularity. Thus, although our analysis of switching two adjacent elements of $\mathcal{L}_k$ is rather long, the total number of computer operations is only $\mathcal{O}(|\mathcal{L}_k|)$. In practice, therefore, it is not expensive to reorder the elements of $\mathcal{L}_k$, which provides good opportunities for saving work when the coefficients $\{\nu_{ki} : i \in \bar{\mathcal{L}}_k\}$ are calculated for every k.

7 Conclusions

Tables 3.1–3.3 show that usually our algorithm requires far fewer than n iterations to satisfy the interpolation equations (1.1) to high accuracy. Indeed, the numbers of iterations are so small that the calculation of all the residuals (4.11) on each iteration can be tolerated. We recall from Section 4, however, that several other actions occur on every iteration. Further, the preliminary work of Sections 5 and 6 is also necessary. Therefore we will consider the amounts of computation of all of these tasks.

Several parts of the computations were timed throughout the experiments of Section 3, the calculations being performed in double precision Fortran on a Sparc 2 workstation. In particular, because the computer operations that generate the residuals (4.11) are independent of the other details of Section 4, we recorded the time of each iteration excluding the residual calculations. These times are given in seconds in the main part of Table 7.1 for the experiments of Tables 3.1 and 3.2, while the times to determine all the residuals (4.11), using the form (4.1) of $s^{(j)}(\cdot)$, are displayed in the last row of the table. We see that the residual calculations dominate the work of each iteration and that the dominance becomes stronger as n increases. Therefore the table corroborates the claim, made at the end of Section 4, that it is suitable to compare the iteration counts of our algorithm with those of the conjugate gradient method.

The updating techniques of Section 6 are included in the preliminary calculations of the experiments of Tables 3.1 and 3.2. Thus we obtained the times that are presented in Table 7.2. The amounts of computation of these calculations seem to be acceptably small when n is large, because all the entries in the last column of Table 7.2 are less than the time to perform two iterations. This conclusion, however, depends on the fact that the last row of Table 7.1 is $\mathcal{O}(n^2)$, but, if Laurent expansions are used to approximate the

$\lvert\mathcal{L}_0\rvert$	Δ	$n=128$	$n=256$	$n=512$	$n=1024$	$n=2048$
8	1.5	0.01	0.02	0.04	0.07	0.14
16	1.5	0.01	0.03	0.06	0.12	0.24
32	1.5	0.02	0.05	0.11	0.22	0.45
8	2.0	0.01	0.02	0.03	0.07	0.15
16	2.0	0.02	0.03	0.06	0.13	0.25
32	2.0	0.03	0.06	0.11	0.22	0.46
8	4.0	0.01	0.02	0.04	0.09	0.18
16	4.0	0.02	0.03	0.06	0.14	0.28
32	4.0	0.02	0.06	0.11	0.24	0.48
Residuals		0.09	0.38	1.57	6.50	26.01

Table 7.1: Times per iteration excluding residuals for Tables 3.1 and 3.2

$\lvert\mathcal{L}_0\rvert$	Δ	$n=128$	$n=256$	$n=512$	$n=1024$	$n=2048$
8	1.5	0.18	0.45	1.11	2.66	6.34
16	1.5	0.23	0.60	1.38	3.37	7.76
32	1.5	0.28	0.74	1.94	4.93	11.24
8	2.0	0.32	0.86	2.08	5.09	12.13
16	2.0	0.35	0.97	2.69	6.03	14.86
32	2.0	0.40	1.17	3.21	8.19	19.88
8	4.0	0.77	2.09	5.72	13.77	35.23
16	4.0	0.73	2.11	5.95	14.99	36.09
32	4.0	0.68	2.28	6.52	17.41	43.99

Table 7.2: Times of the preliminary calculations of Tables 3.1 and 3.2

residuals (4.11), then usually the growth is $\mathcal{O}(n \log n)$. Therefore we expect to achieve much better efficiency by using Laurent expansions. Further, if this better efficiency is achieved, then we should give careful attention not only to the numbers of iterations but also to the complexity of the preliminary work.

Good motitivation for such investigations is provided by a comparison of our algorithm with the solution of the interpolation equations by a direct method. Indeed, we deduce from Tables 3.1, 7.1 and 7.2 that, when $\lvert\mathcal{L}_0\rvert=32$, $\Delta=4.0$ and $n=2048$ for example, the running time of our algorithm has the value

$$43.99 + (0.48 + 26.01) \times 6 \text{ seconds} \approx 3.4 \text{ minutes}. \tag{7.1}$$

On the other hand, when $n=2048$, the Cholesky factorization of the matrix Q^TAQ of the system (1.7) takes about 17 minutes. Further, Table 1 in [8] suggests that the Laurent method would reduce the time to calculate the residuals in expression (7.1) from 26.01 to about 8 seconds. Thus the dependence of the amount of preliminary work on $|\mathcal{L}_0|$ and Δ, which is shown in Table 7.2, becomes important. Moreover, the advantages of our algorithm over direct methods become much stronger as n increases.

The number of operations in the preliminary calculations of Section 5 is $\mathcal{O}(|\mathcal{L}_0|^3 + \sum_{k=1}^{n}\{|\mathcal{L}_k|-|\mathcal{L}_0|\}\,|\mathcal{L}_k|^2)$, but the $|\mathcal{L}_0|^3$ part of this expression is negligible. Further, if the structures of the sets $\{\mathcal{L}_k : k=1,2,\ldots,n\}$ provide suitable intersection properties, then the updating techniques of Section 6 can reduce the work to $\mathcal{O}(\sum_{k=1}^{n}|\mathcal{L}_k|^2)$ operations. Thus the explicit dependence of the amount of preliminary work on the value of $|\mathcal{L}_0|$ becomes slight when the updating techniques can be used extensively, which explains the rather small differences between the entries in the last three rows of Table 7.2. On the other hand, $|\mathcal{L}_0|$ is highly relevant to the number of coefficients that have to be stored in order to apply formula (4.3). Indeed, we have found that the structure (2.7) allows us to retain only $\sum_{k=1}^{n}\{|\mathcal{L}_k|-|\mathcal{L}_0|\}$ coefficients.

It is suggested at the end of Section 2 that it may be suitable to compensate for edge effects by including enough points in the set $\{\underline{x}_k : k\in\mathcal{L}_0\}$. Therefore the previous paragraph exposes both good and bad news. It is helpful that increases in $|\mathcal{L}_0|$ do not demand more storage space for coefficients, but the times of the preliminary calculations may become much longer, due to the $\mathcal{O}(\sum_{k=1}^{n}|\mathcal{L}_k|^2)$ overhead that occurs when the factorization (5.11) is required for every k. Therefore we note that the following change to the structure (2.7) may allow a substantial reduction in the preliminary work.

We ask whether each of the interpolation points $\{\underline{x}_k : k=1,2,\ldots,n\}$ is so close to the edge of the data that $|\mathcal{L}_k|$ has to be relatively large to accommodate edge effects. Thus we partition the data points into two disjoint subsets, $\{\underline{x}_k : k\in\mathcal{I}\}$ and $\{\underline{x}_k : k\in\mathcal{E}\}$ say, where, if the distribution of the data points is fairly even, the number of points in the "edge set" $\mathcal{E}$ may typically be less than $20\sqrt{n}$. Then we replace expression (2.7) by the structure

$$\left.\begin{array}{ll}\mathcal{L}_k = \mathcal{L}_0^{(\mathcal{I})}\cup\bar{\mathcal{L}}_k, & k\in\mathcal{I}\\ \mathcal{L}_k = \mathcal{L}_0^{(\mathcal{E})}\cup\bar{\mathcal{L}}_k, & k\in\mathcal{E}\end{array}\right\}. \tag{7.2}$$

In other words, we now choose one set $\mathcal{L}_0$ for the "interior" points and another set $\mathcal{L}_0$ for the "edge" points, because usually, for large n, we expect the value of $|\mathcal{L}_0^{(\mathcal{I})}|$ to be much less than the old $|\mathcal{L}_0|$. We allow $\mathcal{L}_0^{(\mathcal{I})}$ and $\mathcal{L}_0^{(\mathcal{E})}$ to have some common elements. It is important that we can continue to store only the parameters $\{\nu_{ik} : i\in\bar{\mathcal{L}}_k\}$ for every k. Indeed, it is explained below that these parameters are sufficient for an extension to the procedure of Section 4 that calculates all the coefficients of the initial approximation (1.12). There is no need to give an extension for the correction formula (1.13), because the difference $s^{(j+1)}(\cdot)-s^{(j)}(\cdot)$ is equivalent to expression (1.12) if the data $\{f_k : k=1,2,\ldots,n\}$ are replaced by the residuals $\{f_k - s^{(j)}(\underline{x}_k) : k=1,2,\ldots,n\}$.

Formula (1.12) shows that $s^{(1)}(\cdot)$ is the sum of the functions

$$s^{(1;\mathcal{I})}(\cdot) = \sum_{k\in\mathcal{I}} f_k\,\ell_k(\cdot\,;\mathcal{L}_k) \quad\text{and}\quad s^{(1;\mathcal{E})}(\cdot) = \sum_{k\in\mathcal{E}} f_k\,\ell_k(\cdot\,;\mathcal{L}_k). \tag{7.3}$$

Further, we see that the procedure of Section 4 would yield the parameters of the function

$$s^{(1;\mathcal{I})}(\underline{x}) = \sum_{i=1}^{n} \lambda_i^{(1;\mathcal{I})} \|\underline{x}-\underline{x}_i\|_2^2 \log \|\underline{x}-\underline{x}_i\|_2 + p^{(1;\mathcal{I})}(\underline{x}), \quad \underline{x}\in\mathcal{R}^2, \tag{7.4}$$

if we replaced $\mathcal{L}_0$ by $\mathcal{L}_0^{(\mathcal{I})}$, if we retained the data $\{f_k : k\in\mathcal{I}\}$ and if we set the remaining function values $\{f_k : k\in\mathcal{E}\}$ to zero. Because ν_{ki} is multiplied by f_k in the first line of formula (4.3), it follows that no value of ν_{ki} is needed when $k\in\mathcal{E}$, so it does not matter that we may not have stored some coefficients that would have been required if f_k were nonzero. Similarly, when we generate the parameters of the expression

$$s^{(1;\mathcal{E})}(\underline{x}) = \sum_{i=1}^{n} \lambda_i^{(1;\mathcal{E})} \|\underline{x}-\underline{x}_i\|_2^2 \log \|\underline{x}-\underline{x}_i\|_2 + p^{(1;\mathcal{E})}(\underline{x}), \quad \underline{x}\in\mathcal{R}^2, \tag{7.5}$$

by the method of Section 4, there is no need for the coefficients $\{\nu_{ki} : k\in\mathcal{I}\}$, because we set the function values $\{f_k : k\in\mathcal{I}\}$ to zero. Further, these zero function values usually imply that most of the multipliers $\{\lambda_i^{(1;\mathcal{E})} : i=1,2,\ldots,n\}$ are also zero, which can save much work when the right hand sides (4.8) are calculated. Of course, $s^{(1)}(\cdot)$ has the thin plate spline coefficients

$$\lambda_i^{(1)} = \lambda_i^{(1;\mathcal{I})} + \lambda_i^{(1;\mathcal{E})}, \quad i=1,2,\ldots,n, \tag{7.6}$$

and its polynomial term is the sum $p^{(1;\mathcal{I})}(\cdot)+p^{(1,\mathcal{E})}(\cdot)$.

We have not yet had time to develop automatic ways of choosing the sets $\{\mathcal{L}_k : k=1,2,\ldots,n\}$. Initially we believed that we should aim to satisfy the conditions (2.5), in order to guarantee the convergence of the iterative procedure. We find in Section 3, however, that the number of iterations is quite small for some very crude choices of the sets. In general, therefore, we expect many choices to be adequate, so we are hopeful that $|\mathcal{L}_k|\leq 100$ can be achieved, at least for the "interior" interpolation points. Thus it may be possible to bound the amount of preliminary work by a constant multiple of n. Further, it may be helpful if most of the sets $\{\mathcal{L}_k : k=1,2,\ldots,n\}$ are independent of the position of the edge of the area that is covered by the interpolation points. Therefore we are studying the idea of relaxing the conditions (1.11) by a least squares method, in order to create some freedom that is used to force the function (1.14) to have suitable decay properties. For example, if $|\mathcal{L}_k|$ is between 20 and 30, then we have found that it is suitable to satisfy $|\ell_k(\underline{x};\mathcal{L}_k)|=\mathcal{O}(\|\underline{x}\|^{-3})$ when $\|\underline{x}\|$ is large.

Our research so far shows that the use of approximations to Lagrange functions provides a very promising iterative procedure for the solution of the thin plate spline interpolation equations. Indeed, the results of the numerical experiments are highly encouraging and the amount of computation compares favourably with the work of preconditioned conjugate gradient methods. Therefore we expect interpolation to tens of thousands of data in two dimensions to become a routine calculation.

References

[1] R.K. Beatson and G.N. Newsam (1992), "Fast evaluation of radial basis functions: I", *Comput. Math. Applic.*, Vol. 24, pp. 7–19.

[2] M.D. Buhmann (1990), "Multivariate cardinal interpolation with radial basis functions", *Constr. Approx.*, Vol. 6, pp. 225–255.

[3] J. Duchon (1977), "Splines minimizing rotation-invariant seminorms in Sobolev spaces", in *Constructive Theory of Functions of Several Variables, Lecture Notes in Mathematics 571*, eds. W. Schempp and K. Zeller, Springer-Verlag (Berlin), pp. 85–100.

[4] N. Dyn, D. Levin and S. Rippa (1986), "Numerical procedures for surface fitting of scattered data by radial functions", *SIAM J. Sci. Statist. Comput.*, Vol. 7, pp. 639–659.

[5] G.H. Golub and C.F. Van Loan (1983), *Matrix Computations*, The Johns Hopkins University Press (Baltimore).

[6] C.A. Micchelli (1986), "Interpolation of scattered data: distance matrices and conditionally positive definite functions", *Constr. Approx.*, Vol. 2, pp. 11–22.

[7] M.J.D. Powell (1981), *Approximation Theory and Methods*, Cambridge University Press (Cambridge).

[8] M.J.D. Powell (1993), "Some algorithms for thin plate spline interpolation to functions of two variables", Report No. DAMTP 1993/NA6, University of Cambridge.

[9] R. Sibson and G. Stone (1991) "Computation of thin-plate splines", *SIAM J. Sci. Statist. Comput.*, Vol. 12, pp. 1304–1313.

R.K. Beatson
Mathematics Department
University of Canterbury
Christchurch 1, New Zealand

M.J.D. Powell
Department of Applied Mathematics and Theoretical Physics
University of Cambridge
Silver Street
Cambridge CB3 9EW, England

I S DUFF

The solution of augmented systems

Abstract We examine the solution of sets of linear equations for which the coefficient matrix has the form

$$\begin{pmatrix} H & A \\ A^T & 0 \end{pmatrix}$$

where the matrix H is symmetric. We are interested in the case when the matrices H and A are sparse.

These augmented systems occur in many application areas, for example in the solution of linear programming problems, structural analysis, magnetostatics, differential algebraic systems, constrained optimization, electrical networks, and computational fluid dynamics. We discuss in some detail how they arise in the last three of these applications and consider particular characteristics and methods of solution.

We then concentrate on direct methods of solution. We examine issues related to conditioning and scaling, and discuss the design and performance of a code for solving these systems.

1 Introduction

The augmented matrix

$$\begin{pmatrix} H & A \\ A^T & 0 \end{pmatrix} \tag{1.1}$$

is ubiquitous in the numerical solution of problems in applied mathematics (see, for example, [36]). Since we assume that H and A are sparse, then clearly the matrix (1.1) is even more so. We review a few simple properties of the matrix (1.1) in Section 2 and discuss, in Section 3, applications which give rise to the associated linear system

$$\begin{pmatrix} H & A \\ A^T & 0 \end{pmatrix} \begin{pmatrix} x \\ y \end{pmatrix} = \begin{pmatrix} b \\ c \end{pmatrix}. \tag{1.2}$$

For historical or technical reasons, different approaches to the solution of system (1.2) are used in different application areas. We examine three major application areas and indicate system characteristics and usual solution techniques in each case. In many cases, H is positive definite and is often even diagonal. Also c is very often zero. This additional structure will also be exploited.

We introduce a direct method for the solution of system (1.2) in Section 4 and, in Section 5, consider issues related to the conditioning and scaling of the system. In Section 6, we describe briefly a new code, based on a multifrontal approach but respecting the structure of (1.1). We study the performance of this new code, MA47 from the Harwell Subroutine Library, in Section 7 and present some concluding remarks in Section 8.

2 Basic properties of augmented systems

There are some important facts that should be mentioned concerning the augmented system (1.2). We will assume throughout this paper that H is a square matrix of order m and that A has dimensions $m \times n$ with $m \geq n$. We will use the term *aspect ratio* to denote the ratio of m to n. We will see in Section 3 that the aspect ratio can have a strong influence on the choice of solution technique.

Clearly the matrix (1.1) is symmetric and indefinite. If A has full rank and H is nonsingular, then (1.1) is clearly nonsingular. Indeed, if H is positive definite, we see from the identity

$$\begin{pmatrix} H & A \\ A^T & 0 \end{pmatrix} = \begin{pmatrix} H & 0 \\ A^T & I \end{pmatrix} \begin{pmatrix} H^{-1} & 0 \\ 0 & -A^T H^{-1} A \end{pmatrix} \begin{pmatrix} H & A \\ 0 & I \end{pmatrix}$$

that, by Sylvester's Law of Inertia, we have m positive eigenvalues and n negative eigenvalues. Indeed, if μ_1 and μ_m are the largest and smallest eigenvalues of H and σ_1 and σ_n the largest and smallest singular values of A, then Rusten and Winther [33] show that the spectrum of the matrix (1.1) is contained in the union of the intervals I^- and I^+, where

$$I^- = [\tfrac{1}{2}(\mu_m - \sqrt{\mu_m^2 + 4\sigma_1^2}), \tfrac{1}{2}(\mu_1 - \sqrt{\mu_1^2 + 4\sigma_n^2})] \text{ and } I^+ = [\mu_m, \tfrac{1}{2}(\mu_1 + \sqrt{\mu_1^2 + 4\sigma_1^2})].$$

It is not necessary for H to be positive definite or even nonsingular for the augmented matrix to be nonsingular. A sufficient (and necessary) condition is that A has full column rank (rank(A)=n) and the columns of $\begin{pmatrix} H \\ A^T \end{pmatrix}$ are linearly independent. In this paper, we will assume that the matrix (1.1) is nonsingular. A good reference for the basic properties of augmented systems, which includes the case where A is not full rank, is given by Wedin [38].

Although we will never want to compute the inverse explicitly, it is useful to display it to help later in the understanding of ill conditioning. If H is nonsingular, the inverse of (1.1) is given by

$$\begin{pmatrix} S & B^T \\ B & -(A^T H^{-1} A)^{-1} \end{pmatrix} \tag{2.1}$$

where B is a left-generalized inverse of A given by $(A^T H^{-1} A)^{-1} A^T H^{-1}$ and S can be written as $H^{-1}(I - AB)$, where $(I - AB)$ is the orthogonal projector onto range$(A)^\perp$ (the null space of A^T).

In many cases, the second component in the right-hand side of system (1.2), c, is zero. Although this does not affect the augmented matrix or the properties that we have just described, it can have a significant effect on solution techniques for (1.2). A simple indication of this is that the (2,2) matrix of (2.1) will have no influence on the solution of the system. Some solution techniques fail to respect this and can suffer the consequences, particularly if the (2,2) block is ill conditioned. We return to this theme in Section 5.

3 Occurrence of augmented systems

Several authors (for example, [22], [35], [36], and [37]) have emphasized the importance of the system (1.2) in applied mathematics, indeed Strang [35] calls this system the "fundamental system" and claims that it is indeed so. Principal areas in which such a system arises naturally are optimization, computational fluid dynamics, and electrical networks. We study these in more detail in Sections 3.1 to 3.3. Other important application areas include structural analysis, differential algebraic systems, heat equilibrium, and magnetostatics.

The language for systems of the form (1.2) is also highly dependent on the application area under consideration. In general, we have used the nomenclature from constrained optimization.

3.1 Optimization

Systems of the form (1.2) abound in applications in optimization. The most obvious example is in the minimization

$$\min(\tfrac{1}{2}(Hx, x) - (b, x)) \text{ subject to } A^T x = c$$

where $(\cdot\ ,\ \cdot)$ denotes the Euclidean inner product. The role of the variable y in (1.2) is more clearly seen if the above constrained minimization is expressed as finding the saddle point of the Lagrangian form

$$\tfrac{1}{2}(Hx, x) - (b, x) + (y, A^T x - c)$$

where the variables y are clearly the Lagrange multipliers.

A common name for system (1.2) in optimization is the KKT system, for three people involved in obtaining necessary optimality conditions for differentiable problems using this system, Karush, Kuhn, and Tucker [29]. More generally, iterative methods for linearly and nonlinearly constrained minimization problems frequently solve a sequence of such quadratic minimization problems. Such techniques are normally known as recursive (RQP) or successive (SQP) quadratic programming methods, see for example Gill, Murray, and Wright [25]. In general, A is a matrix of constraint normals while H is a second derivative matrix or an approximation to it. In many cases H is diagonal, for example when solving linear programming problems by interior point methods [28] or when solving weighted least-squares problems using the augmented system approach. Since we are normally solving a nonlinear system and the solution to the quadratic minimization gives the direction along which the current estimate is adjusted, we force this direction to be in the subspace orthogonal to the constraint normals so that c is zero. We will assume that this is so for the remainder of this subsection. Usually A has full column rank and H is symmetric semi-definite. The aspect ratio can occasionally be quite large but usually it is less than 4. Because optimizers like to use A as the Jacobian of the constraint matrix, they perversely use a notation for (1.1) with the transpose on the (1,2) block rather than on the (2,1) block.

There are two main methods for solving these systems in optimization, the range-space method and the null-space method. In range-space methods, variables corresponding to the (1,1) block are eliminated and the variables y are obtained from the equation

$$(A^T H^{-1} A) y = (A^T H^{-1}) b \tag{3.1}$$

which, of course requires H to be nonsingular. The equations (3.1) are then solved by either an iterative method (for example, conjugate gradients) or a direct method (for example, sparse Cholesky). The main problems with this approach are that the system (3.1) can be much more ill conditioned than the original system and the coefficient matrix can be much denser than the augmented system.

In the null-space method use is made of the fact that, since $A^T x = 0$, x lies in the null space of A^T and so can be written as a linear combination of basis vectors of this null space. If these basis vectors are denoted by the matrix Z, then x can be written as Zs, for some vector s. If we then substitute in the equations in the first block of (1.2) we get the system

$$(Z^T H Z) s = Z^T b. \tag{3.2}$$

The coefficient matrix of (3.2) is called the reduced or projected Hessian. Potential problems with this approach are to obtain a good basis Z and the lack of sparsity in the reduced Hessian.

There are two cases where augmented systems arise in optimization that are so important that it is worth mentioning them explicitly. The first is in the subproblem at each stage of an interior point method for the solution of linear programming problems ([28]). The matrix in this case has the form

$$\begin{pmatrix} D^{-2} & A \\ A^T & 0 \end{pmatrix}$$

where the diagonal matrix D has entries $\{x_1, x_2, \ldots, x_n\}$ which are all greater than zero. However, as we approach a solution, many of these components (which are the current estimate of the solution) will tend to zero so that the scaling of the D matrix can become very bad, although work of Stewart [34] shows that the overall system is not so ill conditioned (see also discussion in [37]). Largely because of the economic importance of linear programming, this application has received great attention and many methods of solution have been proposed; including methods based on normal equations, QR, and conjugate gradients, as well as augmented systems. The use of the latter approach in this context is discussed by [11] and [23] inter alios.

Another special case is (unweighted) linear least-squares where the (1,1) matrix is the identity. This case and that of the weighted least-squares just discussed in the interior point subproblem are extensively discussed in works by Björck (for example, [4], [5]). We will consider this special case further in our discussions in Sections 5–8.

Iterative methods are used widely in solving linear systems arising in optimization although they have not been used so often on system (1.1) because it is indefinite. Recent work, for example [2], has investigated extensions of conjugate gradients for this case.

3.2 Computational fluid dynamics

Although we concentrate in this section on applications from computational fluid dynamics, there are many other areas in differential equations giving rise to augmented systems. They occur in elasticity and heat equilibrium problems and indeed whenever second order elliptic equations are solved using a mixed formulation. There are several ways in which system (1.1) arises in computational fluid dynamics. The most obvious way is in the solution of the Stokes' equations for incompressible fluid flow

$$\begin{aligned} -\nabla^2 u + \nabla p &= f \\ \nabla u &= 0 \end{aligned}$$

with appropriate boundary conditions. The matrix H corresponds to the Laplacian and A comes from the discretization of the convective term that also appears in the continuity equations. The variables x are the velocity variables (u) and y are the pressure variables (p). Since in the finite element discretization the velocity function bases are typically quadratic over each element, while the pressure function bases are usually linear, the aspect ratio is sometimes as much as 20, particularly for three-dimensional problems (see, for example [31]). Another characteristic of CFD equations is that the matrix H has very large order, often in the hundreds of thousands.

In these applications, in contrast to the general optimization case, the projected matrix $A^T H^{-1} A$ is not only well conditioned but has a condition number independent of the mesh size used in the discretization. The reduced system (3.1), used in pressure correction methods, is commonly solved using an iterative method like conjugate gradients. One problem with this approach is that it involves solving systems with H as coefficient matrix at every iteration of the conjugate gradient iteration. Since the system in H is often itself solved using an iterative method, we obtain an inner-outer iteration. If conjugate gradients is used for the outer iteration then an accurate answer is required from the inner iteration. However, recently Elman and Golub [19] view the so-called Uzawa method as a Richardson iteration on the reduced system and show that there is no need for an accurate solution of the inner iteration. Because the reduced system is not formed explicitly (it would involve too much work and would normally be dense), it is hard to obtain a good preconditioner. Elman and Golub use diagonal and tridiagonal approximations to the factors of the mass matrix associated with the pressures. Another possibility is to use the matrix $A^T A$ as a preconditioner. Since this matrix is symmetric but otherwise not easy to solve, another iteration is normally used to implement this preconditioner. Recently, Ng et al. [31] have advocated using a direct method to solve the preconditioning equations since, unlike the reduced systems themselves, they can be very ill conditioned. A nice discussion of preconditioning the full augmented system is presented in [33]. Ramage and Wathen [32] have experimented with solving the complete augmented system using the method of conjugate residuals with diagonal scaling.

3.3 Electrical networks

We comment on the application of electrical networks more briefly because they can produce augmented systems with quite different characteristics from those of the previous applications. We consider the case of purely resistive networks and show that augmented systems arise quite simply through an application of Ohm's Law and Kirchhoff's Law. The first Law relates resistances to current and voltages and the corresponding equations can be written

$$Dx - Au = b,$$

while the second Law, which ensures current conservation at the nodes, can be written

$$A^T x = 0,$$

where the variables x, u, and b are the currents in the edges, voltages at the nodes, and external voltages (batteries) respectively, D is a diagonal matrix of resistances, and A is the adjacency matrix for circuit connectivity ($a_{ij} = 1(-1)$ indicates that edge i originates (ends) at node j).

There are several differences between the augmented system consisting of the above equations and those we have considered earlier. The matrix A has only -1 or 1 as possible nonzero entries and its sparsity structure is determined by the physical connectivity of the network. Usually, the matrix D is very badly scaled (if an open circuit is included the corresponding value of D would be infinite while components corresponding to short circuits would be zero (or nearly so)).

Both Strang [36] and Vavasis [37] use electrical networks to introduce augmented systems and the latter develops a variant of his method particularly designed for the numerical character of A. Traditional solution methods are to use an LDL^T factorization with pivoting, but this can be very unstable if the scaling in D is bad enough.

4 Direct methods of solution

As we have seen earlier, particularly in Section 3.2, sometimes the systems can be so large that a direct solution technique by itself is impractical. However, there are many problems, even in computational fluid dynamics, where direct methods are competitive and additionally it is possible to combine such methods with iterative ones when problem sizes become huge. We restrict the rest of our discussion to direct methods of solution.

The most common approach is to form and solve the reduced equations (3.1), using a Cholesky or LDL^T factorization on H to generate the Schur complement matrix $-A^T H^{-1} A$. Cholesky or LDL^T factorization is then used to factorize the Schur complement matrix. This method is only applicable if H is nonsingular and even then can be unstable. Additionally the Schur complement is usually quite dense (and would be full if A had a full row), so that sparse techniques may be inefficient in the second factorization. It is, however, the basis for many of the methods discussed earlier and sometimes the dimensions of the Schur complement are small so the fill-in to the Schur complement is not a problem.

Another approach is to ignore symmetry and use an unsymmetric sparse solver. At first glance this may appear daft but it looks more attractive if the structure of the overall system is considered. In particular, if we are solving a (weighted) least-squares problem, c is zero and we normally wish to solve (1.2) only for the variables y. If the residual, corresponding to variables x, is required we would normally obtain it from the direct computation $b - Ay$. Note that if we choose early pivots in Gaussian elimination from rows corresponding to zero right-hand side components and later pivots from columns corresponding to the variables y then much of the factorization need not be used to obtain the solution. Unfortunately, taking advantage of this in our pivoting algorithm may conflict with preserving sparsity so a compromise must be achieved. In some early experiments [15] this method did show promise but was generally inferior to symmetric factorization of augmented systems or normal equations.

QR based methods will normally suffer from too much fill-in but they are of interest for cases with particular structure in A, for example if A is block diagonal. This class has been examined recently by [27] who have developed a modified form of QR factorization although they have not been concerned with sparsity.

The system (1.2) is indefinite so it would be possible to use a modified Cholesky scheme similar to those proposed for such systems in the optimization context (for example, [20]). However, these methods do not produce an accurate solution of (1.2) and do not respect the structure.

We conclude this section by introducing the approach that we will consider in the remainder of this paper. That is to use an LDL^T factorization of (1.1) where the matrix D is block diagonal (blocks of order 1 and 2) and L is block lower triangular. Our method will inherit much of the stability of the scheme for full systems proposed by Bunch [7], [8]. Clearly the use of 2×2 pivots is necessary, as can be seen from the matrix

$$\begin{pmatrix} 0 & \times \\ \times & 0 \end{pmatrix}. \tag{4.1}$$

The above papers show also that 2×2 pivots are sufficient to ensure stability assuming numerical pivoting is used. As usual with sparse factorizations, we relax the numerical pivoting controls in order to preserve sparsity better. If the augmented matrix (1.1) is denoted by the matrix M with entries m_{ij}, we choose a 1×1 pivot if

$$|m_{kk}| \geq u . \max_{j \neq k} |m_{kj}|, \tag{4.2}$$

and a 2×2 pivot

$$\left| \begin{pmatrix} m_{kk} & m_{k\,k+1} \\ m_{k+1\,k} & m_{k+1\,k+1} \end{pmatrix}^{-1} \right| \begin{pmatrix} \max_{j \neq k,k+1} |m_{kj}| \\ \max_{j \neq k,k+1} |m_{k+1j}| \end{pmatrix} \leq \begin{pmatrix} u^{-1} \\ u^{-1} \end{pmatrix}, \tag{4.3}$$

where u is a threshold parameter ($0 \leq u \leq 0.5$), subject to the entry being suitable on sparsity grounds. We will discuss some aspects of sparsity pivoting later in Section 7.

5 Conditioning and Scaling

Clearly the use of the augmented system should be better for stability than the normal equations. We can see this trivially by observing that the pivot choice of selecting pivots down the diagonal of H first gives the normal equations; so normal equations are a particular case of using the augmented systems. On another level, we note that, although the normal equations appear in the inverse of the augmented system (the (2, 2) block in (2.1)), it was already observed that this block is not involved in the solution if c is zero. We now study the conditioning and error estimation for systems (1.1) where we assume H is diagonal and c is zero. We would like any condition number or error estimate to reflect the structure, and we find that the work in [3] is ideally suited since most of the second block of equations will lie in their category 2 because often the residual is close to zero and we have assumed that c is. If we then define two scaled residuals ω_1 and ω_2, where ω_1 corresponds to the category 1 equations and ω_2 to the others, with corresponding condition numbers K_{ω_1} and K_{ω_2} then the estimate of the error in the solution is given by $\omega_1 K_{\omega_1} + \omega_2 K_{\omega_2}$.

As an example, we compare this estimate on a sequence of three problems where H is γI, $\gamma = 1$, 10^3, 10^5 and A is the matrix WELL1850 from the Harwell-Boeing test set [14]. We have the results shown in Table 5.1, where K_∞ is the classical ∞-norm condition number.

Table 5.1: Comparison of error estimation

K_∞	$\omega_1 K_{\omega_1} + \omega_2 K_{\omega_2}$	actual error
3.10^{18}	9.10^{-1}	1.10^{-1}
1.10^{16}	2.10^{-10}	8.10^{-12}
2.10^{10}	1.10^{-10}	7.10^{-12}

The scaling of such systems that arise in the solution of sparse least-squares problems was addressed at length by Björck at the last Dundee meeting [6]. He uses an α scaling of the form

$$\begin{pmatrix} I & 0 \\ 0 & \alpha^{-1} I \end{pmatrix} \begin{pmatrix} I & A \\ A^T & 0 \end{pmatrix} \begin{pmatrix} \alpha I & 0 \\ 0 & I \end{pmatrix}$$

to give the scaled system

$$\begin{pmatrix} \alpha I & A \\ A^T & 0 \end{pmatrix} \begin{pmatrix} r/\alpha \\ x \end{pmatrix} = \begin{pmatrix} b \\ 0 \end{pmatrix}.$$

The main unresolved issue is how to choose *a priori* a suitable value for α. We are currently experimenting with some ideas for this and presently favour a scheme that uses a crude method to choose an initial α which is then refined if necessary using information from the first attempted solution. We will report on this separately [12] but should mention here that we have found it very beneficial to first scale the matrix (1.1) using the Harwell Subroutine Library scaling routine MC30 (a symmetric variant of the algorithm described in [9]) before applying the α scaling.

For example, we consider the problem where H is the identity matrix and A is the matrix FFFFF800 from the test set of LP problems distributed by David Gay [24]. FFFFF800 is of dimensions 1028×524 and has 6401 nonzero entries. The resulting augmented matrix is scaled to create 5 matrices which are increasingly ill conditioned, with classical condition numbers shown in Table 5.2. We then ran a range of scaling options on these matrices. We denote these five strategies by:

1. Just α scaling.
2. Just scaling by MC30.
3. α scaling followed by MC30.
4. Row scaling of A followed by α scaling.
5. Scaling by MC30 and then α scaling.

Note that when we row scale A in strategy 4, the augmented matrix has the form

$$\begin{pmatrix} D^{-2} & C \\ C^T & 0 \end{pmatrix}$$

where C is the row-equilibrated A ($A = DC$, D diagonal). This is the strategy recommended by Björck [6] when A is badly scaled. We obtain the results in Table 5.2 which clearly indicate that it is vital to scale A prior to using the α scaling and illustrates the advantage of strategy 5. We study various scaling strategies and their effect on the solution of augmented systems in [12].

Table 5.2: Error in solution for various scalings

Matrix	Condition number	Scaling					
	K_∞	None	1	2	3	4	5
FFFFF_02	1.10^{09}	4.10^{-13}	3.10^{-13}	5.10^{-13}	4.10^{-13}	1.10^{-12}	5.10^{-13}
FFFFF_04	1.10^{13}	1.10^{-08}	6.10^{-13}	4.10^{-12}	1.10^{-12}	5.10^{-13}	5.10^{-13}
FFFFF_06	1.10^{16}	4.10^{-05}	5.10^{-13}	2.10^{-06}	2.10^{-05}	3.10^{-13}	1.10^{-13}
FFFFF_08	1.10^{20}	6.10^{-02}	2.10^{-11}	5.10^{-03}	3.10^{-03}	6.10^{-13}	2.10^{-13}
FFFFF_10	1.10^{23}	5.10^{+00}	9.10^{-05}	4.10^{-03}	9.10^{-02}	2.10^{-11}	2.10^{-13}

6 MA47, a sparse structured symmetric indefinite solver

We will use a multifrontal approach in designing our algorithm for the LDL^T factorization of sparse symmetric indefinite matrices. The details of such an approach are not necessary to understand this paper but further background can be obtained from the original papers by Duff and Reid [17], [18]. We now describe the features of multifrontal methods that are needed to follow the subsequent discussion. As is common in sparse elimination, the factorization is split into a symbolic phase, which performs an analysis using only the sparsity pattern of the matrix, and a numerical factorization phase.

In a multifrontal method, the sparse factorization proceeds by a sequence of factorizations on small dense matrices, called frontal matrices. The frontal matrices and a partial ordering for the sequence are determined by a computational tree where each node represents a full matrix factorization and each edge the transfer of data from child to parent node. This tree, which can be obtained from the elimination tree ([10], [30]), is determined solely by the sparsity pattern of the matrix and an initial pivot ordering that can be obtained by standard sparsity preserving orderings such as minimum degree. When using the tree to drive the numerical factorization, eliminations at any node can proceed as soon as those at the child nodes have completed, giving added flexibility for issues such as exploitation of parallelism. Normally, the complete frontal matrix cannot be factorized but only a few steps of Gaussian elimination are possible, after which the remaining reduced matrix (the Schur complement) needs to be summed (assembled) with other data at the parent node before further factorizations can take place. Thus the frontal matrices can be written

$$\begin{pmatrix} F_{11} & F_{12} \\ F_{12}^T & F_{22} \end{pmatrix}$$

where the matrix F_{11} is factorized and the Schur complement $F_{22} - F_{12}^T F_{11}^{-1} F_{12}$ is assembled with other Schur complements and original data at the parent node. Note that the elimination is done using full linear algebra and direct addressing while all the indirect addressing is confined to the assembly.

Since the symbolic phase chooses pivots only on the basis of the structure of the system, it is quite possible that they cannot be used during the numerical factorization because of the failure of the pivots to satisfy the criterion (4.2) or (4.3). If there are pivots rejected in this way during the numerical factorization, all that happens is that the corresponding block F_{11} is smaller than forecast by the analysis and the Schur complement correspondingly larger. Thus the amount of work and storage for the factorization can rise although the factorization can always be performed since, at the root node, any entry in the frontal matrix can be chosen as pivot and all the matrix can be factorized respecting inequalities (4.2) or (4.3).

In our original code [17], the symbolic phase used a minimum degree ordering that assumed any diagonal entry could be chosen as pivot. This did not cause many problems with the numerical factorization in our early experiments with indefinite systems, the overheads because of delayed pivots being seldom more than 10%. However, the situation with augmented systems is quite different since the symbolic phase might choose a sequence of pivots from the (2,2) block early in the factorization. If there has not been fill-in to these positions, then they will be zero during numerical factorization and the effect on subsequent frontal matrix sizes can be very significant. We were alerted to this problem by Gill, Murray, and Saunders (personal communication, 1989) and designed an analysis and factorization scheme that took account explicitly of the zero block [13]. In this work, we were concerned not only to avoid choosing zero diagonal entries as pivots during the analysis but also to take into account the structure of the matrix during subsequent factorization. Based on the work in [13], we have recently developed a new code which has been released in the Harwell Subroutine Library [1] under the name MA47.

The earlier code, MA27 [16] in the Harwell Subroutine Library, was run on the matrix

$$\begin{pmatrix} I_{m-n} & & A_1 \\ & 0_n & A_2 \\ A_1^T & A_2^T & 0_n \end{pmatrix} \tag{6.1}$$

where the matrix A_2 is a nonsingular $n \times n$ submatrix of the matrix FFFFF800 that was used in the experiments in Section 5. The symbolic phase of MA27 forecasts that 1.5 million floating-point operations are needed for the numerical factorization. However, the subsequent numerical factorization of MA27 requires 16.5 million floating-point operations when the threshold parameter u of (4.2) and (4.3) has the value 0.1.

In MA47, not only do we avoid pivoting on zeros in the symbolic phase but we also take account of the block structure. For example, in the matrix

$$\begin{pmatrix} \times & \times & \times & \times & \times & \times & \times \\ \times & 0 & 0 & 0 & 0 & 0 & 0 \\ \times & 0 & & & & & \\ \times & 0 & & & & & \\ \times & 0 & & & & & \\ \times & 0 & & & & & \\ \times & 0 & & & & & \end{pmatrix},$$

if the (1,1) entry is chosen as pivot, the whole matrix is thereafter full. However, if the top right block of order two is chosen first (mathematically and structurally equivalent to pivoting on the (1,2) entry followed by the (2,1) entry), the remaining Schur complement is completely zero. Notice that this pivot choice and that of (1,1) followed by (2,2) are mathematically equivalent although the sparse code will not recognize the cancellation caused by the (2,2) pivot. In our new code, we want to choose 2×2 pivots of this form, called *tile* pivots and pivots of the form (4.1), called *oxo* pivots.

In general, the Schur complement from these structured systems will not be either zero or full but they can all be represented by the block form

$$\begin{pmatrix} 0 & X & X \\ X & X & X \\ X & X & 0 \end{pmatrix}. \tag{6.2}$$

Our new code thus allows for non-full frontal matrices of this type and chooses pivots in symmetric pairs based on an unsymmetric Markowitz count.

One consequence of our strategy of allowing non-full Schur complements is that a front matrix is not always absorbed by the parent node and can propagate up the tree. For example, if an *oxo* pivot is chosen at the parent node, where the (1,1) entry is not in the Schur complement but the (2,2) entry is in the (1,1) block, then the resulting Schur complement will comprise two overlapping matrices of the form (6.2). Our alternatives are to store the Schur complement as a general sparse matrix or to exploit structure but then have multiple fronts. We choose the second alternative but notice that either causes our new code to be much more complicated than the old. A second issue, the effect of which we will observe in the next section, is that our sparsity criterion will seldom choose structured pivots when H is diagonal and the aspect ratio is high because normally the Markowitz cost for an off-diagonal will be significantly higher than for the entry on the diagonal. Thus we would not expect many *tiles* to be chosen in such cases. Since, in such examples, MA27 is likely for similar arguments to pivot first in the (1,1) block of (1.1), we would expect little improvement for the new code and in fact possibly worse performance because of the much more complicated structure of the MA47 code.

We now examine this performance briefly in the following section.

7 Preliminary experiments with MA47

There are some good reasons for favouring the selection of *tiles* or *oxo* pivots because the structure of the augmented system is unchanged.

We adapted the symbolic phase of MA47 to force the selection of more *tiles* and show the performance on the problem

$$\begin{pmatrix} I & A \\ A^T & 0 \end{pmatrix}, \tag{7.1}$$

where the matrix A is the aforementioned FFFFF800 matrix.

Table 7.1: Forcing of *tiles*

number *tiles* Forecast	Actual	Ops (×1000)	Storage (Kwords)	Time (SPARC-1 secs)
89	77	6.8	69.3	10.1
150	137	8.7	78.6	12.3
230	217	6.7	66.2	10.0
249	234	7.0	69.2	11.0
371	302	12.1	85.7	21.9

We can see from the results in Table 7.1 that the performance is very flat over a wide range of forcing *tiles*. Also, although not noted in the table, the numerical accuracy was essentially unchanged. From this run and several others, we conclude that there is little point in modifying MA47 but that approaches that favour selecting *tiles* (for example, [26]) should not be too disadvantaged from a point of view of sparsity unless they severely overdo the amount of forcing.

Finally, when we ran MA47 on the example (6.1) of Section 6, only 7,954 floating-point operations were forecast by the symbolic phase, nearly a two-hundred fold improvement. The subsequent numerical factorization of MA47 also required only 7,954 floating-point operations, now an improvement of over two thousand fold! Additionally, the accuracy was slightly better with MA47 although, because of the more complicated code, the factorization times for MA27 and MA47 on a SPARC-10 were 3.15 and 0.07 seconds respectively, for a still creditable ratio of 45.

We show in Table 7.2, results comparing MA47 with MA27 on a range of matrices of this form, where the threshold parameter for both codes has been set to 0.001. These results emphasize that significant gains can be obtained with the new code. However, unfortunately sometimes a penalty is paid for the extra complications in the new code and, on unconstrained problems of the form of (7.1) with the same LP matrices as in Table 7.2, MA27 usually outperforms MA47. For example, on problem (7.1) where A is equal to the FFFFF800 problem, the numerical factorization of MA47 requires .95 seconds on a SPARC-10 while MA27 needs only .57 seconds. It was results of this kind that prevented us making MA27 an obsolescent routine in the Harwell Subroutine Library. I should stress that, although not recorded in the table, the solution was obtained with a similar satisfactory accuracy for all the runs of both MA27 and MA47.

Table 7.2: Comparison between MA27 and MA47 on constrained least-squares problems. Time in secs on Sun SPARC-10.

		CPU time			No. Operations		Nonzeros
matrix A	Method	Symb	Fact	Total	Forecast	Actual	in factors
ADLITTLE	MA47	0.020	0.010	0.030	771	771	619
	MA27	0.010	0.020	0.030	4356	12533	1303
AFIRO	MA47	0.010	0.000	0.010	180	180	180
	MA27	0.000	0.000	0.000	518	1297	283
ISRAEL	MA47	0.140	0.020	0.160	3129	3129	2961
	MA27	0.070	0.120	0.190	103528	333566	8247
CAPRI	MA47	0.080	0.020	0.100	2612	2612	2604
	MA27	0.050	0.140	0.190	134972	415996	11188
SHARE1B	MA47	0.070	0.030	0.100	4487	4482	1743
	MA27	0.020	0.040	0.060	32741	58287	3834
VTPBASE	MA47	0.040	0.020	0.060	1471	1471	1471
	MA27	0.030	0.060	0.090	16566	60992	3742
E226	MA47	0.110	0.030	0.140	3463	3463	3463
	MA27	0.080	0.210	0.290	189126	583512	12913
BEACONFD	MA47	0.150	0.030	0.180	3876	3876	3876
	MA27	0.130	0.160	0.290	152572	438060	10941
FFFFF800	MA47	0.290	0.060	0.350	7954	7954	7954
	MA27	0.380	1.030	1.410	1512220	4336104	44377

8 Conclusions

We have reviewed the use of augmented systems illustrating application areas in which they appear and commenting on a range of solution techniques used. We have described a general approach for their solution based on a direct method using 2×2 pivoting and have illustrated that it could prove useful in certain cases. Sometimes our new code yields a substantial improvement over an earlier code that did not respect the structure.

References

[1] Anon. *Harwell Subroutine Library A Catalogue of Subroutines (Release 11).* Theoretical Studies Department, AEA Industrial Technology, 1993.

[2] M. Arioli, T. F. Chan, I. S. Duff, N. I. M. Gould, and J. K. Reid. Computing a search direction for large-scale linearly-constrained nonlinear optimization calculations. Technical Report RAL 93-066, Central Computing Department, Rutherford Appleton Laboratory, 1993.

[3] M. Arioli, J. Demmel, and I. S. Duff. Solving sparse linear systems with sparse backward error. *SIAM J. Matrix Anal. Appl.*, 10:165–190, 1989.

[4] Å. Björck. Least squares methods. In P. G. Ciarlet and J. L. Lions, editors, *Handbook of Numerical Analysis*, volume I: Finite Difference Methods-Solution of Equations in R^n. Elsevier/North Holland, Amsterdam, 1990.

[5] Å. Björck. Error analysis of least squares algorithms. In G. H. Golub and P. Van Dooren, editors, *Numerical Linear Algebra, Digital Signal Processing and Parallel Algorithms*, NATO ASI Series, pages 41–73, Berlin, 1991. Springer.

[6] Å. Björck. Pivoting and stability in the augmented system method. In Griffiths and Watson, editors, *Numerical Analysis Proceedings, Dundee 91*, pages 1–16, London, 1992. Longman.

[7] J. R. Bunch. Analysis of the diagonal pivoting method. *SIAM J. Numer. Anal.*, 8:656–680, 1971.

[8] J. R. Bunch and B. N. Parlett. Direct methods for solving symmetric indefinite systems of linear systems. *SIAM J. Numer. Anal.*, 8:639–655, 1971.

[9] A. R. Curtis and J. K. Reid. On the automatic scaling of matrices for Gaussian elimination. *J. Inst. Maths. Applics.*, 10:118–124, 1972.

[10] I. S. Duff. Full matrix techniques in sparse Gaussian elimination. In G.A. Watson, editor, *Numerical Analysis Proceedings, Dundee1981*, Lecture Notes in Mathematics 912, pages 71–84, Berlin, 1981. Springer-Verlag.

[11] I. S. Duff. The solution of large-scale least-squares problems on supercomputers. *Annals of Operations Research*, 22:241–252, 1990.

[12] I. S. Duff, N. I. M. Gould, and J. Patricio. Experiments in scaling augmented systems. Technical Report (to appear), Rutherford Appleton Laboratory, 1993.

[13] I. S. Duff, N. I. M. Gould, J. K. Reid, J. A. Scott, and K. Turner. Factorization of sparse symmetric indefinite matrices. *IMA Journal of Numerical Analysis*, 11:181–204, 1991.

[14] I. S. Duff, R. G. Grimes, and J. G. Lewis. Users' guide for the Harwell-Boeing sparse matrix collection (Release I). Technical Report RAL 92-086, Rutherford Appleton Laboratory, 1992.

[15] I. S. Duff and J. K. Reid. A comparison of some methods for the solution of sparse overdetermined systems of linear equations. *J. Inst. Maths. Applics.*, 17:267–280, 1976.

[16] I. S. Duff and J. K. Reid. MA27—a set of Fortran subroutines for solving sparse symmetric sets of linear equations. Technical Report R.10533, AERE, Harwell, England, 1982.

[17] I. S. Duff and J. K. Reid. The multifrontal solution of indefinite sparse symmetric linear systems. *ACM Trans. Math. Software*, 9:302–325, 1983.

[18] I. S. Duff and J. K. Reid. The multifrontal solution of unsymmetric sets of linear systems. *SIAM J. Scient. Statist. Comput.*, 5:663–641, 1984.

[19] H.C. Elman and G.H. Golub. Inexact and preconditioned uzawa algorithms for saddle point problems. Technical Report CS-TR-3075, Compuetr Science Department, University of Maryland, 1993.

[20] E. Eskow and R. B. Schnabel. A preconditioned iterative method for saddlepoint problems. *SIAM Journal on Matrix Analysis and Applications*, 13:887–904, 1992.

[21] A. Forsgren and W. Murray. Newton methods for large-scale linear equality-constrained minimization. *SIAM Journal on Matrix Analysis and Applications*, 14:560–587, 1993.

[22] M. Fortin and R. Glowinski. *Augmented Lagrangian Methods: Applications to the Numerical Solution of Boundary-Value Problems. Volume 15 of Studies in Mathematics and its Applications.* North-Holland, 1983.

[23] R. Fourer and S. Mehrota. Solving symmetric indefinite systems in an interior-point method for linear programming. Technical Report Technical Report 92-01, Department of Industrial Engineering and Management Sciences, Northwestern University, 1992.

[24] D. M. Gay. Electronic mail distribution of linear programming test problems. Mathematical Programming Society. COAL Newsletter, 1985.

[25] P. E. Gill, W. Murray, and M. H. Wright. *Practical Optimization.* Academic Press, 1981.

[26] N. I. M. Gould. KKTSOL/MA49: A set of Fortran subroutines for solving sparse symmetric sets of linear equations which arise in constrained optimization calculations. Technical Report (to appear), CERFACS, Toulouse, France, 1993.

[27] M. Gulliksson and P.-Å. Wedin. Modifying the QR-decomposition to constrained and weighted linear least squares. *SIAM Journal on Matrix Analysis and Applications*, 13:1298–1313, 1992.

[28] N. Karmarkar. A new polynomial-time algorithm for linear programming. *Combinatorica*, 4:373–395, 1984.

[29] H. W. Kuhn and A. W. Tucker. Nonlinear programming. In J. Neyman, editor, *Proceedings of Second Berkeley Symposium on Mathematical Statistics and Probability*, pages 481–492, California, 1951. University of California, Press.

[30] J. H. W. Liu. The role of elimination trees in sparse factorization. *SIAM J. Matrix Anal. Appl.*, 11:134–172, 1990.

[31] E. Ng, B. Nitrosso, and B. Peyton. On the solution of Stokes' pressure system within N3S using supernodal Cholesky factorization. Technical Report (to appear), Electricité de France, 1993.

[32] A. Ramage and A.J Wathen. Iterative solution techniques for the Navier-Stokes equations. Technical Report AM-93-01, School of Mathematics, University of Bristol, 1993.

[33] T. Rusten and R. Winther. A preconditioned iterative method for saddlepoint problems. *ACM Transactions on Mathematical Software*, 17:306–312, 1991.

[34] G. W. Stewart. On scaled projections and pseudoinverses. *Linear Algebra and its Applications*, 112:189–193, 1989.

[35] G. Strang. *Introduction to Applied Mathematics.* Wellesley-Cambridge Press, 1986.

[36] G. Strang. A framework for equilibrium equations. *SIAM Review*, 30:283–297, 1988.

[37] S. A. Vavasis. Stable numerical algorithms for equilibrium systems. *SIAM Journal on Matrix Analysis and Applications*, 12:000–000, 1993.

[38] P-Å. Wedin. Perturbation theory and condition numbers for generalized and constrained linear least squares problems. Technical Report UMINF 125.85, Inst Inf Processing, University of Umeå, 1985.

Acknowledgement

I would like to thank my colleagues at CERFACS and RAL for their assistance. Ali Bouaricha and João Patricio for helping obtain some of the numerical results and Nick Gould, John Reid, and Jennifer Scott for commenting on the manuscript. I also thank Åke Björck for discussions on Section 5.

Iain S Duff
Rutherford Appleton Laboratory
Didcot
Oxon OX11 0QX
England

C M ELLIOTT AND A R GARDINER

One dimensional phase field computations

Abstract The phase field equations arise in the modelling of the solidification of undercooled liquids. They are the coupled parabolic equations:-

$$c\theta_t + \frac{l}{2}u_t = k\Delta\theta + f$$

$$\tau u_t = \gamma\Delta u - \psi'(u) + \alpha\theta$$

where ψ is a double well potential and f is a volumetric heat source or sink. By choosing an appropriate scaling of τ, γ and α with respect to a small parameter ϵ the phase field equations are a regularisation of a moving boundary problem for the heat equation $c\theta_t = k\Delta\theta$. We consider numerical issues for the problem in one space dimension for both a quartic and a double obstacle potential. Model problems are considered to enable comparisons of the two choices of potential as $\epsilon \to 0$.

1 Introduction

In this paper we present some numerical computations of two variants of the phase-field equations with a scaling of parameters corresponding to approximations of moving boundary problems for the one dimensional heat equation. The phase field equations (PF)

$$c\theta_t + \frac{l}{2}u_t = k\theta_{xx} + f, \tag{1.1a}$$

$$\tau u_t = \gamma u_{xx} - \psi'(u) + \alpha\theta, \tag{1.1b}$$

are known to describe liquid/solid phase transitions. Here $\theta(x,t)$ is the temperature, $u(x,t)$ is a phase variable, x is the spatial variable and t denotes time. The constants c, l and k are respectively, the specific heat, latent heat and conductivity. The prescribed function f is a volumetric heat source or sink. The constants τ, γ and α are constants which we scale with a small parameter ϵ in order to approximate, as $\epsilon \to 0$, moving boundary problems for the heat equation

$$c\theta_t = k\theta_{xx} + f.$$

The homogeneous free energy function ψ is a symmetric double well with minimum value 0 so that $\psi(-1) = \psi(+1) = 0$. A typical function is

$$\psi(r) := \frac{1}{4}(r^2 - 1)^2, r \in I\!R. \tag{1.2}$$

These equations have been much studied. We refer to Caginalp [3], Caginalp and Fife [4], Stoth [14] and the references cited therein.

A new variant (PFO) introduced by Blowey and Elliott [1, 2] is defined by the homogeneous energy function

$$\psi(r) = \frac{1}{2}(1-r^2) + I_{[-1,1]}(r), \qquad r \in I\!R \tag{1.3}$$

where I is the indicator function of the interval $[-1,1]$ so that $I_{[-1,1]}$ vanishes for $|r| \leq 1$ and is $+\infty$ for $|r| > 1$. The equation (1.1b) is then interpreted as the inclusion

$$-\tau u_t + \gamma u_{xx} + u + \alpha\theta \in \partial I_{[-1,1]}(u) \tag{1.4}$$

or equivalently the complimentary conditions

$$(\tau u_t - \gamma u_{xx} - u - \alpha\theta)(|u| - 1) = 0 \tag{1.5a}$$

$$(\tau u_t - \gamma u_{xx} - u - \alpha\theta)\text{sign}(u) \leq 0, \quad |u| \leq 1 \tag{1.5b}$$

holding almost everywhere in the space and time.

Choosing the scaling

$$\tau = \sigma_1\epsilon^2, \quad \gamma = \epsilon^2, \quad \alpha = \sigma_2\epsilon, \tag{1.6}$$

one expects, from formal and rigorous analysis, that solutions of (1.1) approximate the moving boundary problem:-

$$c\theta_t = k\theta_{xx} + f \qquad x \in (0, s(t)) \cup (s(t), L) \tag{1.7a}$$

$$k\theta_x(s(t)+0,t) - k\theta_x(s(t)-0,t) = [k\theta_x]_-^+ = -l\frac{\mathrm{d}s}{\mathrm{d}t} \tag{1.7b}$$

$$\theta(s(t),t) = -\delta_v\frac{\mathrm{d}s}{\mathrm{d}t} \tag{1.7c}$$

$$u(x,t) = \text{sign}(x - s(t)) \tag{1.7d}$$

$$\delta_v = C_\psi\frac{\sigma_1}{\sigma_2} \tag{1.7e}$$

where $x = s(t)$ denotes the position of the interface of the moving boundary. The solid and liquid regions correspond to $u = -1$ and $u = +1$ respectively. As formulated (1.7a,b,c,d) allow supercooling/superheating i.e. the temperature can take negative/positive values in the liquid/solid phases.

The constant C_ψ in (1.7e) is defined by

$$C_\psi := \frac{1}{2}\int_{-\infty}^{\infty}[U'(z)]^2\mathrm{d}z$$

where $U(z)$ is the unique monotone standing wave solution of

$$-U'' + \psi'(U) = 0, \quad \forall z \in I\!R, \tag{1.8a}$$

$$U(0) = 0, \quad U'(z) \geq 0, \quad \forall z \in I\!R, \tag{1.8b}$$

$$\lim_{z \to \pm\infty} U(z) = \pm 1, \tag{1.8c}$$

so that in the two cases

$$U(z) = \tanh\left(\frac{z}{\sqrt{2}}\right) \quad z \in I\!R, \quad C_\psi = \frac{\sqrt{2}}{3}, \qquad \text{(PF)} \tag{1.9a}$$

$$U(z) = \begin{cases} 1 & z \geq \pi/2 \\ \sin z & |z| < \pi/2 \\ -1 & z \leq -\pi/2 \end{cases}, \quad C_\psi = \frac{\pi}{4}, \qquad \text{(PFO)}. \tag{1.9b}$$

A related sub-problem is obtained by taking $\theta(x,t)$ to be prescribed in (1.1b), setting $\sigma_1 = 1$ and $\sigma_2 = C_\psi$ in the scaling set out by (1.6), this yields the reaction diffusion equation

$$\epsilon^2 u_t = \epsilon^2 u_{xx} - \psi'(u) + C_\psi \epsilon \theta \tag{1.10}$$

We refer to (1.10) with ψ being (1.2) or (1.3) as (RD) and (RDO) respectively. In these cases the limit problem is

$$u(x,t) = \text{sign}(x - s(t)), \quad \frac{ds}{dt} = -\theta(s(t),t). \tag{1.11}$$

2 Convergence as $\epsilon \to 0$

Throughout the paper we use the scaling (1.6) and label the solutions of (1.1a,b) or (1.10) as $\{\theta^\epsilon, u^\epsilon, s^\epsilon\}$ or $\{u^\epsilon, s^\epsilon\}$ where $s^\epsilon(t)$ denotes the zero of $u^\epsilon(x,t)$. Formal asymptotics indicate that in the neighbourhood of the interface $u^\epsilon(x,t)$ has an inner expansion whose zero order term is $U((x - s(t))/\epsilon)$ where U solves (1.8).

This suggests the choice of initial condition for the phase variable u^ϵ to be

$$u^\epsilon = U\left(\frac{x - s_0}{\epsilon}\right), \tag{2.1}$$

which yields for (PF) and (RD)

$$u^\epsilon(x,0) = \tanh\left(\frac{x - s_0}{\sqrt{2}\epsilon}\right), \tag{2.2a}$$

and for (PFO) and (RDO)

$$u^\epsilon(x,0) = \begin{cases} 1 & x > s_0 + \pi\epsilon/2 \\ \sin((x - s_0)/\epsilon) & |x - s_0| < \pi\epsilon/2 \\ -1 & x < s_0 - \pi\epsilon/2 \end{cases}. \tag{2.2b}$$

where s_0 denotes the initial position of the interface $s(0)$.

A rigorous analysis of convergence for (PF) with this initial condition has been carried out by Stoth [14]. Rigorous analysis for (RD) has been carried out by Evans, Soner and Souganidis [10] and Chen [7]. Only a formal analysis exists for (PFO) by Blowey and Elliott [1]. However a precise analysis for (RDO) has been performed by Chen and Elliott [8] (with $\theta = 0$) and Nochetto, Paolini and Verdi [12]. In particular for (RDO) it is known that , [12],

$$|s^\epsilon(t) - s(t)| \leq C\epsilon^2. \tag{2.3}$$

Since for (PFO) and (RDO) the problem (1.5) or (1.10) is a double obstacle problem it holds that $|u^\epsilon(x,t)| = 1$ apart from the region $\Gamma^\epsilon_x(t) = \{x : |u^\epsilon(x,t) < 1\}$ which has width $O(\epsilon)$. This means that provided $s(t)$ lies in the interior of $(0, L)$ imposing either the Dirichlet conditions

$$u^\epsilon(0,t) = -1, \quad u^\epsilon(L,t) = +1 \tag{2.4a}$$

or the Neumann conditions

$$\frac{\partial u^\epsilon}{\partial x}(0,t) = \frac{\partial u^\epsilon}{\partial x}(L,t) = 0, \tag{2.4b}$$

yields the same solution $u^\epsilon(x,t)$.

However the choice of boundary conditions for u^ϵ in the (PF) and (RD) variants is more problematic. The interior layer centered on $s^\epsilon(t)$ is of width $O(\epsilon|\ln\epsilon|)$ and accurately computing the exponential variation of $u^\epsilon(x,t)$ in x away from the interface appears to be of importance in yielding the correct limit as $\epsilon \to 0$. Possible boundary conditions to impose are (2.4) but other boundary conditions have been suggested by Caginalp and Socolovsky [5] for example; see section 6.

3 Numerical Solution of the Phase Field Equations

In this paper we use a uniform spatial mesh with constant time steps. We set L/N to be the spatial mesh size and Δt to be the time step. Employing backward Euler time discretisation and central spatial differences yields the algebraic problem for each $n \geq 1$

$$(cI + k\Delta tA)\underline{\theta}^n + \frac{l}{2}\underline{u}^n = \underline{b}^n, \tag{3.1a}$$

$$(\tau I + \gamma\Delta tA)\underline{u}^n - \alpha\Delta t\underline{\theta}^n + \Delta t\underline{g}(\underline{u}^n) = \underline{d}^n, \tag{3.1b}$$

for the vectors of nodal values $\{\underline{\theta}^n, \underline{u}^n\}$ corresponding to the time levels $n\Delta t$. Here A is the central difference approximation to the operator $-\partial^2 \cdot /\partial x^2$. The right hand sides $\underline{b}^n$ and $\underline{d}^n$ contain the contributions of inhomogeneous boundary conditions and heat source as well as the terms $(c\underline{\theta}^{n-1} + \frac{l}{2}\underline{u}^{n-1})$ and $\tau\underline{u}^{n-1}$. The notation $\underline{g}(\underline{u})$ is used for the vector $\{\psi'(u^n_j)\}$ where j is the index for the jth node of the grid.

The nonlinear equations (3.1) are solved by the following successive relaxation procedure. The matrix A is decomposed as

$$A = D - L - L^T, \tag{3.2}$$

where D is diagonal and L is lower triangular. Then we consider the iteration

$$\begin{bmatrix} cI + \Delta t k(D-L) & \frac{l}{2}I \\ -\alpha\Delta t I & (\tau I + \gamma\Delta t(D-L)) \end{bmatrix} \begin{bmatrix} \underline{\theta}_k^n \\ \underline{u}_k^n \end{bmatrix} + \begin{bmatrix} 0 \\ \Delta t \underline{g}(\underline{u}_k^n) \end{bmatrix}$$

$$= \begin{bmatrix} \underline{b}^n \\ \underline{d}^n \end{bmatrix} + \begin{bmatrix} \Delta t k L & 0 \\ 0 & \gamma\Delta t L^T \end{bmatrix} \begin{bmatrix} \underline{\theta}_{k-1}^n \\ \underline{u}_{k-1}^n \end{bmatrix}. \tag{3.3}$$

Thus each iteration consists of a sweep through the the mesh points and involves the solution of the two by two system

$$(c + \Delta t k a_{jj})\theta_j + \frac{l}{2}u_j = \tilde{b}_j \tag{3.4a}$$

$$-\Delta t \alpha \theta_j + (\tau + \Delta t \gamma a_{jj})u_j + \Delta t g(u_j) = \tilde{d}_j \tag{3.4b}$$

where a_{jj} is a diagonal element of A. Substituting (3.4a) into (3.4b) yields the nonlinear equation

$$\left(\tau + \Delta t \gamma a_{jj} + \frac{\Delta t \alpha l}{2(c + \Delta t k a_{jj})} - \Delta t\right) u_j + \Delta t \beta(u_j) = \tilde{d}_j + \frac{\tilde{b}_j \Delta t \alpha}{c + \Delta t k a_{jj}}, \tag{3.5}$$

where

$$\beta(u) = u^3, \qquad \text{(PF)}, \tag{3.6a}$$

$$\beta(u) = \partial I_{[-1,1]}(u), \qquad \text{(PFO)}. \tag{3.6b}$$

Since $\beta(\cdot)$ is monotone increasing it follows that (3.5) has a unique solution provided

$$\Delta t < \tau + \left(\gamma a_{jj} + \frac{\alpha l}{2(c + \Delta t k a_{jj})}\right) \Delta t.$$

With β defined by (3.6a) Newtons method is used to solve (3.5). In the case of (3.6b) we easily obtain the solution by noting that the unique solution of ($\delta > 0$)

$$\delta u + \beta(u) = q \tag{3.7a}$$

is given by

$$u = \begin{cases} q/\delta & |q/\delta| < 1 \\ \text{sign}(q), & |q/\delta| > 1 \end{cases}. \tag{3.7b}$$

¿From the above discussion it is clear that the same program can be used for each of (PF) and (PFO) provided one has the two options for the scalar nonlinear equation (3.5) and the appropriate choices of initial and boundary conditions are made.

A similar relaxation iteration is employed for (RD) and (RDO). Clearly the procedure is identical to that described above except $\underline{\theta}^n$ is prescribed in (3.1b) and equation (3.4b) is immediately set in the form (3.7a).

The relaxation can be speeded up by using an overrelaxation parameter $\omega \in [1, 2]$. In the double obstacle iteration the overrelaxed iterate is projected onto the interval $[-1, 1]$ as described in Elliott and Ockendon [9] for example.

4 Convergence of the Gauss-Seidel Iteration for a Linear system

We consider the linear system

$$\begin{bmatrix} cI + k\Delta tA & \frac{l}{2}I \\ -\alpha\Delta tI & \tau I + \gamma\Delta tA \end{bmatrix} \begin{bmatrix} \underline{\theta} \\ \underline{u} \end{bmatrix} = \begin{bmatrix} \underline{b} \\ \underline{d} \end{bmatrix}$$

where $\underline{b}$ and $\underline{d}$ are prescribed vectors of dimension M, k, l, γ and α are positive constants, c and τ are non-negative constants and A is an $M \times M$ positive semi-definite matrix. Furthermore we assume that

$$A = D - L - U, \quad U = L^T$$

where D is a positive diagonal matrix and L is lower triangular with a zero diagonal. The iterative method we analyse is :-

$$\begin{bmatrix} cI + k\Delta t(D-L) & \frac{l}{2}I \\ -\alpha\Delta tI & \tau I + \gamma\Delta t(D-L) \end{bmatrix} \begin{bmatrix} \underline{\theta}^n \\ \underline{u}^n \end{bmatrix} = \Delta t \begin{bmatrix} kU & 0 \\ 0 & \gamma U \end{bmatrix} \begin{bmatrix} \underline{\theta}^{n-1} \\ \underline{u}^{n-1} \end{bmatrix} + \begin{bmatrix} \underline{b} \\ \underline{d} \end{bmatrix}.$$

This decouples into M two by two systems with coefficient matrix

$$\begin{bmatrix} c + \Delta tka_{jj} & \frac{l}{2} \\ -\alpha\Delta t & \tau + \gamma\Delta ta_{jj} \end{bmatrix}.$$

Since this matrix has determinant

$$(\tau + \gamma\Delta ta_{jj})(c + \Delta tka_{jj}) + \frac{l}{2}\alpha\Delta t$$

the iteration is well defined under our assumptions on the parameters τ, γ, c, l and α.

The iteration matrix has eigenvalues μ corresponding to eigenvectors decomposed as $(\underline{z}_\theta, \underline{z}_u)$ satisfying:-

$$\Delta t \begin{pmatrix} kU & 0 \\ 0 & \gamma U \end{pmatrix} \begin{pmatrix} \underline{z}_\theta \\ \underline{z}_u \end{pmatrix} = \mu \begin{pmatrix} cI + \Delta tk(D-L) & \frac{l}{2}I \\ -\alpha\Delta tI & \tau I + \gamma\Delta t(D-L) \end{pmatrix} \begin{pmatrix} \underline{z}_\theta \\ \underline{z}_u \end{pmatrix}.$$

For convergence we require the magnitude of the eigenvalues to be less than 1. Clearly

$$(cI + k\Delta t(D - L - \frac{1}{\mu}U))\underline{z}_\theta + \frac{l}{2}\underline{z}_u = 0$$

$$(\tau I + \gamma\Delta t(D - L - \frac{1}{\mu}U))\underline{z}_u - \alpha\Delta t\underline{z}_\theta = 0$$

so that $\underline{z}_\theta$ solves

$$(\tau I + \gamma \Delta t(D - L - \frac{1}{\mu}U))(cI + k\Delta t(D - L - \frac{1}{\mu}U))\underline{z}_\theta + \alpha \Delta t \frac{l}{2} \underline{z}_\theta = 0$$

and $(\mu, \underline{z}_\theta)$ is determined by the eigenvalue problem

$$(D - L - \frac{1}{\mu(\kappa)}U)\underline{z}_\theta = \kappa \underline{z}_\theta,$$

with κ given by the roots of

$$q(\kappa) := (\tau + \gamma \Delta t \kappa)(c + k \Delta t \kappa) + \alpha \frac{l}{2} \Delta t = 0.$$

We now suppose that

$$D = aI$$

and that $D^{-1}(L + U)$ is consistently ordered and is 2-cyclic, [Varga [15]]. It follows that $\mu(\kappa)$ satisfies

$$0 = \det[(a - \kappa)I - (L + \frac{1}{\mu}U)] = \det[(a - \kappa)I - \frac{1}{\mu^{1/2}}(L + U)].$$

Denoting by λ the eigenvectors of the Jacobi iteration matrix $D^{-1}(L + U)$ it follows that

$$\mu(\kappa) = \frac{\lambda^2 a^2}{(a - \kappa)^2}$$

and

$$|\mu| = \frac{|\lambda|^2 a^2}{(a - \kappa_r)^2 + \kappa_i^2}, \qquad \kappa = \kappa_r + i\kappa_i.$$

However since $q(0) = \tau c + \alpha \Delta t \frac{l}{2} > 0$ and $q'(0) = \Delta t(\gamma c + \tau k) \geq 0$ it follows that either $\kappa_r = 0$ and $\kappa_i \neq 0$ or $\kappa_r < 0$. Thus

$$|\mu| < |\lambda|^2.$$

Hence we have convergence of the iteration provided the Jacobi method converges for the matrix A. This is known to be true when A arises as a central difference approximation to $\partial^2 \cdot /\partial x^2$ on a uniform grid with Dirichlet boundary conditions.

5 Model Problems

Problem I

We consider (RD) and (RDO) with $\theta(x,t) \equiv -1$ so that the limit problem is

$$u(x,t) = \text{sign}(x - s(t)), \quad \frac{ds}{dt} = 1, \quad s(0) = s_0 \tag{5.1a}$$

with solution

$$s(t) = s_0 + t. \tag{5.1b}$$

Problem II

We consider (RD) and (RDO) with $\theta(x,t) = -x$ so that the limit problem is

$$u(x,t) = \text{sign}(x - s(t)), \quad \frac{\mathrm{d}s}{\mathrm{d}t} = s, \quad s(0) = s_0 \tag{5.2a}$$

with solution

$$s(t) = s_0 e^t. \tag{5.2b}$$

Problem III

We consider a travelling wave solution of (1.7a,b,c,d) with $c = k = l = 0$ and $f(x) = 0$ given by

$$\theta(x,t) = \begin{cases} e^{-V(x-s(t)-vt)} - (1 + \delta_v V) & x > s(t) \\ -\delta_v V & x < s(t) \end{cases} \tag{5.3a}$$

$$\frac{\mathrm{d}s}{\mathrm{d}t} = V, \quad s(t) = s_0 + Vt, \quad u(x,t) = \text{sign}(x - s(t)). \tag{5.3b}$$

Problem IV

We consider (1.7a,b,c,d) with $f(x) = 4\delta_v/x^2$ and $c = l = k = 1$. A solution of these equations is

$$s(t) = 2t^{1/2} \tag{5.4a}$$

$$\theta(x,t) = -2\frac{\delta_v}{x} + A\left(\text{erf}\left(\frac{x}{2t^{1/2}}\right) - \text{erf}1\right) \qquad x < s(t) \tag{5.4b}$$

$$\theta(x,t) = -2\frac{\delta_v}{x} + B\left(\text{erfc}\left(\frac{x}{2t^{1/2}}\right) - \text{erfc}1\right) \qquad x < s(t) \tag{5.4c}$$

where

$$A + B = e^1 l\pi^{1/2}. \tag{5.4d}$$

6 Numerical Experiments

Computations were performed for the model problems outlined in Section 5 using the discretisation described in Section 3. For model problems I and II computations were performed over the range $x \in [0, 0.5]$, with $s_0 = 0.2$ and with a time step given by $\Delta t = \frac{1}{2}h^2$. The initial condition was given by (2.2a) for the (RD) problem and (2.2b) for the (RDO) problem. Zero Neumann boundary conditions were used.

In Figures 1 and 2 we show the initial profiles of the phase parameter for the values of ϵ used in computations with (RD) and (RDO) respectively. Figure 1 shows the profiles

$$\cdots\cdots \equiv \epsilon = 0.04, \qquad \cdot - \cdot - \equiv \epsilon = 0.02, \qquad — \equiv \epsilon = 0.01,$$

and Figure 2 the profiles

$$\cdots\cdots \equiv \epsilon = 0.1, \qquad \cdot - \cdot - \equiv \epsilon = 0.05, \qquad — \equiv \epsilon = 0.025.$$

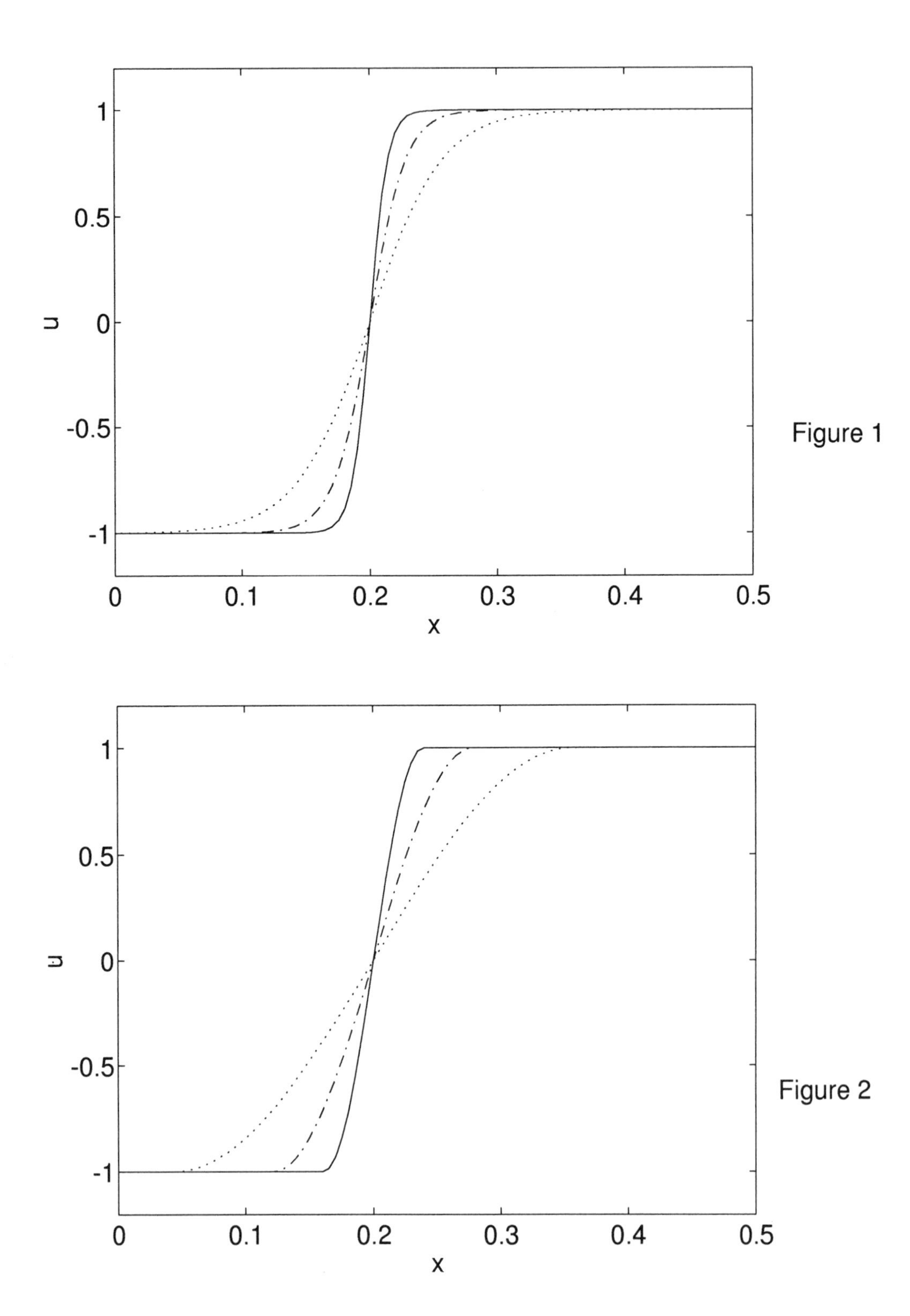
1
0.5
0
-0.5
-1
u
0
0.1
0.2
0.3
0.4
0.5
x
Figure 1
1
0.5
0
-0.5
-1
u
0
0.1
0.2
0.3
0.4
0.5
x
Figure 2

For these values of ϵ it can be seen that the spreading of the interfacial region is a lot greater for the (RD) problem ($\approx \epsilon|\ln \epsilon|$) than the (RDO) problem ($\pi\epsilon$).

Problem I

In Figures 3 and 4 the average error in the location of the free boundary over the time interval $0 \leq t \leq 0.1$ is plotted against $1/(2h)$ for several values of ϵ.

Figure 1 shows results for the (RDO) problem when $\epsilon = 0.1, 0.05$ and 0.025. It can be clearly seen that as $h \to 0$ the error becomes $O(\epsilon^2)$, as expected. Figure 2 shows results when $\epsilon = 0.04$ and 0.02 for the (RD) problem with the error again of $O(\epsilon^2)$.

Problem II

Computations were performed for the (RDO) and (RD) problems, with the average error in the location of the free boundary over the interval $0 \leq t \leq \ln(1.5)/5$ being calculated. Figure 5 shows results for (RDO) when $\epsilon = 0.1$,0.05 and 0.025. In contrast with the behaviour of the error for (RDO) in Problem I, if sufficient grid points lie in the interfacial region then the error can take, for h large, values below that of its asymptotic value as $h \to 0$. As $h \to 0$ then the error is again observed to be of $O(\epsilon^2)$. For the (RD) problem we again take $\epsilon = 0.04$ and 0.02, the results shown in Figure 6 show that the behaviour of the error is analogous to that for (RD) in Problem I with the error once more of $O(\epsilon^2)$ as $h \to 0$. Observe that we always measure the error in the computed approximation with respect to the free boundary problem ($\epsilon = 0$) rather than the 'ϵ' problem as $h \to 0$.

Problem III

We consider the moving boundary problem (1.7a,b,c,d) for $t \in [0, 0.1]$. We take $V = 1$, $\delta_v = 0.05$, $s_0 = 0.2$ and solve on the interval $x \in [0, 0.5]$. Initial conditions for u^ϵ are again given by (2.2) while those for θ^ϵ are given by (4.3a). Two sets of boundary conditions are considered :-

A: Zero Neumann conditions for the phase parameter and exact Neumann conditions given by (4.3a) for the temperature.

B: Dirichlet boundary conditions are imposed for both the phase and temperature variables. For the temperature we use the exact values given by (4.3a)

$$\theta^\epsilon(0,t) = \theta_a(t), \qquad \theta^\epsilon(0.5,t) = \theta_b(t).$$

For the (PFO) problem we take

$$u^\epsilon(0,t) = -1, \qquad u^\epsilon(0.5,t) = +1.$$

While for the (PF) problem we take

$$u(0,t) = u_a(t), \qquad u(0.5,t) = u_b(t),$$

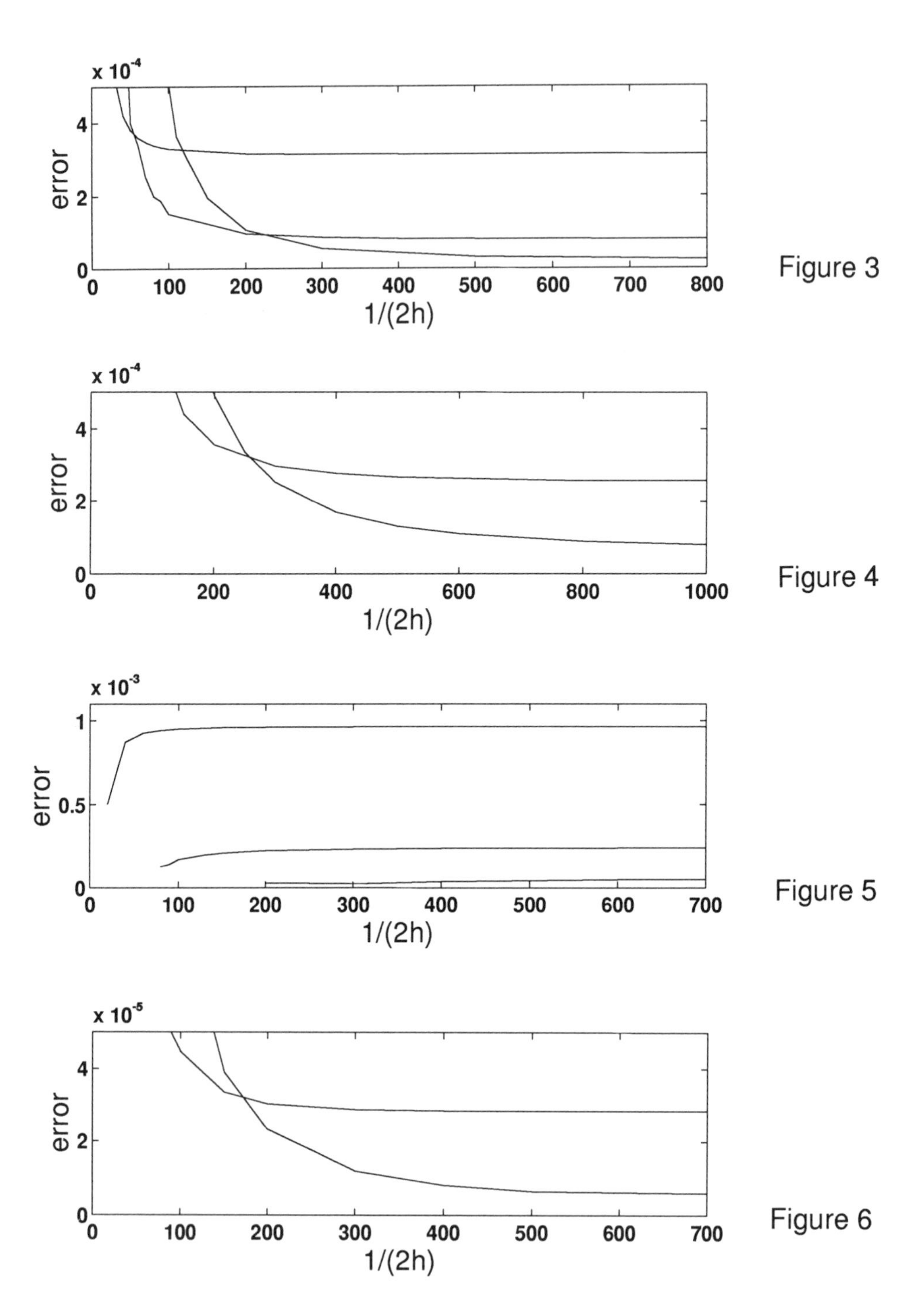

Figure 3

Figure 4

Figure 5

Figure 6

where u_a and u_b are the roots of

$$-\psi'(u_a) + \alpha\theta_a \equiv u - u^3 + \alpha\theta_a = 0 \quad \text{and} \quad u - u^3 + \alpha\theta_b = 0,$$

closest to -1 and $+1$ respectively. The Dirichlet boundary conditions are chosen to approximate local equilibrium in (1.1b) far from the interface. (c.f. Caginalp and Socolovsky [6])

The time step is taken to be $\Delta t = h^2$. We again monitor the average error in the position of the free boundary over the time interval [0,0.1]. In addition we measure the discrete L_2 error in the temperature at $t = 0.1$, given by

$$\left(\sum_{i=1}^{n+1} h(\theta(x,t) - \theta_i)^2 \right)^{\frac{1}{2}}. \tag{6.1}$$

For the (PFO) problem Figures 7 and 8 show the average error in the interface and the L_2 error (5.1) respectively. The solid lines show the results for computations with Neumann boundary conditions and the broken lines those using Dirichlet boundary conditions. In these Figures we have

$$1: \quad \epsilon = 0.1\ , \qquad 2: \quad \epsilon = 0.05, \qquad 3: \quad \epsilon = 0.025.$$

The results of computations for the (PF) problem are shown in Figures 9 and 10. Figure 10 showing the average error in the position of the interface and Figure 10 the L_2 error (5.1), again the solid lines show the results for Neumann boundary conditions and the broken lines results for Dirichlet boundary conditions. In these Figures we have

$$1: \quad \epsilon = 0.04, \qquad 2: \quad \epsilon = 0.02, \qquad 3: \quad \epsilon = 0.01.$$

From Figures 7—10 it can be seen that for both (PF) and (PFO) that the behaviour of the error as $h \to 0$ appears to be of the form $O(\epsilon)$.

To provide a comparison between results for the (PF) and (PFO) problems Table 6.1 contains results when the two problems were solved, using both Neumann and Dirichlet boundary conditions, over the interval $t \in [0, 0.1]$ with 300 spatial grid points for identical values of ϵ.

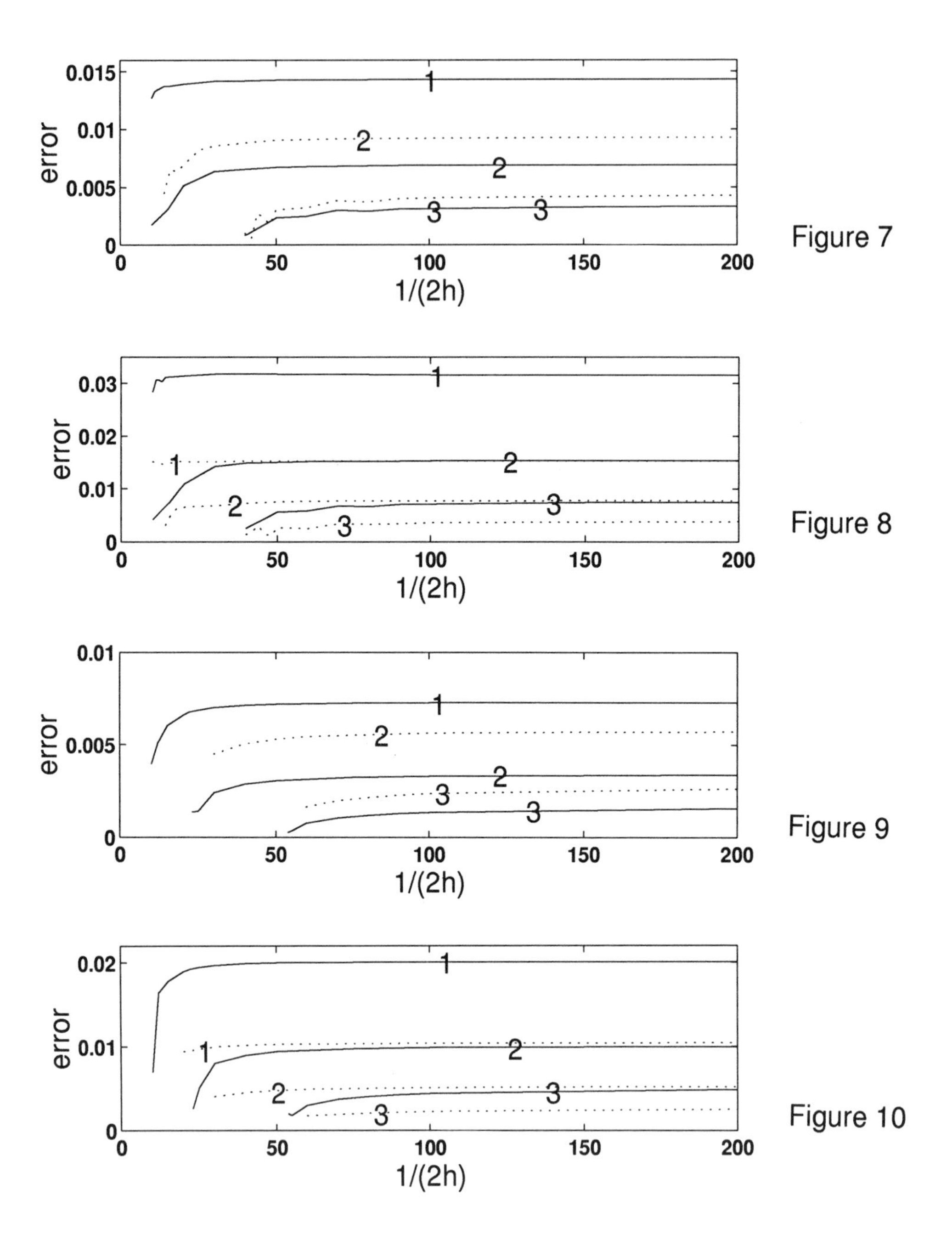

0.015
0.01
0.005
0
error
0
50
100
150
200
1/(2h)
1
2
2
3
3

Figure 7

0.03
0.02
0.01
0
error
0
50
100
150
200
1/(2h)
1
1
2
2
3
3

Figure 8

0.01
0.005
0
error
0
50
100
150
200
1/(2h)
1
2
2
3
3

Figure 9

0.02
0.01
0
error
0
50
100
150
200
1/(2h)
1
1
2
2
3
3

Figure 10

ϵ	Average error in location of interface for $t \in [0, 0.1]$ (PF)	(PFO)	Discrete L_2 error in temperature at $t = 0.1$ (PF)	(PFO)
	Neumann Boundary Conditions			
0.02	3.3949×10^{-3}	2.6742×10^{-3}	1.0010×10^{-2}	5.9597×10^{-3}
0.01	1.6089×10^{-3}	1.1555×10^{-3}	4.9542×10^{-3}	2.6146×10^{-3}
	Dirichlet Boundary Conditions			
0.02	5.7333×10^{-3}	3.3898×10^{-3}	5.2300×10^{-3}	3.0554×10^{-3}
0.01	2.6836×10^{-3}	1.4403×10^{-3}	2.6031×10^{-3}	1.3259×10^{-3}

Table 6.1

The results indicate, for this problem, that the double obstacle formulation provides the more accurate results.

It is observed for both the average error in the interface and the error (6.1) in both (PF) and (PFO) that, as observed in error for (RD) in Problem II, for fixed ϵ with sufficient spatial grid points the error reduces rapidly before before rising and tending toward an asymptotic value as $h \to 0$. This behaviour is illustrated in Figure 11 for both problems when $\epsilon = 0.02$. For the (PF) problem we have a much 'smoother' variation in the error with the number of grid points at which the minimum occurs easy to determine. In contrast the error for the (PFO) problem 'oscillates' about its minimum value before rising up to its asymptotic value, with the number of grid points at which the minimum error occurs being a lot harder to determine. It is possible to determine relationships between ϵ and h for which this minimum h^* occurs from experimental data for both problems. For the (PF) problem we obtain

$$h^* = 1.9241\epsilon^{1.1559},$$

and for (PFO)

$$h^* = 9.2325\epsilon^{1.7876},$$

although the relation for (PFO), for the reasons giving above, should be treated with scepticism. These relationships for (PF) and (PFO) are plotted in Figures 12 and 13 respectively, where the 'o's mark points obtained from numerical experiments.

Figure 14 shows the initial temperature field and the computed temperature field at $t = 0.1$ when $\epsilon = 0.0125$ and with 200 spatial grid points. The dotted lines indicate the position of the free boundary $s(t)$ at $t = 0$ and $t = 0.1$.

In computations we find that the profiles of the phase parameter in both (PF) and (PFO) over the time interval remain very close to the standing wave profiles (2.2). Consequently the number of spatial grid points in the interfacial region remains approximately constant in time.

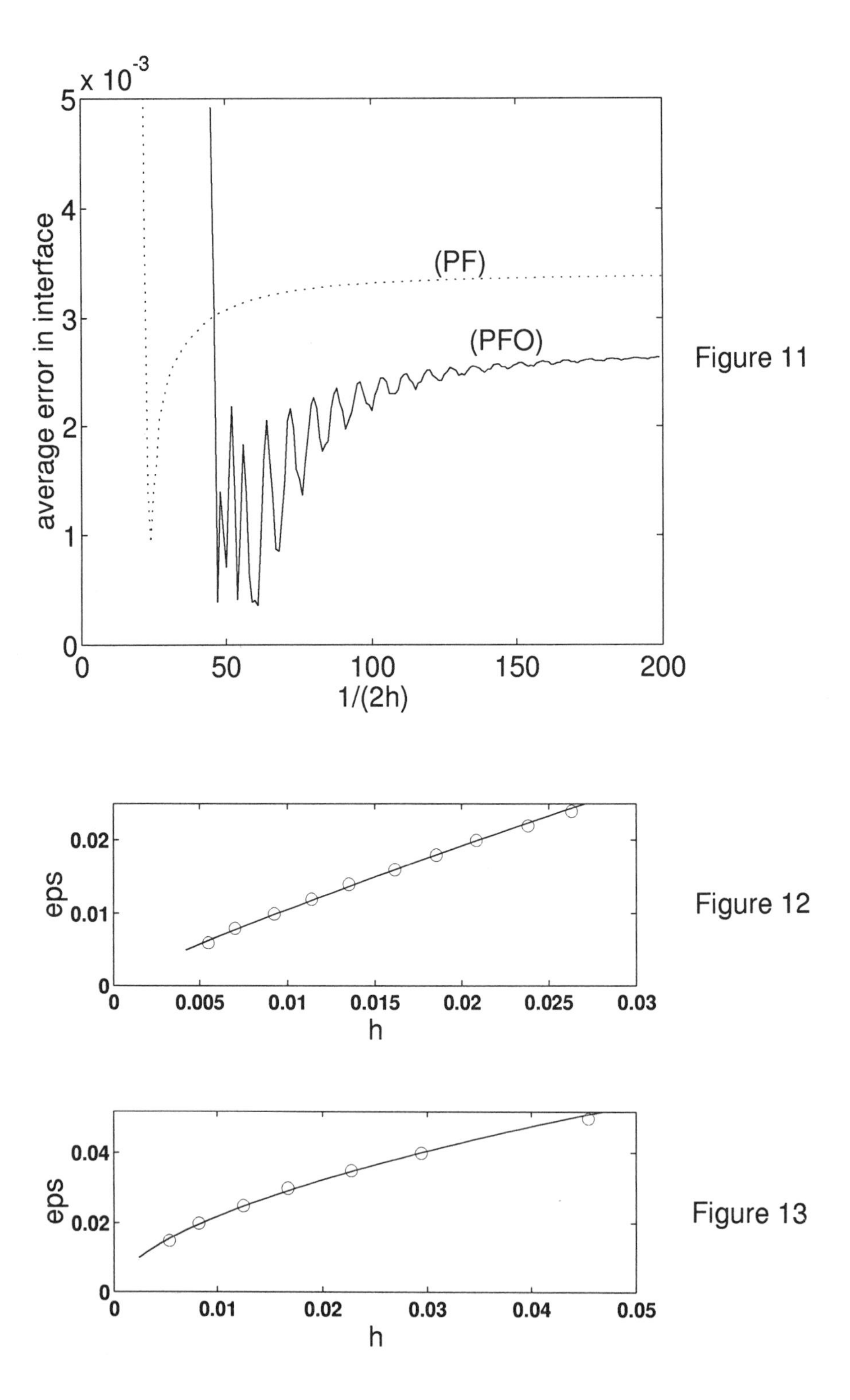

Figure 11

Figure 12

Figure 13

Problem IV

We consider the problem in the interval $x \in [1,2]$ with $t_0 = 0.3$ and $s_0 = 2(0.3)^{1/2}$. We take $\delta_v = 0.05$ and

$$B = -\frac{\theta_\infty}{\mathrm{erfc}1}, \qquad A = e^1\sqrt{\pi} + \frac{\theta_\infty}{\mathrm{erfc}1},$$

where $\theta_\infty = 0.2$. Taking 800 spatial grid points so that the discretisation error is negligible and computing the average error in the location of the free boundary in the interval $t \in [0.3, 0.3572]$, and the discrete L_2 error in the temperature at $t = 0.3572$ we obtain, for three values of ϵ, the results in Table 6.2.

ϵ	Average error in location of interface for $t \in [0.3, 0.3572]$		Discrete L_2 error in temperature at $t = 0.3572$	
	(PF)	(PFO)	(PF)	(PFO)
0.05	—	6.2611×10^{-3}	—	7.7128×10^{-3}
0.025	5.5850×10^{-3}	3.2471×10^{-3}	6.8771×10^{-3}	4.0624×10^{-3}
0.0125	2.7684×10^{-3}	1.6109×10^{-3}	3.6615×10^{-3}	2.0375×10^{-3}

Table 6.2

¿From these results it can again be seen that the reduction in the error as ϵ is reduced is of the $O(\epsilon)$ in both the position of the interface and the temperature. Again it can be seen that the (PFO) formulation provides the better results when the same parameters are used in both problems.

Finally, in Figure 15, we show the profile of the computed temperature at $t = 0.3572$, along with its initial profile $t = 0.3$. The dotted lines indicate the position of the free boundary at the two times.

7 Summary

1. The numerical experiments described in section 6 illustrate the convergence of the phase field models (PF, PFO, RD, RDO) to moving boundary problems. For Problems I and II the rate of convergence (2.3), $O(\epsilon^2)$ is confirmed for (RDO) and the experiments suggest the same rate holds for (RD). For the full phase field model the rate of convergence for both the moving boundary location and the temperature appears to be $O(\epsilon)$.

2. The attraction of the phase field field methodology for computing the solutions of moving boundary problems lies in the implicit way in which the interface is determined as the zero level surface of a phase variable. In principle one can use standard

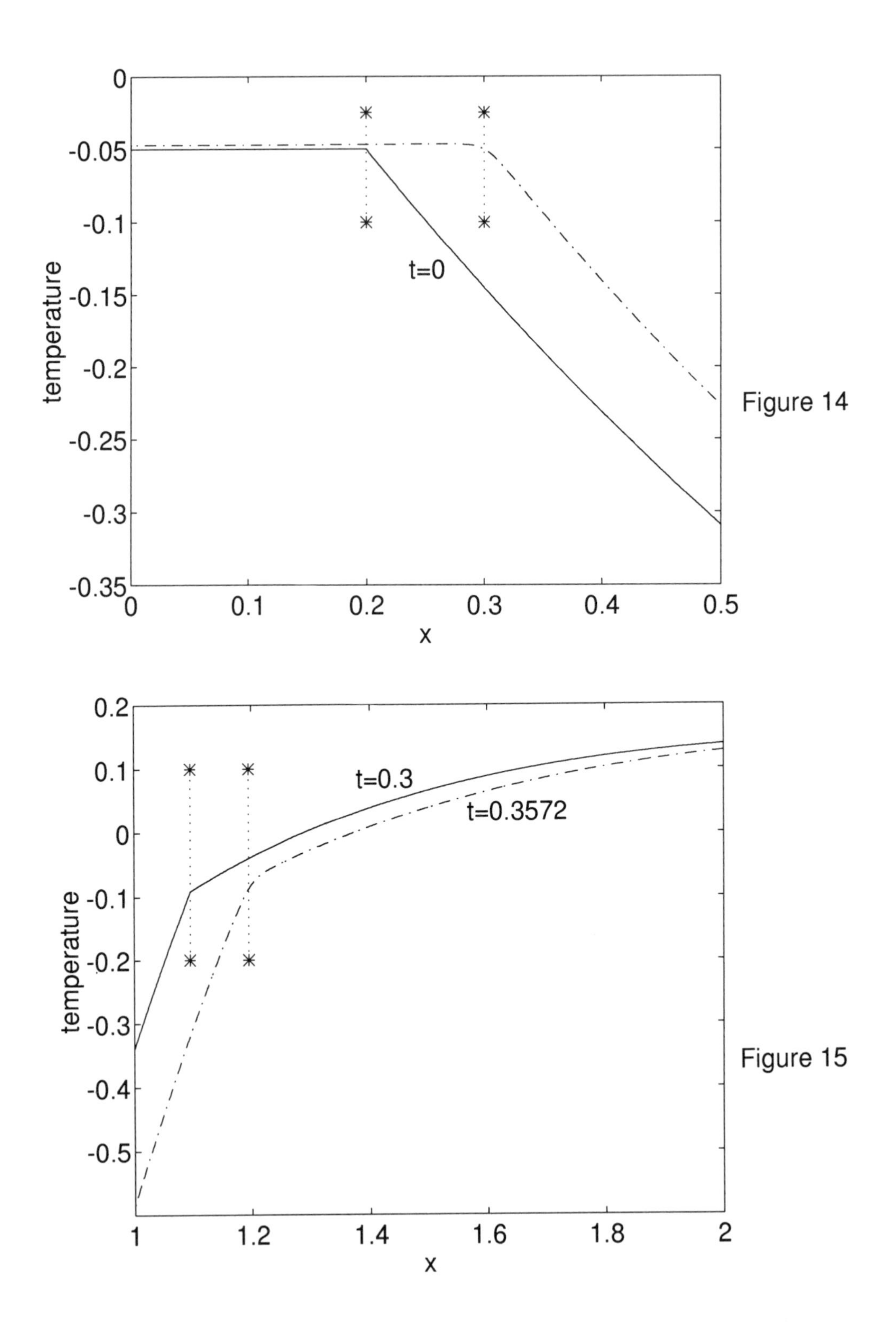

Figure 14

Figure 15

methods for solving nonlinear parabolic equations. (Actually this is particularly important in higher dimensions where the instability of interfaces between a solid and undercooled liquid leads to a complex morphology involving tip splitting of growing fingers and dendrite growth: (see Kobayashi [11], Wheeler, Murray and Schaefer [16]).) However one sees from these simple one dimensional computations that the phase variable u varies rapidly in a narrow region. This leads to a requirement that the spatial step size be small in order for the interfacial region to be resolved. Indeed it is natural to conjecture that letting h and ϵ tend to zero in the discretisation one requires h to go to zero faster than ϵ in order to 'fill up' the interfacial region with grid points. Indeed the optimal experimental h^* has this property for model problem III.

3. The double obstacle models (PFO, RDO) have the following clear advantages over their smooth counterparts (PF, RD) :-
   - The interface thickness is $O(\epsilon)$ compared with $O(\epsilon|\ln\epsilon|)$
   - Since there is no variation of u^ϵ outside the interfacial region the choice of boundary conditions poses no difficulty whereas for the smooth free energy an optimal choice of boundary condition for u^ϵ is unclear.
   - In principle u^ϵ need only be computed in a narrow region of width $O(\epsilon)$ whereas for the smooth model u^ϵ has to be computed everywhere. This is likely to be particularly useful in multi-dimensions and when adaptive spatial meshes are employed. (c.f. Paolini and Verdi [13]).
4. The smooth potential model (PF) has the advantage over the double obstacle model (PFO) in that it may be a good physical model for $\epsilon > 0$.

References

[1] J. F. Blowey, C. M. Elliott, "Curvature dependent phase boundary motion and parabolic double obstacle problems." Proceedings of the (IMA) (Minneapolis) workshop :- 'Degenerate Diffusions' ed. Wei-Ming Ni, L.A. Peletier and J. L. Vazquez. Springer Verlag, New York. **47**, 19 - 60, (1993).

[2] J.F. Blowey, C. M. Elliott, " A phase-field model with a double obstacle potential." To appear in 'Surface tension and motion by mean curvature' ed G. Buttazzo and A. Visintin, Gruyter, Berlin. (1993)

[3] G. Caginalp, "An analysis of a phase field model of a free boundary." Arch. Rat. Mech. Anal. **92**, 205-245 (1986)

[4] G. Caginalp, P. Fife, "Dynamics of layered interfaces arising from phase boundaries." SIAM J. Appl. Math. **48**, 506—518 (1988).

[5] G. Caginalp, E. Socolovsky, "Computation of sharp phase boundaries by spreading: the planar and spherically symmetric case." J. Comp Phys. **95**, 85—100 (1991).

[6] G. Caginalp, E. Socolovsky, "Phase field computations of single-needle crystals, crystal growth and motion by mean curvature" (to appear).

[7] X. Chen, "Generation and propagation of interface in reaction diffusion equations", J. Diff. Eqns. **96**, 116—141, (1992).

[8] X. Chen, C. M. Elliott, "Asymptotics for a parabolic double obstacle problem." Preprint submitted for publication.

[9] C.M. Elliott, J.R. Ockendon, "Weak and variational methods for free and moving boundary problems," Pitman, London, 1982.

[10] L.C. Evans, H. M, Soner, P. E. Souganidis, "Phase transitions and generalized motion by mean curvature" Comm. Pure. Appl. Math. **XLV**, 1097 — 1123, (1992).

[11] R. Kobayashi, " Modeling and numerical simulations of dendritic growth", Physica (D), **63**, 410 — 423, (1993).

[12] R. H. Nochetto, M. Paolini, C. Verdi, "Sharp error analysis for curvature dependent evolving fronts", M^3AS (to appear).

[13] M. Paolini, C. Verdi, "Asymptotic and numerical analysis of the mean curvature flow with a space dependent relaxation parameter." Asymptotic Anal. **5**, 553 — 594, (1992).

[14] B. E. E. Stoth, "The coupled Stefan problem with the Gibbs-Thomson law as a singular limit of the phase-field equations in the radial case." (preprint)

[15] R. S. Varga, "Matrix Iterative Analysis." Prentice Hall, Englewood Cliffs, N.J., (1962).

[16] A.A. Wheeler, B.T. Murray, R. J. Schaefer, "Computation of dendrites using a phase field model" Physica (D), (to appear).

Acknowledgement

This work was supported by the SERC by the grants GR/F85659 and GR/H61445.

Professor C. M. Elliott
Dr A. R. Gardiner
Centre for Mathematical Analysis and its Applications
University of Sussex
Brighton
BN1 9QH
UK

J DE FRUTOS AND J M SANZ-SERNA

Erring and being conservative

1 Introduction

The classical analysis of numerical methods for time-dependent, ordinary or partial differential equations is based on the ideas of stability, consistency and convergence. Roughly speaking, consistency means small local errors and stability means that errors do not build up catastrophically. Together, consistency and stability yield convergence: small (global) errors. However it is clear that there are useful theoretical properties of a method beyond its consistency, stability and convergence. Here we are interested in conserved quantities: the differential equations being integrated may possess one or several quantities (mass, energy, etc.) that are conserved in the true evolution and it is reasonable to demand that the numerical scheme also preserves those quantities. Several reasons are usually invoked for using schemes with such conservation properties. In a recent paper [6], C.W. Gear writes "In some cases, failure to maintain certain invariants leads to physically impossible solutions". In other cases conservation quantities are deemed important to avoid spurious blow-up of the numerical solution. In a classical paper [1], Arakawa writes "If we can find a finite difference scheme which has constraints analogous to the integral constraints of the differential form, the solution will not show the false 'noodling', followed by computational instability".

Since, in general, a quantitatively accurate solution cannot be unphysical, the preceding and similar arguments in favour of conservative schemes only apply to long-time integrations. Here the term long-time refers to integrations on time intervals $0 < t < t_{max}$ that are so long, relatively to the step-size Δt being used, that errors become large and the numerical solutions deviate significantly from the theoretical solutions. Therefore the preceding arguments *seem* to imply that *while errors are small* conservation properties are not too important; they become important when t_{max} is so large that the integration goes very wrong quantitatively. In the latter, long-time regime, conservative methods are quantitatively wrong but qualitatively acceptable, while nonconservative numerical solutions may be unacceptable from both the quantitative and the qualitative viewpoints.

However such an assessment of the merits of conservative schemes is too severe. In actual fact, in many cases, conservative schemes have better error propagation mechanisms that render them superior from a quantitative point of view and they should be preferred even for computations where the numerical solution remains close to the theoretical solution. An instance is presented in [3]. It is shown there that, when integrating the two-body problem with some conservative schemes, the leading term of the global error grows linearly with t, while for 'general' schemes the growth is quadratic. This makes conservative methods more efficient than general methods when accurate solutions are needed.

In the present paper we use the Korteweg-de Vries (KdV) equation as a model case. Only soliton solutions are considered, but this particular solution is particularly relevant because other solutions asymptotically give rise to solitons. After presenting the differential equation (Section 2) and the numerical methods considered (Section 3), we describe the behaviour of the numerical solutions by means of soliton perturbation theory (Section 4). It turns out that schemes that preserve the integrals of the solution and the solution squared behave much better than 'general' schemes. Numerical illustrations are presented in Section 5. In the final Section 6 we consider the same issue by using functional analytic techniques. It is shown rigoroursly that, for conservative schemes, the leading term of the global error consists of an error in the soliton phase, plus errors that are uniformly bounded for all positive times.

The main observation in the paper is that, if we look at the local error of a numerical method as a vector in a suitable phase space, then conservation properties imply constraints for the direction of the local error. When local errors build up to give rise to the global error, their directions are not irrelevant: there are harmful directions that lead to faster error accumulation. In many instances, the local error of a conservative scheme has a *direction* that renders it relatively harmless and this gives the scheme an advantage. These features are not captured by standard convergence analyses, which just take into account the *size* of the local error.

2 The Korteweg-de Vries equation

We write the KdV equation in the form

$$u_t - 6uu_x + u_{xxx} = 0, \quad -\infty < x < \infty, \quad t > 0. \tag{2.1}$$

The symbol Φ_t, $t \geq 0$, will denote the t-flow of the equation: if $u_0 = u_0(x)$ is a function of the spatial variable x, then $\Phi_t(u_0)$ is the function of x given by $u(\cdot, t)$, where $u(\cdot,\cdot)$ is the solution of (2.1) with initial condition $u(x, 0) = u_0(x)$, $-\infty < x < \infty$.

Among the many remarkable properties of (2.1) we focus on two: the existence of solitons and the existence of conservation laws.

A function $\phi(\xi)$, $\xi = x - ct$ provides a solution of (2.1) if ($' \equiv d/d\xi$)

$$-c\phi' - 6\phi\phi' + \phi''' = 0,$$

or, if ϕ and its derivatives vanish at ∞,

$$-c\phi - 3\phi^2 + \phi'' = 0. \tag{2.2}$$

After an elementary integration of this second-order ordinary differential equation, it is found that for $c > 0$ and real d, the function

$$\phi(\xi; c, d) = -\frac{c}{2} \operatorname{sech}^2 \frac{\sqrt{c}}{2}(\xi + d), \tag{2.3}$$

provides, through the recipe

$$u(x, t) = \phi(x - ct; c, d), \tag{2.4}$$

a soliton (solitary travelling wave solution) of (2.1) moving at velocity c and initially located at $x = -d$. Obviously, varying the value of d results in a translation of (2.4) along the x-axis. Note that the amplitude or depth of (2.3) is $-c/2$; thus deeper solitons travel faster than shallower solitons.

We now turn our attention to conservation laws for (2.1). There is an infinite number of these, but we only need the first two. For smooth solutions of (2.1) the following quantities do not vary with t:

$$I_1(u) = \int_{-\infty}^{\infty} u(x,t)\,dx, \tag{2.5}$$

$$I_2(u) = \int_{-\infty}^{\infty} u^2(x,t)\,dx. \tag{2.6}$$

3 Numerical methods

We consider semidiscrete (discrete t, continuous x) numerical methods for (2.1). Fully discrete and continuous t, discrete x algorithms may also be treated by the techniques in this paper; for brevity they are not included. If Δt denotes the time step and U^n is the numerical solution at time level $t_n = n\Delta t$, then we write the (one-step) method in the symbolic form

$$U^{n+1} = \Psi_{\Delta t}(U^n). \tag{3.1}$$

The local error (at a function $u_0 = u_0(x)$) is, by definition,

$$L_{\Delta t}(u_0) = \Psi_{\Delta t}(u_0) - \Phi_{\Delta t}(u_0). \tag{3.2}$$

If p denotes the order of the method, then $L_{\Delta t}(u_0) = O(\Delta t^{p+1})$ for suitably smooth u_0. We also assume the existence, for smooth u_0, of an asymptotic expansion

$$L_{\Delta t}(u_0) = \Delta t^{p+1}\ell_{p+1}(u_0) + o(\Delta t^{p+1}), \tag{3.3}$$

where ℓ_{p+1} is an operator independent of Δt and involving differentiation with respect to x. This requirement is fulfilled for all methods of practical interest.

If I is a conserved quantity of (2.1), then $\Psi_{\Delta t}$ conserves exactly I if, for all Δt and u_0,

$$I(\Psi_{\Delta t}(u_0)) = I(u_0).$$

From here and $I(\Phi_{\Delta t}(u_0)) = I(u_0)$, it follows, via Taylor expansion, that

$$0 = I(\Psi_{\Delta t}(u_0)) - I(\Phi_{\Delta t}(u_0)) = \Delta t^{p+1} I'_{u_0}(\ell_{p+1}(u_0)) + o(\Delta t^{p+1}),$$

where I'_{u_0} is the first variation of I evaluated at u_0 and acting on $\ell_{p+1}(u_0)$. Hence

$$I'_{u_0}(\ell_{p+1}(u_0)) = 0. \tag{3.4}$$

For instance, for the functional I_2 in (2.6)

$$I(u_0 + \epsilon v) = I(u_0) + \epsilon \int_{-\infty}^{\infty} 2u_0(x)v(x)\,dx + o(\epsilon),$$

so that

$$I'_{u_0}(v) = \int_{-\infty}^{\infty} 2u_0(x)v(x)\,dx$$

and (3.4) means that for methods that conserve I_2 exactly

$$\forall u_0, \quad \int_{-\infty}^{\infty} 2u_0\ell_{p+1}(u_0)\,dx = 0. \tag{3.5}$$

As pointed out in the introduction it is useful to think of this as an orthogonality relation between $\ell_{p+1}(u_0)$ and u_0.

For I_1 in (2.5) obviously $I'_{u_0} = I_1$ and conservation implies

$$\forall u_0, \quad \int_{-\infty}^{\infty} \ell_{p+1}(u_0)\,dx = 0. \tag{3.6}$$

This relation implies geometrically that $\ell_{p+1}(u_0)$ is contained in the hyperplane of functions with vanishing integral.

It is clear that (3.4) holds not only for methods that conserve I exactly, but for all methods for which

$$I(\Psi_{\Delta t}(u_0)) - I(u_0) = o(\Delta t^{p+1}). \tag{3.7}$$

In what follows, it should be understood that the properties that we prove by using (3.4) for methods that exactly conserve I also hold for methods that satisfy the weaker requirement (3.7).

4 Perturbation theory

Following the well-known methodology of modified equations [14], [7], we now introduce the (modified) equation

$$u_t - 6uu_x + u_{xxx} = \Delta t^p \ell_{p+1}(u), \quad -\infty < x < \infty, \quad t > 0. \tag{4.1}$$

The flow $\tilde{\Phi}_{\Delta t}$ of this equation differs from the method mapping $\Psi_{\Delta t}$ in terms $o(\Delta t^{p+1})$; the right hand side of (4.1) counterbalances the leading term of the difference $\Psi_{\Delta t} - \Phi_{\Delta t}$. Since $\tilde{\Phi}$ and Ψ coincide to higher order than Φ and Ψ, it is expected that solutions of (4.1) describe the behaviour of the numerical solution U with higher accuracy than solutions of the original KdV equation (2.1) being integrated.

Perturbation theory [10], [11] can be used to describe the solution of (4.1) when the initial condition is a soliton profile $\phi(x; c, d)$. It turns out that, to leading order in the perturbation parameter Δt^p, the modified solution is of the form

$$\sigma(x,t) + w(x,t) \tag{4.2}$$

where w is a function to be discussed later and

$$\sigma(x,t) = \phi(x - \mu(t); 4\kappa^2(t), d) = -2\kappa^2(t)\,\mathrm{sech}^2\,\kappa(t)[x - \mu(t) + d],$$

with κ and μ given by the differential equations

$$\frac{d\kappa}{dt} = \frac{M(\kappa)\Delta t^p}{8\kappa^2}, \tag{4.3}$$

$$\frac{d\mu}{dt} = 4\kappa^2 - \frac{N(\kappa)\Delta t^p}{8\kappa^2}, \tag{4.4}$$

with initial conditions $\kappa(0) = \sqrt{c}/2$, $\mu(0) = 0$. In turn, M, N in (4.3)–(4.4) are functions of κ given by

$$M(\kappa) = \int_{-\infty}^{\infty} \ell_{p+1}(\phi(x; 4\kappa^2, 0))\, \phi(x; 4\kappa^2, 0)\, dx,$$

$$N(\kappa) = \int_{-\infty}^{\infty} \ell_{p+1}(\phi(x; 4\kappa^2, 0)) \left[\kappa x \operatorname{sech}^2 \kappa x + \tanh\ \kappa x + \tanh^2 \kappa x\right] dx.$$

Note that σ in (4.2) has, for each fixed value of t, the shape of the soliton of depth $-2\kappa^2(t)$. For the unperturbed problem with $\Delta t = 0$, (4.3) shows that $-2\kappa^2$ remains equal to its initial value $-c/2$. Furthermore, in the unperturbed case, $d\mu/dt = c$, $\mu(t) = ct$, so that σ in (4.2) reproduces the soliton solution $\phi(x - ct; c, d)$. When the perturbation is present, the perturbed soliton depth $-2\kappa^2$ evolves according to (4.3) and furthermore the phase velocity $d\mu/dt$ also undergoes changes.

For a scheme that satisfies (3.5), $M \equiv 0$ and the depth of σ does not vary with t. Furthermore, from (4.4) we see that the phase velocity of σ is of the form $c + m\Delta t^p$, with $m = -N(\sqrt{c}/2)/(2c)$, a constant. Hence, for schemes that conserve I_2 exactly, the σ contribution to (4.2) is a soliton of the correct depth travelling at a perturbed, constant velocity.

On the other hand when (3.5) does not hold, then, in general, $M \neq 0$, and solving the equations (4.3)–(4.4) to leading order in Δt, we find that the perturbed soliton depth is given by

$$-2\kappa^2(t) = -\left(\frac{c}{2} + \frac{M(\sqrt{c}/2)\Delta t^p}{\sqrt{c}} t\right) + O(\Delta t^{2p}),$$

with a linear variation with t at leading order. Substitution in (4.4) reveals that then the perturbed phase velocity differs from the unperturbed c in terms $t\Delta t^p$, which in turn implies that the phase $x - \mu(t)$ in (4.2) differs from the unperturbed $x - ct$ in terms $t^2\Delta t$. Thus, *in the 'general' case and to leading order in Δt, depth errors grow linearly with t and phase errors grow quadratically with t. This is to be compared with the case of conserved I_2 where there is no depth error and phase errors grow linearly.*

Note that in the perturbation formulae, the *direction* of ℓ_{p+1} is important; while in conventional error analysis only the size of the local error is taken into account.

The function w in (4.2) has not been discussed so far. This is also $O(\Delta t^p)$ and describes, on the one hand, the change in the soliton shape due to the perturbation and, on the other, a *tail* induced by the perturbation. Since the tail gets in general longer and longer as t increases (see the numerical experiments in the next section), w grows with t. However, perturbation theory shows that there is no tail formation when

$$\int_{-\infty}^{\infty} \ell_{p+1}(\phi(x; 4\kappa^2, 0))\ \tanh^2 \kappa x\, dx = 0. \tag{4.5}$$

We now observe that in view of the relation $\tanh^2 z = 1 - \operatorname{sech}^2 z$, (4.5) holds provided that $M \equiv 0$ and

$$\int_{-\infty}^{\infty} \ell_{p+1}(\phi(x; 4\kappa^2, 0))\, dx = 0,$$

i.e. if the numerical method exactly preserves I_1 and I_2. In this case, w merely represents an $O(\Delta t^p)$ distortion in the soliton shape.

5 Numerical experiments

We first consider the standard backward Euler rule. For an equation $du/dt = A(u)$ this is given by

$$U^{n+1} - U^n = \Delta t A(U^{n+1}).$$

It is well known that the leading term of the truncation error for this rule is $(1/2)\Delta t^2 u_{tt}$. After replacing time derivatives by x-derivatives, we find that (3.3) holds with $p = 1$ and

$$\ell_2(u_0) = \partial_x \left[18u_0^2 \partial_x u_0 - 6u_0 \partial_x^3 u_0 - 9\partial_x u_0 \partial_x^2 u_0 + \frac{1}{2}\partial_x^5 u_0 \right].$$

To implement the method, we introduce a computational domain $-20 < x < 20$ and replace ∂_x by its standard pseudospectral approximation on a grid consisting of 128 equally spaced points. The errors introduced by this spatial discretization are negligible when compared with the time integration errors. For implementation details see [4].

The initial condition is chosen to be $\phi(x; 4, 10)$, i.e. we are dealing with a soliton of velocity $c = 4$ (depth -2) initially located at $x = -10$. The integration is carried out for $0 < t < t_{max} = 6$, so that the soliton travels a distance of 24 units.

For the dissipative backward Euler rule I_2 is not conserved. Actually $M < 0$, and perturbation theory predicts a decrease in the soliton depth as t increases. Furthermore (4.5) does not hold and a tail is expected.

Fig. 5.1 shows the solution at t_{max} when $\Delta t = 1/160$. The continuous line depicts the true position of the soliton ϕ and the crosses give the numerical solution U. It is clear that, in agreement with (4.2) the computed solution consists of a soliton profile plus a tail. The dotted line is the modified soliton σ computed by the the formulae (4.3)–(4.4): the agreement with the numerical solution is excellent. This shows both the ability of the modified equation to approximate the numerical solution and the success of the perturbation theory in describing the behaviour of the solutions of the modified equation (4.1).

Fig. 5.2 gives, for $\Delta t = 1/160, 1/320, 1/640$ the maximum norm of the error $u(\cdot, t_n) - U^n$ as a function of t. While the errors are not too large, say below 0.6, they grow quadratically, as predicted by the perturbation theory. When the phase of the soliton is completely wrong, so that the computed and true soliton do not overlap, the maximum norm of the error is 2. Thus as $t \to \infty$ the errors saturate.

We consider next the nondissipative implicit midpoint rule ($p = 2$)

$$U^{n+1} - U^n = \Delta t A\left(\frac{1}{2}[U^n + U^{n+1}]\right),$$

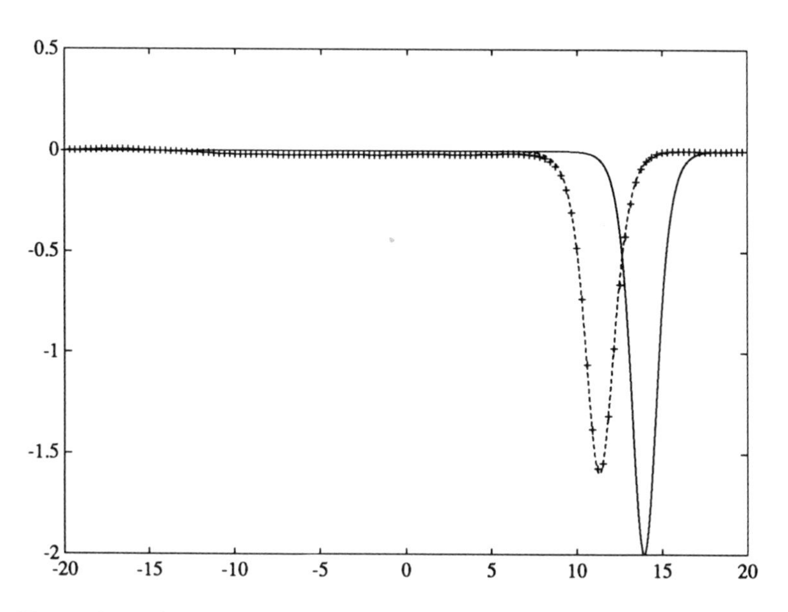

Figure 5.1: True soliton (solid line), modified soliton σ (dotted line) and numerical solution (crosses) for the backward Euler rule, $t = 6$, $\Delta t = 1/160$

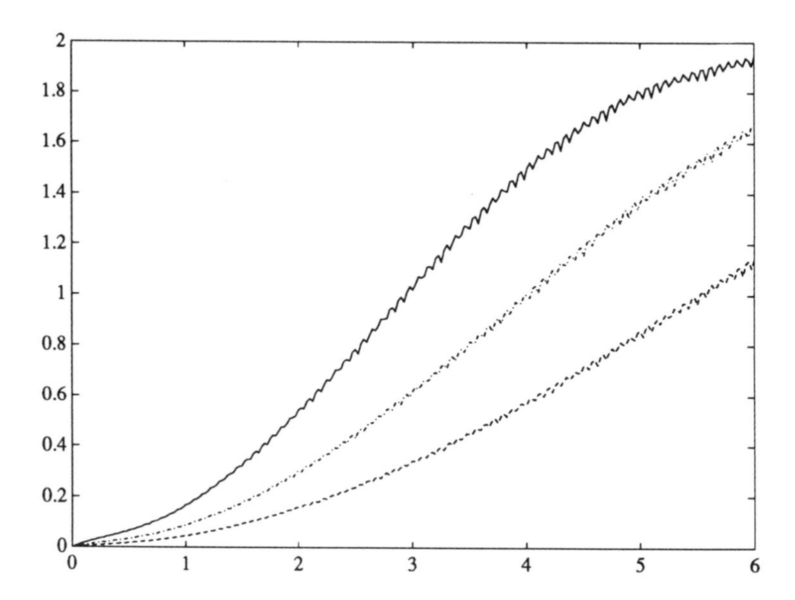

Figure 5.2: Maximum norm of the error as a function of time for the backward Euler rule, $\Delta t = 1/160, 1/320, 1/640$

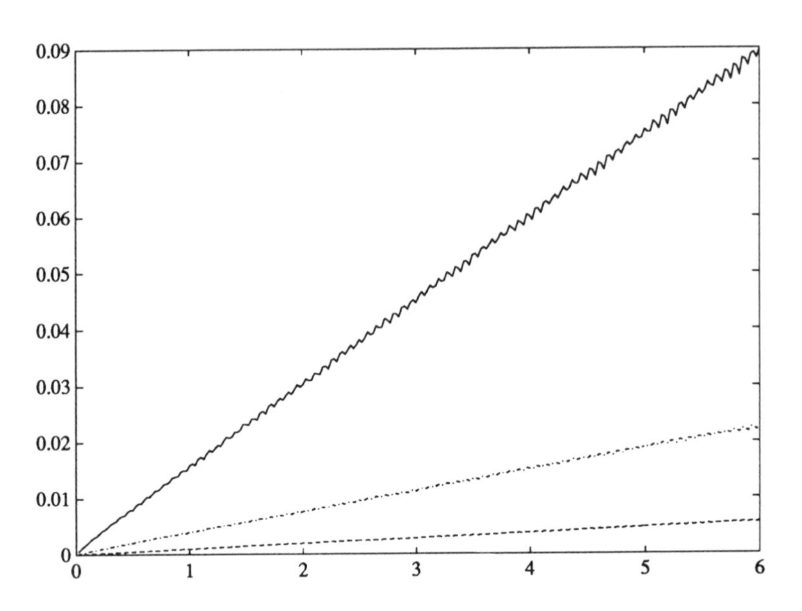

Figure 5.3: Maximum norm of the error as a function of time for the midpoint rule, $\Delta t = 1/40, 1/80, 1/160$

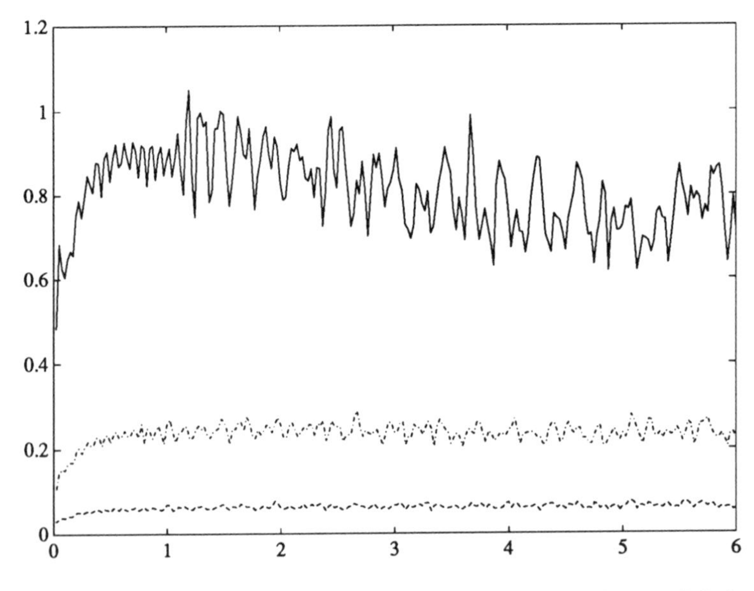

Figure 5.4: Maximum norm $\times 10^3$ of the error with respect to the modified soliton as a function of time for the midpoint rule, $\Delta t = 1/40, 1/80, 1/160$

Table 5.1: Midpoint rule errors with respect to the true solution u and modified soliton σ

	$\|U^n - u(\cdot,t_n)\|_\infty$		$\|U^n - \sigma(\cdot,t_n)\|_\infty$	
Δt	$t_n = 3$	$t_n = 6$	$t_n = 3$	$t_n = 6$
1/40	4.46E-2	8.98E-2	8.60E-4	7.22E-4
1/80	1.12E-2	2.27E-2	2.39E-4	2.22E-4
1/160	2.79E-3	5.69E-3	6.67E-5	5.68E-5

which conserves linear and quadratic invariants [12], [13] and in particular I_1 and I_2. Perturbation theory predicts no tail formation, correct depth and linear growth of phase errors. Experiments (not all of them are shown here) confirm these expectations. Fig. 5.3 gives the maximum norm of the error as a function of t when $\Delta t = 1/40, 1/80, 1/160$: errors clearly grow linearly with t. For the same runs, we have given in Fig. 5.4, the maximum norm of the difference between the computed U^n and the perturbed soliton σ in (4.2). Table 5.1 provides at $t = 3$ and $t = 6$ the maximum norm of the (true) error $U^n - \phi(\cdot - ct_n; c, d)$ with respect the true solution and of the error $U^n - \sigma(\cdot, t_n)$ with respect the perturbed soliton. In the table we see that the errors with respect to σ are much smaller than the true errors with respect to ϕ. In other words, the bulk of the true error consists of the phase error that is removed by comparing with the relocated $\sigma = \phi(x - \mu(t_n); c, d)$ rather than with the KdV solution $\phi(x - ct; c, d)$. However the error with respect to σ still shows an $O(\Delta t^2)$ behaviour: this is due to the w contribution to (4.2). On the other hand note, both in the table and in Fig. 5.4, that errors with respect to σ do not grow with t: with no tail formation w only accounts for a distortion in soliton shape and does not experiment secular growth. To sum up, in this case, experiments show that the true error consists of a phase error, with a $t\Delta t^2$ behaviour, plus further $O(\Delta t^2)$ errors that remain bounded as t increases.

For another application of soliton perturbation theory to numerical methods see [9].

Of course, the aim of the experiments above was not to show that the dissipative, first-order backward Euler rule is not suitable for the problem at hand. This fact is universally appreciated. Our goal was to point out that methods that behave differently with respect to conservation laws also possess different error propagation mechanisms.

6 Analytic results

6.1 Preliminaries

The results mentioned so far are not rigorous. To begin with, asymptotic expansions have been used, without paying attention to the norms in which they may be valid. The combinations of values of t and Δt for which the perturbation formulae are valid have not been spelled out. Furthermore, even though the technique of modified equations may be used *in some cases* in a rigourous way [7], this would require, if at all feasible, a detailed mathematical analysis of the (method dependent) modified equation (4.1). In this section we present a different, but related approach. The key idea is to note that for most practical methods the global error $U^n - u(\cdot, t_n)$ possesses an asymptotic expansion, whose leading term is a solution of a variational equation, see e.g. [8]. We therefore start by studying the variational equation of the KdV equation.

We use the standard Sobolev spaces H^k, k an integer, with norm $\|\cdot\|_k$.

6.2 The homogeneous variational equation around a soliton solution

In what follows, we fix a velocity $c_0 > 0$ and consider the moving coordinates $X = x - c_0 t$, $T = t$, in which the KdV equation becomes

$$u_T - c_0 u_X - 6uu_X + u_{XXX} = 0. \tag{6.1}$$

In the new variables the soliton (2.4) with velocity c_0 and $d = 0$ becomes a T-independent (equilibrium) solution $\psi(X) := \phi(X; c_0, 0)$ of (6.1). Similarly, the soliton of velocity $c \neq c_0$ of (2.1) becomes a soliton of velocity $c - c_0$ of (6.1).

We study perturbations of the equilibrium ψ. If $e(X,0)$ is a small perturbation, then, to the first order of small quantities, the solution of (6.1) whith initial condition $\psi(X)+e(X,0)$ is $\psi(X)+e(X,T)$, where $e(X,T)$ satisfies the (linear) variational equation

$$e_T - c_0 e_X - 6\psi(X)e_X - 6\psi'(X)e + e_{XXX} = 0, \tag{6.2}$$

or

$$e_T = \frac{\partial}{\partial X} L_2 e,$$

with the second-order operator L_2 given by

$$L_2 e = (c_0 + 6\psi)e - e_{XX}. \tag{6.3}$$

Alternatively, the same variational equation may be obtained by first linearizing in (2.1) and then changing variables $(x,t) \to (X,T)$.

Let us find some particular solutions of (6.2). We start from the easily proved relations

$$L_2 \psi' = 0 \tag{6.4}$$

and

$$L_2 \chi = -\psi, \tag{6.5}$$

where $\chi(X) = \phi_c(X; c_0, 0)$.

From (6.4) we see that $\psi'(X)$ is an equilibrium (i.e. T-independent) solution of (6.2). This is readily interpreted: $\psi(X) + \epsilon\psi'(X)$ coincides except for $o(\epsilon)$ terms with $\psi(X+\epsilon)$, i.e. with the soliton of velocity 0 located at $-\epsilon$, a solution of (6.1). Hence $\epsilon\psi'(X)$ is a solution of the variational equation.

On the other hand, by using (6.5)-(6.4), it is a trivial task to prove that,

$$\chi(X) - T\psi'(X) \tag{6.6}$$

also provides a solution of (6.2). The interpretation of this fact is as follows: Consider the initial condition $\psi(X) + \epsilon\chi(X)$ for (6.1). This is, except for order $o(\epsilon)$ terms, the initial profile of the soliton of velocity ϵ located at the origin. At time $T > 0$, this has travelled a distance ϵT so that the solution of (6.1) with initial condition $\psi(X) + \epsilon\chi(X)$ is $\psi(X - \epsilon T) + \epsilon\chi(X - \epsilon T) + o(\epsilon)$. Hence

$$\epsilon(\chi(X) - T\psi'(X)) = [\psi(X - \epsilon T) + \epsilon\chi(X - \epsilon T)] - \psi(X) + o(\epsilon)$$

has to satisfy the variational equation (6.2).

6.3 Stability of the homogeneous variational equation

The expression (6.6) shows that the variational equation (6.2) possesses solutions that *grow unboundedly* with T. This corresponds to the fact that ψ is not a Lyapunov-stable equilibrium of (6.1): if the initial profile is changed from ψ into that of a soliton with a small velocity the difference between ψ and the new soliton solution grows unboundedly. However, Benjamin [2] proved that ψ is stable in the sense that if $u(X,T)$ is a solution of (6.1) with $u(X,0) - \psi(X)$ small, then, for $T > 0$, a suitable translation $u(X - \mu(T), T)$ of $u(X,T)$ remains close to $\psi(X)$ (uniformly in T). Thus the Lyapunov instability only manifests itself as a phase error. The techniques used by Benjamin can be applied to prove the following result [5].

Theorem 6.1 *There exists a constant $C > 0$ such that if $e(X,T)$ is a solution of (6.2) then, for all $T > 0$,*

$$\|e(\cdot,T) - \mu(T)\psi'(\cdot)\|_1 \leq C\,\|e(\cdot,0)\|_1 \qquad (6.7)$$

with $\mu(T) = (e(\cdot,T),\psi')/(\psi',\psi')$.

The theorem shows that, even though the solutions $e(X,T)$ may grow unboundedly with T, they only do so in the direction of ψ', cf. (6.6). This of course corresponds to secular growth of *phase errors* in the KdV equation (6.1).

6.4 The nonhomogeneus variational equation

We now look at the nonhomogeneous initial-value problem

$$e_T = \frac{\partial}{\partial X} L_2 e + s, \quad T > 0, \qquad (6.8)$$

$$e(T = 0) = 0, \qquad (6.9)$$

where the source s is assumed to be independent of T. The following result holds.

Theorem 6.2 *Assume that the (T-independent) source term s is in L^2, is the (distributional) derivative of an L^2 function S and satisfies*

$$(s,\psi) = 0. \qquad (6.10)$$

Then, for a constant C independent of s and T,

$$\left\| e(\cdot,T) - \frac{(e(\cdot,T),\psi')}{(\psi',\psi')}\psi' \right\|_1 \leq C\,\|s\|_{-1}, \quad T > 0. \qquad (6.11)$$

Proof. We first look for a T-independent solution e_1 of (6.8). This satisfies

$$0 = e_{1,T} = \frac{\partial}{\partial X} L_2 e_1 + s = \frac{\partial}{\partial X} L_2 e_1 + \frac{\partial}{\partial X} S,$$

or

$$L_2 e_1 = -S. \tag{6.12}$$

The kernel of the operator L_2 mapping H^2 in L^2 is spanned by ψ'. Hence (Fredholm's alternative) (6.12) has a solution e_1 provided that $(-S, \psi') = 0$, a condition that follows from (6.10). Furthermore e_1 is uniquely defined under the additional condition $(e_1, \psi') = 0$ and with e_1 defined in this way

$$\|e_1\|_2 \leq C \|S\|_0 = C \|s\|_{-1}, \tag{6.13}$$

with C a constant associated with L_2

We now set $e_2 = e - e_1$, where e_2 has to solve

$$e_{2,T} = \frac{\partial}{\partial X} L_2 e_2,$$
$$e_2(T = 0) = -e_1.$$

By (6.7) and (6.13)

$$\left\| e_2(\cdot, T) - \frac{(e_2(\cdot, T), \psi')}{(\psi', \psi')} \psi' \right\|_1 \leq \|e_1\|_1 \leq C \|s\|_{-1}.$$

From this bound and after noticing that $(e_2(\cdot, T), \psi') = (e(\cdot, T), \psi')$, we conclude

$$\left\| e(\cdot, T) - \frac{(e(\cdot, T), \psi')}{\psi', \psi')} \psi' \right\|_1 \leq \|e_1\|_1 + C \|s\|_{-1}$$

and a new application of (6.13) concludes the proof. □

We see that, once more, the theorem ensures that the component of e orthogonal to ψ' remains uniformly bounded for all $T > 0$. However the source term has to satisfy some qualifications: to possess an antiderivative in L^2 and to satisfy (6.10). The solution

$$T\chi - \frac{T^2}{2} \psi'$$

corresponding to the source $s = \chi$ reveals that when $(s, \psi) \neq 0$, the growth with T is not confined to the direction of ψ'. (This particular solution clearly corresponds to quadratic growth of phase errors along with linearly growing amplitude errors, as we found in our discussion in Section 4.)

Before closing this subsection it is important to observe that if s is in $L^2 \cap L^1$ then it has an antiderivative in L^2,

$$S(x) = \int_{-\infty}^{x} s(\bar{x})\, d\bar{x},$$

if and only if $I_1(s) = 0$ (so that the integral decays as $x \to \infty$). Hence for reasonably smooth sources s, the hypotheses of the theorem read $I_1(s) = 0$, $(s, \psi) = 0$; these are essentially the geometric constraints we found in Section 4. These constraints again identify directions in phase-space that are relatively harmless.

6.5 Conclusion

We now revert to the original (x, t) variables. The nonhomogeneous variational equation (6.8) becomes

$$e_t - 6\psi(x - c_0t)e_x - 6\psi'(x - c_0t)e + e_{xxx} = s(x - c_0t), \qquad (6.14)$$

with s a real function of a real variable ξ. The bound (6.11) becomes

$$\left\| e(\cdot, t) - \frac{(e(\cdot, t), \psi')}{(\psi', \psi')}\psi' \right\|_1 \leq C \|s\|_{-1}, \quad t > 0.$$

where now ψ' is to be interpreted as the function $\psi'(\cdot - c_0t)$.

Assume that for a numerical method $\Psi_{\Delta t}$ satisfying (3.3) it holds that the global error when integrating $\phi(x - c_0t; c_0, 0)$ has an asymptotic expansion

$$U^n = u(\cdot, t_n) + \Delta t^p e(\cdot, t_n) + R(\cdot, t),$$

where R is a residual with $\|R(\cdot, t)\|_1 = o(\Delta t^p)$ uniformly in bounded t intervals and e is the solution of the variational equation (6.14) with source term

$$s(\xi) = \ell_{p+1}(\psi(\xi)).$$

We can conclude from Theorem 2 that *the leading error term* $\Delta t^p e$ *consists of a phase error in the direction of* ψ' *plus errors that are uniformly bounded for all* t, *provided* that the local error satisfies $(\ell_{p+1}(\psi), \psi) = 0$ and $I_1(\ell_{p+1}(\psi)) = 0$. These conditions on the local error are once more satisfied for methods that exactly preserve I_1 and I_2.

Of course to apply rigorously the last result it is necessary to first check that the numerical method converges in H^1 and has the indicated asymptotic expansion for the global error. While this has not been done for the methods considered in the preceding section, we strongly feel that it is feasible by using routine techniques. The verification of the existence of such an asymptotic expansion would however take us too far away from the main purpose of this article: to show additional evidence for the fact that numerically preserving conservation laws not only implies better qualitative behaviour, but may also lead to better error bounds.

References

[1] A. Arakawa, *Computational design for long-term numerical integration of the equations of fluid motion: two-dimensional incompressible flow. Part I,* J. Comput. Phys., 1 (1966), 119–143.

[2] T. B. Benjamin, *The stability of solitary waves,* Proc. R. Soc. Lond. A., 328 (1972), 153–183.

[3] M. P. Calvo and J. M. Sanz-Serna, *The development of variable-step symplectic integrators, with application to the two-body problem,* SIAM J. Sci. Comput., (1993), to appear.

[4] J. de Frutos and J. M. Sanz-Serna, *An easily implementable fourth-order method for the time integration of wave problems,* J. Comput. Phys., 103 (1992), 160–168.

[5] J. de Frutos and J. M. Sanz-Serna, *Error growth and invariant quantities in numerical methods: a case study,* Applied Mathematics and Computation Reports, Report 1993/4, Universidad de Valladolid.

[6] C. W. Gear, *Invariants and numerical methods for ODEs,* Physica D, 60 (1990), 303–310.

[7] D. F. Griffiths and J. M. Sanz-Serna, *On the scope of the method of modified equations,* SIAM J. Sci. Comput., 7 (1986), 994–1008.

[8] E. Hairer, S. P. Nørsett and G. Wanner, *Solving Ordinary Differential Equations I, Nonstiff problems,* Springer, Berlin, 1987.

[9] R. L. Herman and C. J. Knickerbocker, *Numerically induced phase shift in the KdV soliton,* J. Comput. Phys., 104 (1993), 50–55.

[10] V. I. Karpman and E. M. Maslov, *Perturbation theory for solitons,* Sov. Phys. JETP, 46 (1977), 281–291.

[11] V. I. Karpman and E. M. Maslov, *Structure of tails produced under the action of perturbations of solitons,* 48 (1978), 252–259.

[12] J. M. Sanz-Serna, *Runge-Kutta schemes for Hamiltonian systems,* BIT, 28 (1988), 877-883.

[13] J. M. Sanz-Serna and M. P. Calvo, *Numerical Hamiltonian Problems,* Chapman and Hall, London, 1993.

[14] R. F. Warming and B. J. Hyett, *The modified equation approach to the stability and accuracy analysis of finite difference methods,* J. Comput. Phys., 14 (1974), 159–179.

Acknowledgement

This research has been supported by project DGICYT PB89-0351.

J. de Frutos and J.M. Sanz-Serna
Departamento de Matemática Aplicada y Computación
Facultad de Ciencias
Universidad de Valladolid
Valladolid, Spain

P E GILL, W MURRAY, D B PONCELEON AND M A SAUNDERS

Solving reduced KKT systems in barrier methods for linear programming

Abstract In barrier methods for constrained optimization, the main work lies in solving large linear systems $Kp = r$, where K is symmetric and indefinite.

For linear programs, these KKT systems are usually reduced to smaller positive-definite systems $AH^{-1}A^Tq = s$, where H is a large principal submatrix of K. These systems can be solved more efficiently, but $AH^{-1}A^T$ is typically more ill-conditioned than K.

In order to improve the numerical properties of barrier implementations, we discuss the use of "reduced KKT systems", whose dimension and condition lie somewhere in between those of K and $AH^{-1}A^T$.

We have implemented reduced KKT systems in a primal-dual algorithm for linear programming, based on the sparse indefinite solver MA27 from the Harwell Subroutine Library. Some features of the algorithm are presented, along with results from the *netlib* LP test set.

1 Introduction

We discuss barrier methods for solving a linear program (LP) expressed in the standard form

$$\begin{array}{ll} \underset{x}{\text{minimize}} & c^Tx \\ \text{subject to} & Ax = b, \quad l \le x \le u, \end{array} \tag{1.1}$$

where A is $m \times n$ $(m \le n)$. We assume that an optimal solution (x^*, π^*) exists, where π^* is a set of Lagrange multipliers for the constraints $Ax = b$.

Implicit within most of the current barrier or interior-point algorithms is a so-called *KKT system* of the form

$$K\begin{pmatrix} \Delta x \\ -\Delta\pi \end{pmatrix} = \begin{pmatrix} w \\ r \end{pmatrix}, \qquad K \equiv \begin{pmatrix} H & A^T \\ A & \end{pmatrix}, \tag{1.2}$$

in which H is positive semidefinite and diagonal. The search direction $(\Delta x, \Delta\pi)$ is used to update the current solution estimate (x, π). In some cases it is obtained by solving the same KKT system with more than one right-hand side. It is critical that such systems be solved quickly and reliably.

If H is nonsingular it is common to use it as a block pivot and reduce (1.2) to a system involving $AH^{-1}A^T$. In general this is an *unstable process* because H usually contains some very small diagonals. The main advantages are that $AH^{-1}A^T$ is much smaller than K and it is positive-definite.

Our aim is to discuss an intermediate strategy in which *part of* H is used as a block pivot. The reduced systems obtained are considerably smaller than K, and typically no more than twice as large as $AH^{-1}A^T$. The approach retains the numerical reliability of factoring the full matrix K, with an efficiency that is closer to that of using $AH^{-1}A^T$. It also provides a convenient way of dealing with *free variables* and *dense columns* (i.e., variables with bounds $-\infty \leq x_j \leq \infty$ and columns of A that have many nonzeros).

The proposed use of reduced KKT systems is motivated by the sensitivity analysis in [27] and by the investigation of preconditioners for KKT systems in [10]. Note from both references that *large* diagonals in H give K a deceptively high condition number, but they do not cause sensitivity in the solution of systems involving K. (Similarly, a system $Dx = b$ with $D = \text{diag}(10^{20}, 10^{10}, 1, 1, 1)$ has a well-defined solution, even though $\text{cond}(D) = 10^{20}$.)

In fact, large diagonals of H are *desirable*, since they are obvious candidates for a block pivot. For example, if $\|A\| \approx 1$, any diagonals H_{jj} significantly larger than one could be included in the block pivot. The associated *reduced* system reflects the true sensitivity of linear systems involving K.

2 A Primal-Dual LP Algorithm

For concreteness we describe the main parts of a primal-dual barrier algorithm for LP.[1] After deriving the KKT systems to be solved, we are able to discuss certain numerical issues.

In order to handle upper and lower bounds symmetrically, we slightly generalize the approach of Lustig *et al.* [15, 16] and restate problem (1.1) as follows:

$$\begin{array}{ll} \underset{x,\, s_1,\, s_2,\, p}{\text{minimize}} & c^Tx + \frac{1}{2}\|\gamma x\|^2 + \frac{1}{2}\|p\|^2 \\ \text{subject to} & \begin{array}{rcl} Ax \qquad\qquad + \delta p & = & b, \\ x - s_1 \qquad\qquad & = & l, \\ x \qquad + s_2 \qquad & = & u, \end{array} \end{array} \tag{2.1}$$

with s_1, $s_2 \geq 0$. The scalars γ and δ are intended to be "small". The dual variables associated with the three sets of equality constraints will be denoted by π, z and $-y$. At a solution, z and y are non-negative.

We assume that $l < u$, since fixed variables ($l_j = u_j$) can be absorbed into b. If $l_j = -\infty$ or $u_j = +\infty$, we omit the corresponding equation in $x - s_1 = l$ or $x + s_2 = u$. (In particular, both equations are omitted for free variables.) Symmetric treatment of the bounds via s_1 and s_2 allows a problem to be treated "as it stands", without converting the bounds to $0 \leq x \leq u$ (say). The latter practice is hazardous in the case of "almost free variables", whose bounds are not large enough to be treated as infinite (e.g., $-10^6 \leq x_j \leq 10^6$).

The perturbations involving γ and δ are included to "regularize" the problem. The term $\frac{1}{2}\|\gamma x\|^2$ ensures that $\|x^*\|$ is bounded, and the term δp allows $Ax = b$ to be satisfied in some least-squares sense if the original constraints have no feasible solution.

[1] Related references for LP are [22, 14, 21, 15, 16, 23, 24, 25].

Objective perturbations of the form $\frac{1}{2}\|\gamma x\|^2$ have been studied by Mangasarian *et al.* [19, 18], who show that the LP solution is not perturbed if γ is sufficiently small. The approach has been successfully pursued in the interior-point context by Setiono [30].

An alternative form of regularization is the proximal-point method of Rockafellar [28], which involves an objective term of the form $\frac{1}{2}\|\gamma(x - x_k)\|^2$ and does not perturb the problem as $x_k \to x^*$ even if γ is not particularly small. Again, this approach has been successfully explored by Setiono [29, 31, 32].

2.1 The Barrier Subproblem

Barrier methods include both "outer" and "inner" iterations. The outer iterations reduce a positive scalar *barrier parameter* μ toward zero, and the inner iterations are those used to solve a *barrier subproblem*, which is a nonlinear minimization problem with only equality constraints. For a given μ, the barrier subproblem for (2.1) is to minimize

$$c^T x + \tfrac{1}{2}\|\gamma x\|^2 + \tfrac{1}{2}\|p\|^2 - \mu \sum_j \ln(s_1)_j - \mu \sum_j \ln(s_2)_j$$

subject to the same equality constraints. Optimality conditions for this subproblem include the equation $p = \delta\pi$. We can therefore eliminate p immediately. The remaining optimality conditions may be stated as the following equations:

$$f_\mu(x, s_1, s_2, \pi, z, y) = \begin{pmatrix} b - Ax - \delta^2\pi \\ l - x + s_1 \\ u - x - s_2 \\ c + \gamma^2 x - A^T\pi - z + y \\ \mu e - S_1 Z e \\ \mu e - S_2 Y e \end{pmatrix} \equiv \begin{pmatrix} r \\ t_1 \\ t_2 \\ t \\ v_1 \\ v_2 \end{pmatrix} = 0,$$

where e is a column of ones and S_1, S_2, Z, Y are diagonal matrices composed from s_1, s_2, z, y.

If Newton's method is applied to the equations $f_\mu = 0$, we obtain the Newton equations $J_\mu d = -f_\mu$, where J_μ is the Jacobian of f_μ, and d is the Newton search direction $(\Delta x, \Delta s_1, \Delta s_2, \Delta\pi, \Delta z, \Delta y)$. The next iterate is found by taking a step along the Newton direction that maintains positive values of the variables s_1, s_2, z, y. In practice, taking different steps in the primal and dual variables leads to fewer iterations [15, 16, 9].

2.2 The Newton Direction

The Newton equations for generating a search direction $(\Delta x, \Delta s_1, \Delta s_2, \Delta\pi, \Delta z, \Delta y)$ are

$$\begin{array}{rrrrrrcl}
A\Delta x & & & + \ \delta^2\Delta\pi & & & = & r \\
\Delta x & -\Delta s_1 & & & & & = & t_1 \\
\Delta x & & +\Delta s_2 & & & & = & t_2 \\
-\gamma^2\Delta x & & & +A^T\Delta\pi & + \ \Delta z & - \ \Delta y & = & t \\
 & Z\Delta s_1 & & & +S_1\Delta z & & = & v_1 \\
 & & Y\Delta s_2 & & & +S_2\Delta y & = & v_2,
\end{array}$$

which may be solved by defining

$$\begin{aligned} H_0 &= S_1^{-1}Z + S_2^{-1}Y, \\ w &= S_1^{-1}(Zt_1 + v_1) + S_2^{-1}(Yt_2 - v_2) - t, \end{aligned}$$

solving the KKT-like system

$$K \begin{pmatrix} \Delta x \\ -\Delta \pi \end{pmatrix} = \begin{pmatrix} w \\ r \end{pmatrix}, \qquad K \equiv \begin{pmatrix} H_0 + \gamma^2 I & A^T \\ A & -\delta^2 I \end{pmatrix}, \tag{2.2}$$

and finally solving the equations

$$\begin{aligned} \Delta s_1 &= \Delta x - t_1, \\ \Delta s_2 &= t_2 - \Delta x, \\ S_1 \Delta z &= v_1 - Z\Delta s_1, \\ S_2 \Delta y &= v_2 - Y\Delta s_2. \end{aligned}$$

It is straightforward to show that any values of x and π give the same search direction $(\Delta s_1, \Delta s_2, \Delta z, \Delta y)$.

The above definition of K shows its dependence on the regularization parameters γ and δ. Later we shall not need H_0 itself, but will work with $H \equiv H_0 + \gamma^2 I$. An important property is that γ and δ help preserve the nonsingularity of K. For example, if A does not have full row rank, K is singular unless we choose $\delta > 0$. Similarly, if some columns of A associated with free variables are linearly dependent; then K is singular unless we choose $\gamma > 0$. (An alternative means for preserving nonsingularity with free variables is given in [8].)

Since $p = \delta\pi$ at a solution, the regularization terms in the objective function are effectively $\frac{1}{2}\|\gamma x\|^2 + \frac{1}{2}\|\delta\pi\|^2$, which has a satisfying symmetry and shows that both x^* and π^* are bounded if γ and δ are nonzero.

3 Reduced KKT Systems

If $H = H_0 + \gamma^2 I$ is nonsingular, (2.2) may be solved using the *range-space equations* of optimization:

$$(AH^{-1}A^T + \delta^2 I)\Delta\pi = r - AH^{-1}w, \qquad H\Delta x = A^T\Delta\pi + w. \tag{3.1}$$

The main benefit is that $AH^{-1}A^T + \delta^2 I$ is much smaller than K and is positive definite, so that well-established sparse Cholesky factorizations can be applied. Some drawbacks are as follows:

1. $AH^{-1}A^T + \delta^2 I$ is normally more ill-conditioned than K.

2. Free variables complicate direct use of (3.1) by making H_0 singular.

3. Relatively dense columns in A degrade the sparsity of $AH^{-1}A^T + \delta^2 I$ and its factors.

To keep H_0 nonsingular in the presence of free variables, some authors have treated each such variable as the difference between two nonnegative variables. Lustig, *et al.* [15, 16] report satisfactory performance on the *netlib* LP test set, which contains a few relevant examples. To deal with problems involving *many* free variables, other authors have introduced moving artificial bounds (e.g. [20] and Vanderbei [34], who cites difficulties with the first approach).

Dense columns have been handled by using Cholesky factors of the sparse part of $AH^{-1}A^T$ to precondition the conjugate-gradient method (e.g. [11]) or to form a certain Schur complement [20, 15]. A difficulty is that the "sparse" Cholesky factor usually becomes even more ill-conditioned than the one associated with all of $AH^{-1}A^T$. More recently, the approach of *splitting* or *stretching* dense columns has been proposed and implemented with success [17, 35]. The increased problem size is perhaps an inconvenience if not a difficulty. (See also Grcar [13], who recommends the term *stretching* and gives an extremely thorough development and analysis of this new approach to solving sparse linear equations.)

In general, small diagonals of H prevent it from being a "good" block pivot from a numerical point of view. That is, if Gaussian elimination were applied to K, not all diagonals of H would be acceptable as pivots. To overcome this difficulty, we note that in practice *most of* H is likely to be acceptable as a block pivot. Thus we partition H and A as

$$H = \begin{pmatrix} H_N & \\ & H_B \end{pmatrix}, \qquad A = \begin{pmatrix} N & B \end{pmatrix},$$

where the diagonals of H_N are a "reasonable" size compared to the nonzeros in the corresponding columns of N. In general, a reordering of the variables is implied. The column dimension of N may be anywhere from 0 to n (and similarly for B). The KKT system (2.2) becomes

$$\begin{pmatrix} H_N & & N^T \\ & H_B & B^T \\ N & B & -\delta^2 I \end{pmatrix} \begin{pmatrix} \Delta x_N \\ \Delta x_B \\ -\Delta\pi \end{pmatrix} = \begin{pmatrix} w_N \\ w_B \\ r \end{pmatrix},$$

which may be solved via the *reduced KKT system*

$$K_B \begin{pmatrix} \Delta x_B \\ -\Delta\pi \end{pmatrix} = \begin{pmatrix} w_B \\ r - NH_N^{-1}w_N \end{pmatrix}, \qquad K_B \equiv \begin{pmatrix} H_B & B^T \\ B & -NH_N^{-1}N^T - \delta^2 I \end{pmatrix}. \tag{3.2}$$

The final step is to solve the diagonal system $H_N \Delta x_N = w_N + N^T \Delta\pi$.

If H_N happens to be all of H, $K_B = -(AH^{-1}A^T + \delta^2 I)$ and the reduced KKT system is equivalent to (3.1). Otherwise, K_B is a symmetric indefinite matrix. Like K it can be processed by a sparse indefinite solver such as MA27 [4, 5] (an implementation of the factorization described in [2, 1]). The above difficulties are resolved as follows:

1. K_B need not be more ill-conditioned than K.

2. Free variables are always included in B.

3. Dense columns of A are also included in B. (They do not necessarily degrade the sparsity of the K_B factors.)

As with current Cholesky solvers, MA27 has an *Analyze* phase (to choose a row and column ordering for K_B and to set up a data structure for its factors) and a *Factor* phase (to compute the factors themselves). Some disadvantages of working with reduced KKT systems are:

1. Whenever the N–B partition is altered, a new *Analyze* is required. This is usually less expensive than the *Factor* phase, often by an order of magnitude, and we expect it to be needed only sometimes (typically the last few iterations). However, it can be costly for certain structures in A.

2. To date, sparse indefinite factorization is more expensive than Cholesky factorization when there is a large degree of indefiniteness (as in KKT systems).

Further recent work on solving large indefinite systems (to avoid the difficulties associated with $AH^{-1}A^T$) appears in [33, 24, 6, 37]. Iterative solution of the full KKT system is explored in [10].

4 Numerical Examples

To illustrate some numerical values arising in the reduced KKT systems of Section 3, we apply the basic primal-dual algorithm to two LP problems of the form

$$\min\ c^T x \quad \text{subject to} \quad Ax = b, \quad x \geq 0,$$

with

$$A = \begin{pmatrix} 1 & 1 & 3 & 3 \\ 1 & 2 & 1 & 2 \end{pmatrix}, \qquad c = \begin{pmatrix} 1 & 1 & 0 & 0 \end{pmatrix}^T,$$

using MATLAB™ [26] with about 16 digits of precision.

4.1 A Non-degenerate LP

We first let the right-hand side and optimal solution be

$$b = \begin{pmatrix} 6 \\ 3 \end{pmatrix}, \qquad x^* = \begin{pmatrix} 0 & 0 & 1 & 1 \end{pmatrix}^T.$$

At the start of the fifth iteration of the primal-dual algorithm, we have
$x = (3.9\text{e-}6,\ 3.2\text{e-}6,\ 1.000005,\ 0.999992)$, $z = (0.70,\ 0.70,\ 2.0\text{e-}6,\ 1.7\text{e-}6)$, and

$$K = \begin{pmatrix} 1.8\text{e+}5 & & & & 1 & 1 \\ & 2.2\text{e+}5 & & & 1 & 2 \\ & & 2.0\text{e-}6 & & 3 & 1 \\ & & & 1.7\text{e-}6 & 3 & 2 \\ 1 & 1 & 3 & 3 & & \\ 1 & 2 & 1 & 2 & & \end{pmatrix} = \begin{pmatrix} H & A^T \\ A & \end{pmatrix},$$

where $H = X^{-1}Z$. Let *Htol* define which diagonals of H are considered large enough to form a block pivot, i.e., the jth column of H is included in H_N (and column a_j of A is included in N) if $H_{jj} \geq Htol$. In terms of conventional error analysis for Gaussian elimination, $Htol = 1$ or 0.1 should be "safe", while $Htol < 10^{-3}$ (say) is likely to be unreliable. Various values of *Htol* give the following reduced KKT systems:

Htol	K_B	cond(K_B)
0.1	$\begin{pmatrix} 2.0\text{e-}6 & & 3 & 1 \\ & 1.7\text{e-}6 & 3 & 2 \\ 3 & 3 & -1\text{e-}5 & -1\text{e-}5 \\ 1 & 2 & -1\text{e-}5 & -2\text{e-}5 \end{pmatrix}$	7.5
1.8e-6	$\begin{pmatrix} 1.7\text{e-}6 & 3 & 2 \\ 3 & -4\text{e+}6 & -1\text{e+}6 \\ 2 & -1\text{e+}6 & -5\text{e+}5 \end{pmatrix}$	5e+6
1e-20	$\begin{pmatrix} -1\text{e+}7 & -5\text{e+}6 \\ -5\text{e+}6 & -3\text{e+}6 \end{pmatrix}$	58

The large diagonals of H make K seem rather ill-conditioned ($\text{cond}(K) = 10^5$), but pivoting on those diagonals ($Htol = 0.1$) gives a very favorable reduced system: $\text{cond}(K_B) = 7.5$. Allowing one small pivot ($Htol = 1.8\text{e-}6$) gives the expected large numbers and high condition: $\text{cond}(K_B) \approx 10^6$. A second small pivot would normally have a similar effect, but here we have $m = 2$. The large numbers arising from m small pivots happen to form a very *well*-conditioned reduced system: $\text{cond}(K_B) = \text{cond}(AH^{-1}A^T) = 58$.

In general, the structure of K is such that pivoting on *any* nonzero diagonals of H should be safe if the following conditions hold:

- There are m *or more* small pivots of similar size (to within one or two orders of magnitude).
- The associated m or more columns of A form a well-conditioned matrix.

Unfortunately, in the presence of primal degeneracy there will be *less than* m small pivots, as the next example shows.

4.2 A Degenerate LP

Now let the right-hand side and optimal solution be

$$b = \begin{pmatrix} 3 \\ 2 \end{pmatrix}, \qquad x^* = \begin{pmatrix} 0 & 0 & 0 & 1 \end{pmatrix}^T.$$

At the start of the sixth iteration of the primal-dual algorithm, we have $x = (3.6\text{e-}7,\ 6.0\text{e-}7,\ 9.3\text{e-}7,\ 0.9999987)$, $z = (0.61,\ 0.32,\ 0.29,\ 2.2\text{e-}7)$, and

$$K = \begin{pmatrix} 1.7\text{e+}6 & & & & 1 & 1 \\ & 5.3\text{e+}5 & & & 1 & 2 \\ & & 3.1\text{e+}5 & & 3 & 1 \\ & & & 2.2\text{e-}7 & 3 & 2 \\ 1 & 1 & 3 & 3 & & \\ 1 & 2 & 1 & 2 & & \end{pmatrix}.$$

Two representative values of *Htol* give the following reduced KKT systems:

Htol	K_B	cond(K_B)
0.1	$\begin{pmatrix} 2.2\text{e-}7 & 3 & 2 \\ 3 & -3\text{e-}5 & -1\text{e-}5 \\ 2 & -1\text{e-}5 & -1\text{e-}5 \end{pmatrix}$	8e+5
1e–20	$\begin{pmatrix} -4\text{e+}7 & -3\text{e+}7 \\ -3\text{e+}7 & -2\text{e+}7 \end{pmatrix}$	1e+13

We see that the partially reduced system has a considerably lower condition than the fully reduced system. After one further iteration, the contrast is even greater (see the second table below).

4.3 Condition Numbers at Each Iteration

To further illustrate the effect of small H pivots, we list the condition of the reduced KKT systems arising at each iteration of the primal-dual algorithm with various values of *Htol*. The barrier parameter μ does not appear in K, but it is listed for reference.

For the first (non-degenerate) problem, the following condition numbers cond(K_B) were obtained:

Itn	μ	*Htol*: 0.1	1e–3	1e–4	1.8e–6	1e–20
1	3e–2	14	14	14	14	14
2	2e–3	72	300	300	300	300
3	2e–4	8	59	59	59	59
4	2e–6	7	7	5e+4	74	74
5	2e–8	7	7	7	5e+6	59
6	2e–10	7	7	7	7	7
7	2e–12	7	7	7	7	7

We see that allowing pivots as small as $H_{jj} = 10^{-p}$ is likely to give cond(K_B) $= 10^p$ at some stage, except in the fortuitous case where are there are m or more small pivots.

For the degenerate problem, the following values of cond(K_B) were obtained:

Itn	μ	*Htol*: 0.1	1e-4	1e-20
1	2e-2	14	14	14
2	2e-4	23	23	23
3	6e-5	11	1e+3	1e+3
4	2e-5	114	114	7e+6
5	2e-7	8e+3	8e+3	1e+9
6	2e-9	8e+5	8e+5	1e+13
7	2e-11	8e+7	8e+7	2e+16
8	2e-13	8e+9	8e+9	∞

We see that very small pivots allow the condition of K_B to deteriorate seriously. We cannot expect a method based on $AH^{-1}A^T$ to make meaningful progress on this example beyond the sixth iteration. Since degeneracy is a feature of most real-life problems, it seems clear that small H pivots must be avoided if stability is to be assured.

To date, implementations based on $AH^{-1}A^T$ [16] or some other "unstable" factorization [37] appear capable of attaining 8 digits of precision on most real-life applications, but occasionally attain only 6 digits or less. This is commendable performance, since 6 digits is undoubtedly adequate for most practitioners. It is the "occasionally less" that we maintain some concern about!

5 Numerical Results from the *netlib* Collection

In this section we present results obtained from the *netlib* collection of LP test problems [7]. One aim is to explore the probable dimensions of the reduced KKT systems that must be solved (and determine how often a new *Analyze* is needed).

5.1 Implementation Details

Various run-time parameters are used to define starting points, stopping conditions, etc. (see Gill *et al.* [9]). We assume that all computations are performed with about 16 digits of precision.

To date we have used the Harwell Subroutine Library package MA27 [4, 5] to solve the reduced KKT systems. This is a multifrontal code designed to perform well on vector machines on matrices that are definite or nearly definite (i.e., most eigenvalues have the same sign).

Let $S = -(NH_N^{-1}N^T + \delta^2 I)$ be the Schur complement appearing in the south-east corner of K_B (3.2). In our implementation, three parameters are used to control the choice of N and the factorization of K_B, with typical values as follows:

$$\begin{aligned} \mathit{Htol} &= 10^{-6}, \\ \mathit{ndense} &= 10, \\ \mathit{factol} &= 0.01. \end{aligned}$$

Note that small diagonals of H_N lead to large entries in S, while "dense" columns in N lead to excessive density in S. We therefore use *Htol* and *ndense* to control the partitioning of K in the following way. The jth column of H is included in H_N (and column a_j of A is included in N) if

1. $H_{jj} \geq Htol\, \|a_j\|$, and

2. a_j has fewer than *ndense* nonzeros.

Since we scale A to give $\|a_j\| \approx 1$ for all j, we avoid storing an n-vector of norms by simplifying the first test to just $H_{jj} \geq Htol$. Including $\|a_j\|$ would give slightly greater reliability.

The third parameter *factol* is used as the stability tolerance u [4, 5] when the *Factor* phase of MA27 is applied to K_B. In the extreme case (N void, $K_B = -AH^{-1}A^T$), *factol* is inoperative since MA27 is then performing Cholesky factorization on a (negative) definite matrix. In all other cases, *factol* affects the stability of the numerical factorization and the fill-in in the factors (beyond that predicted by the MA27 *Analyze*).

The regularization values $\gamma = 10^{-5}$ and $\delta = 10^{-5}$ were used, with seemingly satisfactory results. Larger values may perturb the solution too much, and smaller values can lead to near-singularity in K and perhaps divergence of the iterates.

Recall that the regularization terms in the objective are $\frac{1}{2}\|\gamma x\|^2 + \frac{1}{2}\|\delta\pi\|^2$, with $\|x\| \approx 1$ and $\|\pi\| \approx 1$ near a solution. For many problems we have observed that $\|\pi\|$ decreases sharply in the final primal-dual iterations, showing that a nonzero δ helps resolve some ambiguity in the dual solution.

In line with the theory of [19, 18], there is no similar decrease in $\|x\|$ when γ is rather small.

5.2 Solving the KKT Systems

When there are no free variables or dense columns, the tolerances are such that $K_B = -AH^{-1}A^T$ for most of the early iterations. The performance should then be similar to other Cholesky-based implementations.

As the optimal solution is approached, many diagonals of H become small and K_B becomes more indefinite as its dimension increases. In some cases the MA27 *Factor* generates substantially more nonzeros than predicted by the *Analyze*, and the iteration time deteriorates significantly.

As always, improvements in computation time will come from speeding up the KKT solves. A major modification of MA27 is being developed as MA47 [3], and we expect that its performance in the KKT context will be considerably improved. A promising alternative is the code described recently in [6].

During the code development, occasional high iteration counts were usually found to be the result of lax tolerances in forming and solving the reduced KKT systems (just as a simplex code could be expected to iterate indefinitely if an unreliable basis package were used). With the current MA27 factorizer, it remains desirable to use lax tolerances tentatively (to enhance sparsity), since they are often adequate. Provision for iterative refinement and tightening of the factorization tolerances seems to provide a reliable safeguard. For further details, see Gill *et al.* [9].

As an example, most of the test problems solved successfully with the tolerances fixed at $Htol = 10^{-8}$ and $factol = 0.001$. This is *extremely* lax in terms of conventional Gaussian elimination, but note that implementations based on $AH^{-1}A^T$ are effectively using $Htol = 0$ and $factol = 0$, with no increase possible. Iterative refinement can again be invoked, but that alone may be unsuccessful.

5.3 Dimension of the Reduced KKT Systems

Let n_k be the dimension of the reduced KKT system K_B (3.2) at iteration k, and let $r_k = n_k/m$. The first graph in Figure 5.1 plots the ratios r_k for a representative selection of problems (using $Htol = 10^{-6}$ and requesting 8 digits of accuracy). The name of each problem appears near the end of the associated plot.

The value $r_k = 1$ implies that $K_B = AH^{-1}A^T$ at iteration k. For example, problems *scsd6* and *ship12l* both give fully reduced systems at the beginning, since all elements of H are of order 1 initially, and there are no dense columns. In contrast, $r_k \approx 2$ for most of the *pilots* iterations, because almost m columns of A contain 10 or more nonzeros and are included in B throughout.

In general, r_k stays almost constant until the final iterations, when many diagonals of H start falling below *Htol*. The dimension of K_B increases as more columns of A are included in B.

Similarly, let nz_k be the number of nonzeros in the MA27 factorization of K_B at iteration k, and let $\bar{r}_k = nz_k/nz_1$. (The minimum number of nonzeros happens to occur at the first iteration.) The second graph in Figure 5.1 plots the ratios $\bar{r}_k$ for the same problems. The values $\bar{r}_k > 2$ represent a serious loss of sparsity in order to preserve stability.

Some observations follow.

- The dimension of K_B changes from its previous value rather more than half of the time. This determines how many times a new *Analyze* is needed. Some statistics are given in Table 5.1.

- The cpu time for an MA27 *Analyze* is usually moderate compared to a *Factor*, but sometimes it can be substantial. Table 5.1 shows how much time is spent in each phase, as a percentage of the total cpu time. (There is normally one *Factor* per iteration, except on rare occasions when the stability tolerances are tightened.)

- There is typically a sharp increase in the MA27 *Factor* nonzeros during the final iterations. In particular, requesting 8 digits of accuracy rather than 6 carries a substantial cost.

These matters reflect the cost of stability compared to implementations based on $AH^{-1}A^T$ (for which $r_k = \bar{r}_k = 1$ throughout).

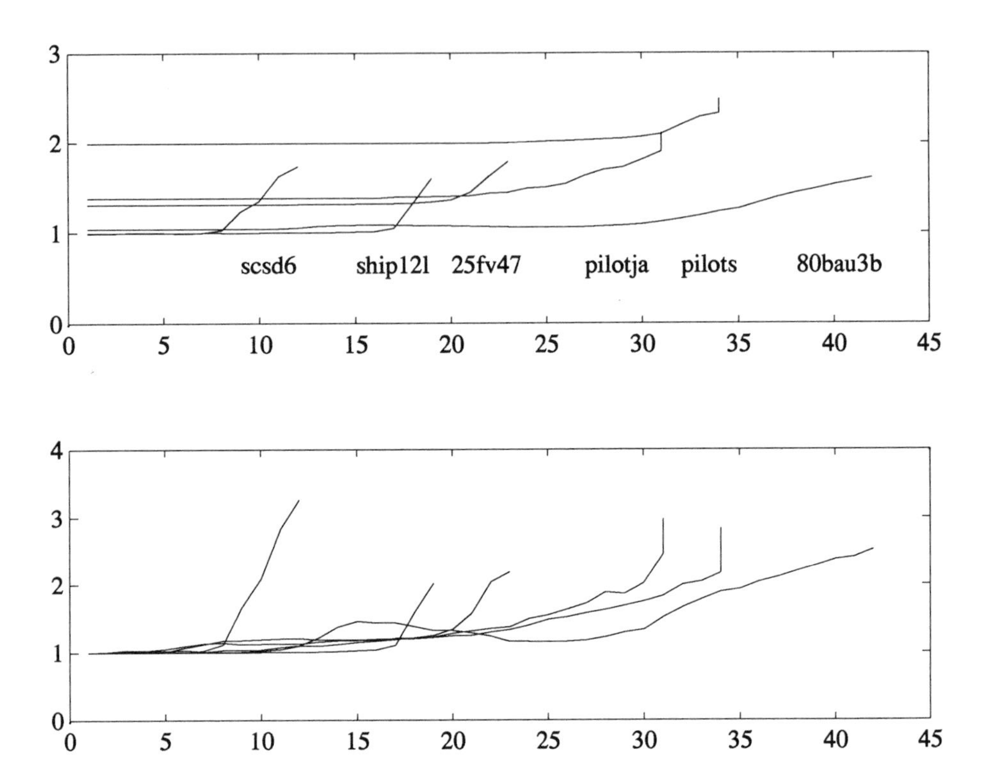

Figure 5.1: The dimension of the reduced KKT systems K_B (relative to m), and the number of nonzeros in the MA27 factors (relative to the first factorization).

6 Conclusions

For various good reasons, most interior-point codes for LP have been based on Cholesky factors of matrices of the form $AH^{-1}A^T$. Excellent performance has been achieved (notably by [16, 25]) and the primary sources of difficulty have been thought to be dense columns and free variables.

Here we have emphasized the fact that the real source of numerical error lies in pivoting on small diagonals of H in the presence of primal degeneracy. Reducing a KKT system K to $AH^{-1}A^T$ is equivalent to pivoting on *all* diagonals of H, regardless of size. We suggest forming "reduced KKT systems" by pivoting on just the diagonals of H that are suitably large. This allows us to avoid factorizing a full KKT system, and often leads to use of $AH^{-1}A^T$ in the early iterations when it is numerically safe.

The primary advantage is intended to be numerical reliability. The drawbacks are that a new *Analyze* is required each time the reduced KKT system changes in dimension, and that we are dependent on the efficiency of a symmetric indefinite factorizer. Some new codes [3, 6] promise to narrow the gap between indefinite and definite solvers. A variant

	Analyze	*Factor*	*Analyze* time	*Factor* time
scsd6	11	12	20%	32%
ship12l	14	19	18%	30%
25fv47	14	23	13%	72%
pilotja	19	32	16%	75%
pilots	20	35	15%	81%
80bau3b	38	42	29%	45%
degen3	9	28	29%	66%

Table 5.1: The number of MA27 *Analyze* and *Factor* calls, and the percentage of time spent in each.

of "reduced KKT systems" has recently been given in [36]. An advantage is that it avoids the need for an indefinite solver, but in its present form it is susceptible to the normal dangers of small pivots in H (see [12]).

In terms of overall strategy, it remains to be seen which of the approaches in [16, 33, 6, 36] and the present paper will offer the most favorable balance between efficiency and reliability.

References

[1] J. R. Bunch and L. Kaufman, "Some stable methods for calculating inertia and solving symmetric linear systems," Mathematics of Computation **31**, 163–179 (1977).

[2] J. R. Bunch and B. N. Parlett, "Direct methods for solving symmetric indefinite systems of linear equations," SIAM J. on Numerical Analysis **8**, 639–655 (1971).

[3] I. S. Duff, N. I. M. Gould, J. K. Reid, J. A. Scott, and K. Turner, "The factorization of sparse symmetric indefinite matrices," Report CSS 236, Computer Science and Systems Division, AERE Harwell, Oxford, England (1989).

[4] I. S. Duff and J. K. Reid, "MA27 – a set of Fortran subroutines for solving sparse symmetric sets of linear equations," Report R-10533, Computer Science and Systems Division, AERE Harwell, Oxford, England (1982).

[5] I. S. Duff and J. K. Reid, "The multifrontal solution of indefinite sparse symmetric linear equations," ACM Transactions on Mathematical Software **9**, 302–325 (1983).

[6] R. Fourer and S. Mehrotra, "Performance of an augmented system approach for solving least-squares problems in an interior-point method for linear programming," Mathematical Programming Society COAL Newsletter **19**, 26–31 (1991).

[7] D. M. Gay, "Electronic mail distribution of linear programming test problems," Mathematical Programming Society COAL Newsletter **13**, 10–12 (1985).

[8] P. E. Gill, W. Murray, D. B. Ponceleón, and M. A. Saunders, "Primal-dual methods for linear programming," Report SOL 91-3, Department of Operations Research, Stanford University (1991).

[9] P. E. Gill, W. Murray, D. B. Ponceleón, and M. A. Saunders, "Solving reduced KKT systems in barrier methods for linear and quadratic programming," Report SOL 91-7, Department of Operations Research, Stanford University (1991).

[10] P. E. Gill, W. Murray, D. B. Ponceleón, and M. A. Saunders, "Preconditioners for indefinite systems arising in optimization," SIAM J. on Matrix Analysis and Applications **13**, 292–311 (1992).

[11] P. E. Gill, W. Murray, M. A. Saunders, J. Tomlin, and M. H. Wright, "On projected Newton methods for linear programming and an equivalence to Karmarkar's projective method," Mathematical Programming **36**, 183–209 (1986).

[12] P. E. Gill, M. A. Saunders, and J. R. Shinnerl, "On the numerical stability of quasi-definite systems," Numerical Analysis Report 93-1, Department of Mathematics, University of California, San Diego, La Jolla, CA (1993).

[13] J. F. Grcar, "Matrix stretching for linear equations," Report SAND90-8723, Sandia National Laboratories, Albuquerque, NM (1990).

[14] M. Kojima, S. Mizuno, and A. Yoshise, "A primal-dual interior point algorithm for linear programming," In N. Megiddo, editor, *Progress in Mathematical Programming: Interior Point and Related Methods*, pages 29–47. Springer Verlag, New York, NY, 1989.

[15] I. J. Lustig, R. E. Marsten, and D. F. Shanno, "Computational experience with a primal–dual interior point method for linear programming," Linear Algebra and its Applications **152**, 191–222 (1991).

[16] I. J. Lustig, R. E. Marsten, and D. F. Shanno, "On implementing Mehrotra's predictor-corrector interior point method for linear programming," SIAM Journal on Optimization **2**, 435–449 (1992).

[17] I. J. Lustig, J. M. Mulvey, and T. J. Carpenter, "The formulation of stochastic programs for interior point methods", Operations Research **39**, 757–770 (1991).

[18] O. L. Mangasarian, "Normal solutions of linear programs," Mathematical Programming Study **22**, 206–216 (1984).

[19] O. L. Mangasarian and R. R. Meyer, "Nonlinear perturbation of linear programs," SIAM J. on Control and Optimization **17**, 745–757 (1979).

[20] A. Marxen, "Primal barrier methods for linear programming," Report SOL 89-6, Department of Operations Research, Stanford University (1989).

[21] K. A. McShane, C. L. Monma, and D. F. Shanno, "An implementation of a primal-dual interior point method for linear programming," ORSA Journal on Computing **1**, 70–83 (1989).

[22] N. Megiddo, "Pathways to the optimal set in linear programming," In N. Megiddo, editor, *Progress in Mathematical Programming: Interior Point and Related Methods*, pages 131–158. Springer-Verlag, NY, 1989.

[23] S. Mehrotra, "Handling free variables in interior methods," Report 91-06, Department of Industrial Engineering and Management Sciences, Northwestern University, Evanston, IL (1991).

[24] S. Mehrotra, "On finding a vertex solution using interior point methods," Linear Algebra and its Applications **152**, 233–253 (1991).

[25] S. Mehrotra, "On the implementation of a primal-dual interior point method," SIAM Journal on Optimization **2**, 575–601 (1992).

[26] C. Moler, J. Little, and S. Bangert, "PRO-MATLAB User's Guide," The MathWorks, Inc., Sherborn, MA, 1987.

[27] D. B. Ponceleón, "Barrier methods for large-scale quadratic programming," Ph.D. thesis, Depertment of Computer Science, Stanford University (1990).

[28] R. T. Rockafellar, "Monotone operators and the proximal-point algorithm," SIAM J. on Control and Optimization **14**, 877–898 (1976).

[29] R. Setiono, "Interior proximal-point algorithms for linear programs," Report 949, Computer Sciences Department, University of Wisconsin, Madison, WI (1990).

[30] R. Setiono, "Interior dual least 2-norm algorithm for linear programs," Report 950, Computer Sciences Department, University of Wisconsin, Madison, WI (1990).

[31] R. Setiono, "Interior dual proximal point algorithm using preconditioned conjugate gradient," Report 951, Computer Sciences Department, University of Wisconsin, Madison, WI (1990).

[32] R. Setiono, "Interior proximal-point algorithm for linear programs," J. Optimization Theory and Applications **74**, 425–444 (1992).

[33] K. Turner, "Computing projections for the Karmarkar algorithm," Linear Algebra and its Applications **152**, 141–154 (1991).

[34] R. J. Vanderbei, "ALPO: Another Linear Program Solver," Technical report, AT&T Bell Laboratories, Murray Hill, NJ (1990).

[35] R. J. Vanderbei, "Splitting dense columns in sparse linear systems," Linear Algebra and its Applications **152**, 107–117 (1991).

[36] R. J. Vanderbei, "Symmetric quasi-definite matrices," Report SOR 91-10, Department of Civil Engineering and Operations Research, Princeton University (1991).

[37] R. J. Vanderbei and T. J. Carpenter, "Symmetric indefinite systems for interior point methods," Mathematical Programming **58**, 1–32 (1993).

Acknowledgement

This research was supported by National Science Foundation Grants DDM-9204547 and DDM-9204208, Department of Energy Grant DE-FG03-92ER25117, and Office of Naval Research Grant N00014-90-J-1242.

Philip E. Gill
Dept. of Mathematics
University of California at San Diego
La Jolla, CA 92093–0112
USA

Walter Murray
Dept. of Operations Research
Stanford University
Stanford, CA 94305-4022
USA

Dulce B. Ponceleón
Apple Computer, Inc.
20525 Mariani Ave, Cupertino
CA 95014
USA

Michael A. Saunders
Dept. of Operations Research
Stanford University
Stanford, CA 94305-4022
USA

G H GOLUB AND GERARD MEURANT

Matrices, moments and quadrature

Abstract In this paper we study methods to obtain bounds or approximations of elements of a matrix $f(A)$ where A is a symmetric positive definite matrix and f is a smooth function. These methods are based on the use of quadrature rules and the Lanczos algorithm for diagonal elements and the block Lanczos or the non-symmetric Lanczos algorithms for the non diagonal elements. We give some theoretical results on the behavior of these methods based on results for orthogonal polynomials as well as analytical bounds and numerical experiments on a set of matrices for several functions f.

1 Definition of the problem

Let A be a real symmetric positive definite matrix of order n. We want to find upper and lower bounds (or approximations, if bounds are not available) for the entries of a function of a matrix. We shall examine analytical expressions as well as numerical iterative methods which produce good approximations in a few steps. This problem leads us to consider

$$u^T f(A) v, \tag{1.1}$$

where u and v are given vectors and f is some smooth (possibly C^∞) function on a given interval of the real line.

As an example, if $f(x) = \frac{1}{x}$ and $u^T = e_i^T = (0, \ldots, 0, 1, 0, \ldots, 0)$, the non zero element being in the i-th position and $v = e_j$, we will obtain bounds on the elements of the inverse A^{-1}.

We shall also consider

$$W^T f(A) W,$$

where W is an $n \times m$ matrix. For specificity, we shall most often consider $m = 2$.

Some of the techniques presented in this paper have been used (without any mathematical justification) to solve problems in solid state physics, particularly to compute elements of the resolvant of a Hamiltonian modeling the interaction of atoms in a solid, see [12], [14], [15]. In these studies the function f is the inverse of its argument.

Analytic bounds for elements of inverses of matrices using different techniques have been recently obtained in [17].

The outline of the paper is as follows. Section 2 considers the problem of characterizing the elements of a function of a matrix. The theory is developed in Section 3 and Section 4 deals with the construction of the orthogonal polynomials that are needed to obtain a numerical method for computing bounds. The Lanczos, non-symmetric Lanczos and block Lanczos methods used for the computation of the polynomials are presented there. Applications to the computation of elements of the inverse of a matrix are described in

Section 5 where very simple iterative algorithms are given to compute bounds. Some numerical examples are given in Section 6, for different matrices and functions f.

2 Elements of a function of a matrix

Since $A = A^T$, we write A as

$$A = Q\Lambda Q^T,$$

where Q is the orthonormal matrix whose columns are the normalized eigenvectors of A and Λ is a diagonal matrix whose diagonal elements are the eigenvalues λ_i which we order as

$$\lambda_1 \le \lambda_2 \le \cdots \le \lambda_n.$$

By definition, we have

$$f(A) = Qf(\Lambda)Q^T.$$

Therefore,

$$\begin{aligned} u^T f(A) v &= u^T Q f(\Lambda) Q^T v \\ &= \alpha^T f(\Lambda)\beta, \\ &= \sum_{i=1}^{n} f(\lambda_i)\alpha_i\beta_i. \end{aligned}$$

This last sum can be considered as a Riemann–Stieltjes integral

$$I[f] = u^T f(A) v = \int_a^b f(\lambda)\, d\alpha(\lambda), \tag{2.1}$$

where the measure α is piecewise constant and defined by

$$\alpha(\lambda) = \begin{cases} 0 & \text{if } \lambda < a = \lambda_1 \\ \sum_{j=1}^{i} \alpha_j\beta_j & \text{if } \lambda_i \le \lambda < \lambda_{i+1} \\ \sum_{j=1}^{n} \alpha_j\beta_j & \text{if } b = \lambda_n \le \lambda \end{cases}$$

When $u = v$, we note that α is an increasing positive function.

The block generalization is obtained in the following way. Let W be an $n \times 2$ matrix, $W = (w_1\ w_2)$, then

$$W^T f(A) W = W^T Q f(\Lambda) Q^T W = \alpha f(\Lambda)\alpha^T,$$

where, of course, α is a $2 \times n$ matrix such that

$$\alpha = (\alpha_1 \ldots \alpha_n),$$

and α_i is a vector with two components. With these notations, we have

$$W^T f(A) W = \sum_{i=1}^{n} f(\lambda_i)\alpha_i\alpha_i^T.$$

This can be written as a matrix Riemann–Stieltjes integral

$$I_B[f] = W^T f(A) W = \int_a^b f(\lambda)\, d\alpha(\lambda).$$

$I_B[f]$ is a 2×2 matrix where the entries of the (matrix) measure α are piecewise constant and defined by

$$\alpha(\lambda) = \sum_{k=1}^{l} \alpha_k \alpha_k^T, \quad \lambda_l \leq \lambda < \lambda_{l+1}.$$

In this paper, we are looking for methods to obtain upper and lower bounds L and U for $I[f]$ and $I_B[f]$,

$$\begin{aligned} L &\leq I[f] \leq U \\ L &\leq I_B[f] \leq U. \end{aligned}$$

In the next section, we review and describe some basic results from Gauss quadrature theory as this plays a fundamental role in estimating the integrals and computing bounds.

3 Bounds on matrix functions as integrals

A way to obtain bounds for the Stieltjes integrals is to use Gauss, Gauss–Radau and Gauss–Lobatto quadrature formulas, see [3],[8],[9]. For 1.1, the general formula we will use is

$$\int_a^b f(\lambda)\, d\alpha(\lambda) = \sum_{j=1}^{N} w_j f(t_j) + \sum_{k=1}^{M} v_k f(z_k) + R[f], \tag{3.1}$$

where the weights $[w_j]_{j=1}^N$, $[v_k]_{k=1}^M$ and the nodes $[t_j]_{j=1}^N$ are unknowns and the nodes $[z_k]_{k=1}^M$ are prescribed, see [4],[5],[6],[7].

3.1 The case $u = v$

When $u = v$, the measure is a positive increasing function and it is known (see for instance [18]) that

$$R[f] = \frac{f^{(2N+M)}(\eta)}{(2N+M)!} \int_a^b \prod_{k=1}^{M} (\lambda - z_k) \left[\prod_{j=1}^{N} (\lambda - t_j) \right]^2 d\alpha(\lambda), \quad a < \eta < b. \tag{3.2}$$

If $M = 0$, this leads to the Gauss rule with no prescribed nodes. If $M = 1$ and $z_1 = a$ or $z_1 = b$ we have the Gauss–Radau formula. If $M = 2$ and $z_1 = a, z_2 = b$, this is the Gauss–Lobatto formula.

Let us recall briefly how the nodes and weights are obtained in the Gauss, Gauss–Radau and Gauss–Lobatto rules. For the measure α, it is possible to define a sequence of polynomials $p_0(\lambda), p_1(\lambda), \ldots$ that are orthonormal with respect to α:

$$\int_a^b p_i(\lambda) p_j(\lambda)\, d\alpha(\lambda) = \begin{cases} 1 & \text{if } i = j \\ 0 & \text{otherwise} \end{cases}$$

and p_k is of exact degree k. Moreover, the roots of p_k are distinct, real and lie in the interval $[a,b]$. We will see how to compute these polynomials in the next Section.

This set of orthonormal polynomials satisfies a three term recurrence relationship (see [20]):

$$\gamma_j p_j(\lambda) = (\lambda - \omega_j)p_{j-1}(\lambda) - \gamma_{j-1}p_{j-2}(\lambda), \quad j = 1,2,\ldots,N \tag{3.3}$$
$$p_{-1}(\lambda) \equiv 0, \quad p_0(\lambda) \equiv 1,$$

if $\int d\alpha = 1$.

In matrix form, this can be written as

$$\lambda p(\lambda) = J_N p(\lambda) + \gamma_N p_N(\lambda) e_N,$$

where

$$p(\lambda)^T = [p_0(\lambda)\ p_1(\lambda) \cdots\ p_{N-1}(\lambda)],$$
$$e_N^T = (0\ 0\ \cdots\ 0\ 1),$$

$$J_N = \begin{pmatrix} \omega_1 & \gamma_1 & & & \\ \gamma_1 & \omega_2 & \gamma_2 & & \\ & \ddots & \ddots & \ddots & \\ & & \gamma_{N-2} & \omega_{N-1} & \gamma_{N-1} \\ & & & \gamma_{N-1} & \omega_N \end{pmatrix}. \tag{3.4}$$

The eigenvalues of J_N (which are the zeroes of p_N) are the nodes of the Gauss quadrature rule (i. e. $M = 0$). The weights are the squares of the first elements of the normalized eigenvectors of J_N, cf. [7]. We note that all the eigenvalues of J_N are real and simple.

For the Gauss quadrature rule (renaming the weights and nodes w_j^G and t_j^G) we have

$$\int_a^b f(\lambda)\, d\alpha(\lambda) = \sum_{j=1}^N w_j^G f(t_j^G) + R_G[f],$$

with

$$R_G[f] = \frac{f^{(2N)}(\eta)}{(2N)!} \int_a^b \left[\prod_{j=1}^N (\lambda - t_j^G) \right]^2 d\alpha(\lambda),$$

and the next theorem follows.

Theorem 3.1 *Suppose $u = v$ in 2.1 and f is such that $f^{(2n)}(\xi) > 0$, $\forall n$, $\forall \xi$, $a < \xi < b$, and let*

$$L_G[f] = \sum_{j=1}^N w_j^G f(t_j^G).$$

Then, $\forall N$, $\exists \eta \in [a,b]$ such that

$$L_G[f] \leq I[f],$$
$$I[f] - L_G[f] = \frac{f^{(2N)}(\eta)}{(2N)!}.$$

Proof: See [18]. The main idea of the proof is to use a Hermite interpolatory polynomial of degree $2N-1$ on the N nodes which allows us to express the remainder as an integral of the difference between the function and its interpolatory polynomial and to apply the mean value theorem (as the measure is positive and increasing). As we know the sign of the remainder, we easily obtain bounds.

To obtain the Gauss–Radau rule ($M = 1$ in 3.1–3.2), we should extend the matrix J_N in 3.4 in such a way that it has one prescribed eigenvalue, see [8].

Assume $z_1 = a$, we wish to construct p_{N+1} such that $p_{N+1}(a) = 0$. From the recurrence relation 3.3, we have

$$0 = \gamma_{N+1} p_{N+1}(a) = (a - \omega_{N+1}) p_N(a) - \gamma_N p_{N-1}(a).$$

This gives

$$\omega_{N+1} = a - \gamma_N \frac{p_{N-1}(a)}{p_N(a)}.$$

We have also

$$(J_N - aI)p(a) = -\gamma_N p_N(a) e_N.$$

Let us denote $\delta(a) = [\delta_1(a), \cdots, \delta_N(a)]^T$ with

$$\delta_l(a) = -\gamma_N \frac{p_{l-1}(a)}{p_N(a)} \quad l = 1, \ldots, N.$$

This gives $\omega_{N+1} = a + \delta_N(a)$ and

$$(J_N - aI)\delta(a) = \gamma_N^2 e_N. \tag{3.5}$$

From these relations we have the solution of the problem as: 1) we generate γ_N by the Lanczos process (see Section 4 for the definition), 2) we solve the tridiagonal system 3.5 for $\delta(a)$ and 3) we compute ω_{N+1}. Then the tridiagonal matrix $\hat{J}_{N+1}$ defined as

$$\hat{J}_{N+1} = \begin{pmatrix} J_N & \gamma_N e_N \\ \gamma_N e_N^T & \omega_{N+1} \end{pmatrix},$$

will have a as an eigenvalue and gives the weights and the nodes of the corresponding quadrature rule. Therefore, the recipe is to compute as for the Gauss quadrature rule and then to modify the last step to obtain the prescribed node.

For Gauss–Radau the remainder R_{GR} is

$$R_{GR}[f] = \frac{f^{(2N+1)}(\eta)}{(2N+1)!} \int_a^b (\lambda - z_1) \left[\prod_{j=1}^{N} (\lambda - t_j) \right]^2 d\alpha(\lambda).$$

Again, this is proved by constructing an interpolatory polynomial for the function and its derivative on the t_js and for the function on z_1.

Therefore, if we know the sign of the derivatives of f, we can bound the remainder. This is stated in the following theorem.

Theorem 3.2 *Suppose $u = v$ and f is such that $f^{(2n+1)}(\xi) < 0$, $\forall n, \forall \xi, a < \xi < b$. Let U_{GR} be defined as*

$$U_{GR}[f] = \sum_{j=1}^{N} w_j^a f(t_j^a) + v_1^a f(a),$$

w_j^a, v_1^a, t_j^a being the weights and nodes computed with $z_1 = a$ and let L_{GR} be defined as

$$L_{GR}[f] = \sum_{j=1}^{N} w_j^b f(t_j^b) + v_1^b f(b),$$

w_j^b, v_1^b, t_j^b being the weights and nodes computed with $z_1 = b$. Then, $\forall N$ we have

$$L_{GR}[f] \le I[f] \le U_{GR}[f],$$

and

$$I[f] - U_{GR}[f] = \frac{f^{(2N+1)}(\eta)}{(2N+1)!} \int_a^b (\lambda - a) \left[\prod_{j=1}^{N} (\lambda - t_j^a)\right]^2 d\alpha(\lambda),$$

$$I[f] - L_{GR}[f] = \frac{f^{(2N+1)}(\eta)}{(2N+1)!} \int_a^b (\lambda - b) \left[\prod_{j=1}^{N} (\lambda - t_j^b)\right]^2 d\alpha(\lambda).$$

Proof : With our hypothesis the sign of the remainder is easily obtained. It is negative if we choose $z_1 = a$, positive if we choose $z_1 = b$.

Remarks :

i) if the sign of the f derivatives is positive, the bounds are reversed.

ii) it is enough to suppose that there exists an n_0 such that $f^{(2n_0+1)}(\eta) < 0$ but, then $N = n_0$ is fixed.

Now, consider the Gauss–Lobatto rule ($M = 2$ in 3.1–3.2), with $z_1 = a$ and $z_2 = b$ as prescribed nodes. Again, we should modify the matrix of the Gauss quadrature rule, see [8]. Here, we would like to have

$$p_{N+1}(a) = p_{N+1}(b) = 0.$$

Using the recurrence relation 3.3 for the polynomials, this leads to a linear system of order 2 for the unknowns ω_{N+1} and γ_N:

$$\begin{pmatrix} p_N(a) & p_{N-1}(a) \\ p_N(b) & p_{N-1}(b) \end{pmatrix} \begin{pmatrix} \omega_{N+1} \\ \gamma_N \end{pmatrix} = \begin{pmatrix} a\, p_N(a) \\ b\, p_N(b) \end{pmatrix}. \tag{3.6}$$

Let δ and μ be defined as vectors with components

$$\delta_l = -\frac{p_{l-1}(a)}{\gamma_N p_N(a)}, \qquad \mu_l = -\frac{p_{l-1}(b)}{\gamma_N p_N(b)},$$

then

$$(J_N - aI)\delta = e_N, \quad (J_N - bI)\mu = e_N,$$

and the linear system 3.6 can be written

$$\begin{pmatrix} 1 & -\delta_N \\ 1 & -\mu_N \end{pmatrix} \begin{pmatrix} \omega_{N+1} \\ \gamma_N^2 \end{pmatrix} = \begin{pmatrix} a \\ b \end{pmatrix},$$

giving the unknowns that we need. The tridiagonal matrix $\hat{J}_{N+1}$ is then defined as in the Gauss–Radau rule.

Having computed the nodes and weights, we have

$$\int_a^b f(\lambda)d\alpha(\lambda) = \sum_{j=1}^N w_j^{GL} f(t_j^{GL}) + v_1 f(a) + v_2 f(b) + R_{GL}[f],$$

where

$$R_{GL}[f] = \frac{f^{(2N+2)}(\eta)}{(2N+2)!} \int_a^b (\lambda - a)(\lambda - b) \left[\prod_{j=1}^N (\lambda - t_j) \right]^2 d\alpha(\lambda).$$

Then, we have the following obvious result.

Theorem 3.3 *Suppose $u = v$ and f is such that $f^{(2n)}(\eta) > 0$, $\forall n$, $\forall \eta, a < \eta < b$ and let*

$$U_{GL}[f] = \sum_{j=1}^N w_j^{GL} f(t_j^{GL}) + v_1 f(a) + v_2 f(b).$$

Then, $\forall N$

$$I[f] \leq U_{GL}[f],$$

$$I[f] - U_{GL}[f] = \frac{f^{(2N+2)}(\eta)}{(2N+2)!} \int_a^b (\lambda - a)(\lambda - b) \left[\prod_{j=1}^N (\lambda - t_j^{GL}) \right]^2 d\alpha(\lambda).$$

We remark that we need not always compute the eigenvalues and eigenvectors of the tridiagonal matrix. Let Y_N be the matrix of the eigenvectors of J_N (or $\hat{J}_N$) whose columns we denote by y_i and T_N be the diagonal matrix of the eigenvalues t_i which give the nodes of the Gauss quadrature rule. It is well known that the weights w_i are given by (cf. [21])

$$\frac{1}{w_i} = \sum_{l=0}^{N-1} p_l^2(t_i).$$

It can be easily shown that

$$w_i = \left(\frac{y_i^1}{p_0(t_i)} \right)^2,$$

where y_i^1 is the first component of y_i.

But, since $p_0(\lambda) \equiv 1$, we have,

$$w_i = (y_i^1)^2 = (e_1^T y_i)^2.$$

Theorem 3.4

$$\sum_{l=1}^N w_l f(t_l) = e_1^T f(J_N) e_1.$$

Proof:

$$\begin{aligned}
\sum_{l=1}^{N} w_l f(t_l) &= \sum_{l=1}^{N} e_1^T y_l f(t_l) y_l^T e_1 \\
&= e_1^T \left(\sum_{l=1}^{N} y_l f(t_l) y_l^T \right) e_1 \\
&= e_1^T Y_N f(T_N) Y_N^T e_1 \\
&= e_1^T f(J_N) e_1.
\end{aligned}$$

The same statement is true for the Gauss–Radau and Gauss–Lobatto rules. Therefore, in some cases where $f(J_N)$ (or the equivalent) is easily computable (for instance, if $f(\lambda) = 1/\lambda$, see Section 5), we do not need to compute the eigenvalues and eigenvectors of J_N.

3.2 The case $u \neq v$

We have seen that the measure in 2.1 is piecewise constant and defined by

$$\alpha(\lambda) = \sum_{k=1}^{l} \alpha_k \delta_k, \quad \lambda_l \leq \lambda < \lambda_{l+1}.$$

For variable signed weight functions, see [19]. We will see later that for our application, u and v can always be chosen such that $\alpha_k \delta_k \geq 0$. Therefore, in this case α will be a positive increasing function.

In the next Section, we will show that there exists two sequences of polynomials p and q such that

$$\begin{aligned}
\gamma_j p_j(\lambda) &= (\lambda - \omega_j) p_{j-1}(\lambda) - \beta_{j-1} p_{j-2}(\lambda), \quad p_{-1}(\lambda) \equiv 0, \quad p_0(\lambda) \equiv 1, \\
\beta_j q_j(\lambda) &= (\lambda - \omega_j) q_{j-1}(\lambda) - \gamma_{j-1} q_{j-2}(\lambda), \quad q_{-1}(\lambda) \equiv 0, \quad q_0(\lambda) \equiv 1.
\end{aligned}$$

Let

$$p(\lambda)^T = [p_0(\lambda)\ p_1(\lambda) \cdots\ p_{N-1}(\lambda)],$$
$$q(\lambda)^T = [q_0(\lambda)\ q_1(\lambda) \cdots\ q_{N-1}(\lambda)],$$

and

$$J_N = \begin{pmatrix}
\omega_1 & \gamma_1 & & & \\
\beta_1 & \omega_2 & \gamma_2 & & \\
& \ddots & \ddots & \ddots & \\
& & \beta_{N-2} & \omega_{N-1} & \gamma_{N-1} \\
& & & \beta_{N-1} & \omega_N
\end{pmatrix}.$$

Then, we can write

$$\begin{aligned}
\lambda p(\lambda) &= J_N p(\lambda) + \gamma_N p_N(\lambda) e_N, \\
\lambda q(\lambda) &= J_N^T q(\lambda) + \beta_N q_N(\lambda) e_N.
\end{aligned}$$

Theorem 3.5

$$p_j(\lambda) = \frac{\beta_j \cdots \beta_1}{\gamma_j \cdots \gamma_1} q_j(\lambda).$$

Proof: The theorem is proved by induction. We have

$$\gamma_1 p_1(\lambda) = \lambda - \omega_1,$$

$$\beta_1 q_1(\lambda) = \lambda - \omega_1,$$

therefore

$$p_1(\lambda) = \frac{\beta_1}{\gamma_1} q_1(\lambda).$$

Now, suppose that

$$p_{j-1}(\lambda) = \frac{\beta_{j-1} \cdots \beta_1}{\gamma_{j-1} \cdots \gamma_1} q_{j-1}(\lambda).$$

We have

$$\begin{aligned} \gamma_j p_j(\lambda) &= (\lambda - \omega_j) p_{j-1}(\lambda) - \beta_{j-1} p_{j-2}(\lambda) \\ &= (\lambda - \omega_j) \frac{\beta_{j-1} \cdots \beta_1}{\gamma_{j-1} \cdots \gamma_1} q_{j-1}(\lambda) - \beta_{j-1} \frac{\beta_{j-2} \cdots \beta_1}{\gamma_{j-2} \cdots \gamma_1} q_{j-2}(\lambda). \end{aligned}$$

Multiplying by $\frac{\gamma_{j-1} \cdots \gamma_1}{\beta_{j-1} \cdots \beta_1}$ we obtain the result.

Hence q_N is a multiple of p_N and the polynomials have the same roots which are also the common eigenvalues of J_N and J_N^T.

We will see that it is possible to choose γ_j and β_j such that

$$\gamma_j = \pm \beta_j,$$

with, for instance, $\gamma_j \geq 0$. Then, we have

$$p_j(\lambda) = \pm q_j(\lambda).$$

We define the quadrature rule as

$$\int_a^b f(\lambda) \, d\alpha(\lambda) = \sum_{j=1}^N f(\lambda_j) s_j t_j + error, \tag{3.7}$$

where λ_j is an eigenvalue of J_N, s_j is the first component of the eigenvector u_j of J_N corresponding to λ_j and t_j is the first component of the eigenvector v_j of J_N^T corresponding to the same eigenvalue, normalized such that $v_j^T u_j = 1$.

We have the following results:

Proposition 3.1 *Suppose that $\gamma_j \beta_j \neq 0$, then the (non-symmetric) Gauss quadrature rule 3.7 is exact for polynomials of degree less than or equal to $N - 1$.*

Proof:

The function f can be written as

$$f(\lambda) = \sum_{k=0}^{N-1} c_k p_k(\lambda),$$

and because of the orthonormality properties

$$\int_a^b f(\lambda)\, d\alpha(\lambda) = c_0.$$

For the quadrature rule, we have

$$\begin{aligned}\sum_{j=1}^{N} f(\lambda_j)s_j t_j q_l(\lambda_j) &= \sum_{j=1}^{N}\sum_{k=0}^{N-1} c_k p_k(\lambda_j)s_j t_j q_l(\lambda_j)\\ &= \sum_{k=0}^{N-1} c_k \sum_{j=1}^{N} p_k(\lambda_j)s_j t_j q_l(\lambda_j).\end{aligned}$$

But $p_k(\lambda_j)s_j$ and $q_l(\lambda_j)t_j$ are respectively the components of the eigenvectors of J_N and J_N^T corresponding to λ_j. Therefore they are orthonormal with the normalization that we chose. Hence,

$$\sum_{j=1}^{N} f(\lambda_j)s_j t_j q_l(\lambda_j) = c_l,$$

and consequently

$$\sum_{j=1}^{N} f(\lambda_j)s_j t_j = c_0,$$

which proves the result.

Now, as in [14], we extend the result to polynomials of higher degree.

Theorem 3.6 *Suppose that $\gamma_j\beta_j \neq 0$, then the (non-symmetric) Gauss quadrature rule 3.7 is exact for polynomials of degree less than or equal to $2N-1$.*

Proof:

Suppose f is a polynomial of degree $2N-1$. Then, f can be written as

$$f(\lambda) = p_N(\lambda)s(\lambda) + r(\lambda),$$

where s and r are polynomials of degree less or equal to $N-1$. Then,

$$\int_a^b f(\lambda)\, d\alpha(\lambda) = \int_a^b p_N(\lambda)s(\lambda)\, d\alpha(\lambda) + \int_a^b r(\lambda)\, d\alpha(\lambda) = \int_a^b r(\lambda)\, d\alpha(\lambda),$$

since p_N is orthogonal to any polynomial of degree less or equal to $N-1$ because of the orthogonality property of the p and q's.

For the quadrature rule, we have

$$\sum_{j=1}^{N} p_N(\lambda_j)s(\lambda_j)s_jt_j + \sum_{j=1}^{N} r(\lambda_j)s_jt_j.$$

But, as λ_j is an eigenvalue of J_N, it is a root of p_N and

$$\sum_{j=1}^{N} p_N(\lambda_j)s(\lambda_j)s_jt_j = 0.$$

As the quadrature rule has been proven to be exact for polynomials of degree less than $N-1$,

$$\int_a^b r(\lambda)\, d\alpha(\lambda) = \sum_{j=1}^{N} r(\lambda_j)s_jt_j,$$

which proves the Theorem.

We will see in the next Section how to obtain bounds on the integral 2.1.

Now, we extend the Gauss–Radau and Gauss–Lobatto rules to the non–symmetric case. This is almost identical (up to technical details) to the symmetric case.

For Gauss–Radau, assume that the prescribed node is a, then, we would like to have $p_{N+1}(a) = q_{N+1}(a) = 0$. This gives

$$(a - \omega_{N+1})p_N(a) - \beta_N p_{N-1}(a) = 0.$$

If we denote $\delta(a) = [\delta_1(a), \ldots, \delta_N(a)]^T$, with

$$\delta_l(a) = -\beta_N \frac{p_{l-1}(a)}{p_N(a)},$$

we have

$$\omega_{N+1} = a + \delta_N(a),$$

where

$$(J_N - aI)\delta(a) = \gamma_N \beta_N e_N.$$

Therefore, the algorithm is essentially the same as previously discussed.

For Gauss–Lobatto, the algorithm is also almost the same as for the symmetric case. We would like to compute p_{N+1} and q_{N+1} such that

$$p_{N+1}(a) = p_{N+1}(b) = 0, \quad q_{N+1}(a) = q_{N+1}(b) = 0.$$

This leads to solving the linear system

$$\begin{pmatrix} p_N(a) & p_{N-1}(a) \\ p_N(b) & p_{N-1}(b) \end{pmatrix} \begin{pmatrix} \omega_{N+1} \\ \beta_N \end{pmatrix} = \begin{pmatrix} ap_N(a) \\ bp_N(b) \end{pmatrix}.$$

The linear system for the q's whose solution is $(\omega_{N+1}, \gamma_N)^T$ can be shown to have the same solution for ω_{N+1} and $\gamma_N = \pm\beta_N$ depending on the signs relations between the p's and the q's.

Let $\delta(a)$ and $\mu(b)$ be the solutions of

$$(J_N - aI)\delta(a) = e_N, \quad (J_N - bI)\mu(b) = e_N.$$

Then, we have

$$\begin{pmatrix} 1 & -\delta(a)_N \\ 1 & -\mu(b)_N \end{pmatrix} \begin{pmatrix} \omega_{N+1} \\ \gamma_N \beta_N \end{pmatrix} = \begin{pmatrix} a \\ b \end{pmatrix}.$$

When we have the solution of this system, we choose $\gamma_N = \pm\beta_N$ and $\gamma_N \geq 0$.

The question of establishing bounds on the integral will be studied in the next Section.

As for the case $u = v$, we do not always need to compute the eigenvalues and eigenvectors of J_N but only the $(1,1)$ element of $f(J_N)$.

3.3 The block case

Now, we consider the block case. The problem is to find a quadrature rule. The integral $\int_a^b f(\lambda) d\alpha(\lambda)$ is a 2×2 symmetric matrix. The most general quadrature formula is of the form

$$\int_a^b f(\lambda) d\alpha(\lambda) = \sum_{j=1}^{N} W_j f(T_j) W_j + error,$$

where W_j and T_j are symmetric 2×2 matrices. In this sum, we have $6N$ unknowns. This quadrature rule can be simplified, since

$$T_j = Q_j \Lambda_j Q_j^T,$$

where Q_j is the orthonormal matrix of the eigenvectors, and Λ_j, the diagonal matrix of the eigenvalues of T_j. This gives

$$\sum_{j=1}^{N} W_j Q_j f(\Lambda_j) Q_j^T W_j.$$

But $W_j Q_j f(\Lambda_j) Q_j^T W_j$ can be written as

$$f(\lambda_1) z_1 z_1^T + f(\lambda_2) z_2 z_2^T,$$

where the vector z_i is 2×1. Therefore, the quadrature rule can be written as

$$\sum_{j=1}^{2N} f(t_j) w_j w_j^T,$$

where t_j is a scalar and w_j is a vector with 2 components. In this quadrature rule, there are also $6N$ unknowns.

In the next Section, we will show that there exists orthogonal matrix polynomials such that

$$\lambda p_{j-1}(\lambda) = p_j(\lambda)\Gamma_j + p_{j-1}(\lambda)\Omega_j + p_{j-2}(\lambda)\Gamma_{j-1}^T,$$
$$p_0(\lambda) \equiv I_2, \quad p_{-1}(\lambda) \equiv 0.$$

This can be written as

$$\lambda[p_0(\lambda),\ldots,p_{N-1}(\lambda)] = [p_0(\lambda),\ldots,p_{N-1}(\lambda)]J_N + [0,\ldots,0,p_N(\lambda)\Gamma_N],$$

where

$$J_N = \begin{pmatrix} \Omega_1 & \Gamma_1^T & & & \\ \Gamma_1 & \Omega_2 & \Gamma_2^T & & \\ & \ddots & \ddots & \ddots & \\ & & \Gamma_{N-2} & \Omega_{N-1} & \Gamma_{N-1}^T \\ & & & \Gamma_{N-1} & \Omega_N \end{pmatrix}, \tag{3.8}$$

is a block tridiagonal matrix of order $2N$ and a banded matrix whose half bandwidth is 2 (we have at most 5 non zero elements in a row).

If we denote $P(\lambda) = [p_0(\lambda),\ldots,p_{N-1}(\lambda)]^T$, we have as J_N is symmetric

$$J_N P(\lambda) = \lambda P(\lambda) - [0,\ldots,0,p_N(\lambda)\Gamma_N]^T.$$

We note that if λ is an eigenvalue, say λ_r, of J_N and if we choose $u = u_r$ to be a two element vector whose components are the first two components of an eigenvector corresponding to λ_r, then $P(\lambda_r)u$ is this eigenvector (because of the relations that are satisfied) and if Γ_N is non singular, $p_N^T(\lambda_r)u = 0$. The difference with the scalar case is that although the eigenvalues are real, it might be that they are of multiplicity greater than 1 (although this is unlikely except in the case of the Gauss-Radau and Gauss-Lobatto rule where this condition is enforced).

We define the quadrature rule as:

$$\int_a^b f(\lambda)\, d\alpha(\lambda) = \sum_{i=1}^{2N} f(\lambda_i)u_i u_i^T + error, \tag{3.9}$$

where $2N$ is the order of J_N, the eigenvalues λ_i are those of J_N and u_i is the vector consisting of the two first components of the corresponding eigenvector, normalized as before. In fact, if there are multiple eigenvalues, the quadrature rule should be written as follows. Let $\mu_i, i = 1,\ldots,l$ be the set of distinct eigenvalues and q_i their multiplicities. The quadrature rule is then

$$\sum_{i=1}^{l} \left(\sum_{j=1}^{q_i} (w_i^j)(w_i^j)^T \right) f(\mu_i). \tag{3.10}$$

We will show in the next Section that the Gauss quadrature rule is exact for polynomials of degree $2N-1$ and how to obtain estimates of the error.

We extend the process described for scalar polynomials to the matrix analog of the Gauss–Radau quadrature rule. Let a be an extreme eigenvalue of A. We would like a to be a double eigenvalue of J_{N+1}. We have

$$J_{N+1}P(a) = aP(a) - [0,\ldots,0,p_{N+1}(a)\Gamma_{N+1}]^T.$$

Then, we need to require $p_{N+1}(a) \equiv 0$. From the recurrence relation this translates into

$$ap_N(a) - p_N(a)\Omega_{N+1} - p_{N-1}(a)\Gamma_N^T = 0.$$

Therefore, if $p_N(a)$ is non singular, we have

$$\Omega_{N+1} = aI_2 - p_N(a)^{-1} p_{N-1}(a) \Gamma_N^T.$$

We must compute the right hand side. This can be done by noting that

$$J_N \begin{pmatrix} p_0(a)^T \\ \vdots \\ p_{N-1}(a)^T \end{pmatrix} = a \begin{pmatrix} p_0(a)^T \\ \vdots \\ p_{N-1}(a)^T \end{pmatrix} - \begin{pmatrix} 0 \\ \vdots \\ \Gamma_N^T p_N(a)^T \end{pmatrix}.$$

Multiplying on the right by $p_N(a)^{-T}$, we get the matrix equation

$$(J_N - aI) \begin{pmatrix} -p_0(a)^T p_N(a)^{-T} \\ \vdots \\ -p_{N-1}(a)^T p_N(a)^{-T} \end{pmatrix} = \begin{pmatrix} 0 \\ \vdots \\ \Gamma_N^T \end{pmatrix}.$$

Thus, we solve

$$(J_N - aI) \begin{pmatrix} \delta_0(a) \\ \vdots \\ \delta_{N-1}(a) \end{pmatrix} = \begin{pmatrix} 0 \\ \vdots \\ \Gamma_N^T \end{pmatrix},$$

and hence

$$\Omega_{N+1} = aI_2 + \delta_{N-1}(a)^T \Gamma_N^T.$$

The generalization of Gauss–Lobatto to the block case is a little more tricky. We would like to have a and b as double eigenvalues of the matrix J_{N+1}. This leads to satisfying the following two matrix equations

$$a p_N(a) - p_N(a) \Omega_{N+1} - p_{N-1}(a) \Gamma_N^T = 0$$

$$b p_N(b) - p_N(b) \Omega_{N+1} - p_{N-1}(b) \Gamma_N^T = 0$$

This can be written as

$$\begin{pmatrix} I_2 & p_N^{-1}(a) p_{N-1}(a) \\ I_2 & p_N^{-1}(b) p_{N-1}(b) \end{pmatrix} \begin{pmatrix} \Omega_{N+1} \\ \Gamma_N^T \end{pmatrix} = \begin{pmatrix} aI_2 \\ bI_2 \end{pmatrix}.$$

We now consider the problem of computing (or avoid computing) $p_N^{-1}(\lambda) p_{N-1}(\lambda)$. Let $\delta(\lambda)$ be the solution of

$$(J_N - \lambda I)\delta(\lambda) = (0 \ldots 0 \; I_2)^T.$$

Then, as before

$$\delta_{N-1}(\lambda) = -p_{N-1}(\lambda)^T p_N(\lambda)^{-T} \Gamma_N^{-T}.$$

We can easily show that $\delta_{N-1}(\lambda)$ is symmetric. We consider solving a 2×2 block linear system

$$\begin{pmatrix} I & X \\ I & Y \end{pmatrix} \begin{pmatrix} U \\ V \end{pmatrix} = \begin{pmatrix} aI \\ bI \end{pmatrix}.$$

Consider the block factorization

$$\begin{pmatrix} I & X \\ I & Y \end{pmatrix} = \begin{pmatrix} I & 0 \\ I & W \end{pmatrix} \begin{pmatrix} I & X \\ 0 & Z \end{pmatrix},$$

thus $WZ = Y - X$.

The solution of the system

$$\begin{pmatrix} I & 0 \\ I & W \end{pmatrix} \begin{pmatrix} U_1 \\ V_1 \end{pmatrix} = \begin{pmatrix} aI \\ bI \end{pmatrix},$$

gives

$$WV_1 = (b - a)I.$$

The next step is

$$\begin{pmatrix} I & X \\ 0 & Z \end{pmatrix} \begin{pmatrix} U \\ V \end{pmatrix} = \begin{pmatrix} U_1 \\ V_1 \end{pmatrix},$$

and we get

$$ZV = V_1 = W^{-1}(b - a)I$$

or

$$(WZ)V = (b - a)I.$$

Therefore

$$V = (b - a)(Y - X)^{-1}.$$

Hence, we have

$$Y - X = p_N^{-1}(b)p_{N-1}(b) - p_N^{-1}(a)p_{N-1}(a) = \Gamma_N(\delta_{N-1}(a) - \delta_{N-1}(b)).$$

This means that

$$\Gamma_N^T = (b - a)(\delta_{N-1}(a) - \delta_{N-1}(b))^{-1}\Gamma_N^{-1},$$

or

$$\Gamma_N^T \Gamma_N = (b - a)(\delta_{N-1}(a) - \delta_{N-1}(b))^{-1}.$$

Then, Γ_N is given as a Cholesky decomposition of the right hand side matrix. The right hand side is positive definite because $\delta_{N-1}(a)$ is a diagonal block of the inverse of $(J_N - aI)^{-1}$ which is positive definite because the eigenvalues of J_N are larger that a and $-\delta_{N-1}(b)$ is the negative of a diagonal block of $(J_N - bI)^{-1}$ which is positive definite because the eigenvalues of J_N are smaller that b.

From Γ_N, we can compute Ω_{N+1}:

$$\Omega_{N+1} = aI_2 + \Gamma_N \delta_{N-1}(a)\Gamma_N^T.$$

As for the scalar case, it is not always needed to compute the weights and the nodes for the quadrature rules.

Theorem 3.7 *We have*

$$\sum_{i=1}^{2N} f(\lambda_i)u_i u_i^T = e^T f(J_N)e,$$

where $e^T = (I_2 \; 0 \ldots 0)$.

Proof:

The quadrature rule is

$$\sum_{i=1}^{2N} u_i f(\lambda_i) u_i^T.$$

If y_i are the eigenvectors of J_N then $u_i = e^T y_i$ and

$$\begin{aligned}
\sum_{i=1}^{2N} u_i f(\lambda_i) u_i^T &= \sum_{i=1}^{2N} e^T y_i f(\lambda_i) y_i^T e \\
&= e^T \left(\sum_{i=1}^{2N} y_i f(\lambda_i) y_i^T \right) e \\
&= e^T Y_N f(T_N) Y_N^T e \\
&= e^T f(J_N) e
\end{aligned}$$

where Y_N is the matrix of the eigenvectors and T_N the diagonal matrix of the eigenvalues of J_N.

Note that bounds for non diagonal elements can also be obtained by considering $e_i^T f(A) e_i$, $e_j^T f(A) e_j$ and $\frac{1}{2}(e_i + e_j)^T f(A)(e_i + e_j)$.

4 Construction of the orthogonal polynomials

In this section we consider the problem of computing the orthonormal polynomials or equivalently the tridiagonal matrices that we need. A very natural and elegant way to do this is to use Lanczos algorithms.

4.1 The case $u = v$

When $u = v$, we use the classical Lanczos algorithm.

Let $x_{-1} = 0$ and x_0 be given such that $\|x_0\| = 1$. The Lanczos algorithm is defined by the following relations,

$$\gamma_j x_j = r_j = (A - \omega_j I) x_{j-1} - \gamma_{j-1} x_{j-2}, \quad j = 1, \ldots$$

$$\omega_j = x_{j-1}^T A x_{j-1},$$

$$\gamma_j = \|r_j\|.$$

The sequence $\{x_j\}_{j=0}^{l}$ is an orthonormal basis of the Krylov space

$$\text{span}\{x_0, A x_0, \ldots, A^l x_0\}.$$

Proposition 4.1 *The vector x_j is given by*

$$x_j = p_j(A) x_0,$$

where p_j is a polynomial of degree j defined by the three term recurrence (identical to 3.3)

$$\gamma_j p_j(\lambda) = (\lambda - \omega_j) p_{j-1} j(\lambda) - \gamma_{j-1} p_{j-2}(\lambda), \quad p_{-1}(\lambda) \equiv 0, \quad p_0(\lambda) \equiv 1.$$

Proof:

$$\gamma_1 x_1 = (A - \omega_1 I)x_0,$$

is a first order polynomial in A. Therefore, the Proposition is easily obtained by induction.

Theorem 4.1 *If $x_0 = u$, we have*

$$x_k^T x_l = \int_a^b p_k(\lambda) p_l(\lambda) d\alpha(\lambda).$$

Proof: As the x_j's are orthonormal, we have

$$\begin{aligned} x_k^T x_l &= x_0^T P_k(A)^T P_l(A) x_0 \\ &= x_0^T Q P_k(\Lambda) Q^T Q P_l(\Lambda) Q^T x_0 \\ &= x_0^T Q P_k(\Lambda) P_l(\Lambda) Q^T x_0 \\ &= \sum_{j=1}^{n} p_k(\lambda_j) p_l(\lambda_j) \hat{x}_j^2, \end{aligned}$$

where $\hat{x} = Q^T x_0$.

Therefore, the p_j's are the orthonormal polynomials related to α that we were referring to in 3.3.

4.2 The case $u \neq v$

We apply the non–symmetric Lanczos algorithm to a symmetric matrix A.

Let $x_{-1} = \hat{x}_{-1} = 0$, and $x_0, \hat{x}_0$ be given with $x_0 \neq \hat{x}_0$ and $x_0^T \hat{x}_0 = 1$. Then we define the iterates for $j = 1, \ldots$ by

$$\gamma_j x_j = r_j = (A - \omega_j I)x_{j-1} - \beta_{j-1} x_{j-2}, \tag{4.1}$$

$$\beta_j \hat{x}_j = \hat{r}_j = (A - \omega_j I)\hat{x}_{j-1} - \gamma_{j-1} \hat{x}_{j-2}, \tag{4.2}$$

$$\omega_j = \hat{x}_{j-1}^T A x_{j-1},$$

$$\gamma_j \beta_j = \hat{r}_j^T r_j.$$

This algorithm generates two sequences of mutually orthogonal vectors as we have

$$x_l^T \hat{x}_k = \delta_{kl}.$$

We have basically the same properties as for the Lanczos algorithm.

Proposition 4.2

$$x_j = p_j(A) x_0, \quad \hat{x}_j = q_j(A) \hat{x}_0,$$

where p_j and q_j are polynomials of degree j defined by the three term recurrences

$$\gamma_j p_j(\lambda) = (\lambda - \omega_j) p_{j-1}(\lambda) - \beta_{j-1} p_{j-2}(\lambda), \quad p_{-1}(\lambda) \equiv 0, \quad p_0(\lambda) \equiv 1,$$

$$\beta_j q_j(\lambda) = (\lambda - \omega_j) q_{j-1}(\lambda) - \gamma_{j-1} q_{j-2}(\lambda), \quad q_{-1}(\lambda) \equiv 0, \quad q_0(\lambda) \equiv 1.$$

Proof: The Proposition is easily obtained by induction.

Theorem 4.2 *If $x_0 = u$ and $\hat{x}_0 = v$, then*

$$x_k^T \hat{x}_l = \int_a^b p_k(\lambda) q_l(\lambda) d\alpha(\lambda) = \delta_{kl}.$$

Proof: As the x_j's and $\hat{x}_j$'s are orthonormal the proof is identical to the proof of Theorem 4.1.

We have seen in the previous Section the relationship between the p and q's. The polynomials q are multiples of the polynomials p.

In this particular application of the non-symmetric Lanczos algorithm it is possible to choose γ_j and β_j such that

$$\gamma_j = \pm \beta_j = \pm \sqrt{|\hat{r}_j^T r_j|},$$

with, for instance, $\gamma_j \geq 0$ and $\beta_j = \text{sgn}(\hat{r}_j^T r_j)\gamma_j$. Then, we have

$$p_j(\lambda) = \pm q_j(\lambda).$$

The main difference with the case of symmetric Lanczos is that this algorithm may break down, e.g. we can have $\gamma_j \beta_j = 0$ at some step.

We would like to use the non-symmetric Lanczos algorithm with $x_0 = e_i$ and $\hat{x}_0 = e_j$ to get estimates of $f(A)_{i,j}$. Unfortunately, this is not possible as this implies $x_0^T \hat{x}_0 = 0$. A way to get around this problem is to set $x_0 = e_i/\delta$ and $\hat{x}_0 = \delta e_i + e_j$. This will give an estimate of $f(A)_{i,j}/\delta + f(A)_{i,i}$ and we can use the bounds we get for the diagonal elements (using for instance symmetric Lanczos) to obtain bounds for the non diagonal entry. An added adantage is that we are able to choose δ so that $\gamma_j \beta_j > 0$ and therefore $p_j(\lambda) = q_j(\lambda)$. This can be done by starting with $\delta = 1$ and restarting the algorithm with a larger value of δ as soon as we find a value of j for which $\gamma_j \beta_j \leq 0$.

Regarding expressions for the remainder, we can do exactly the same as for symmetric Lanczos. We can write

$$R(f) = \frac{f^{(2N)}(\eta)}{(2N)!} \int_a^b p_N(\lambda)^2 \, d\alpha(\lambda).$$

However, we know that $p_N(\lambda) = \pm q_N(\lambda)$ and

$$\int_a^b p_N(\lambda) q_N(\lambda) \, d\alpha(\lambda) = 1.$$

This shows that the sign of the integral in the remainder can be computed using the algorithm and we have the following result.

Theorem 4.3 *Suppose f is such that $f^{(2n)}(\eta) > 0$, $\forall n$, $\forall \eta$, $a < \eta < b$. Then, the quadrature rule 3.7 gives a lower bound if*

$$\prod_{j=1}^{N} sgn(\hat{r}_j^T r_j) = 1,$$

and an upper bound otherwise. In both cases, we have

$$|R(f)| = \frac{f^{(2N)}(\eta)}{(2N)!}.$$

For Gauss–Radau and Gauss–Lobatto we cannot do the same thing. Bounds can be obtained if we choose the initial vectors (e.g. δ) such that the measure is positive and increasing. In this case we are in exactly the same framework as for the symmetric case and the same results are obtained. Note however that it is not easy to make this choice a priori. Some examples are given in Section 6. A way to proceed is to start with $\delta = 1$ and to restart the algorithm 4.2 with a larger value of δ whenever we have $\gamma_j \beta_j \leq 0$.

4.3 The block case

Now, we consider the block Lanczos algorithm, see [10],[16]. Let X_0 be an $n \times 2$ given matrix, such that $X_0^T X_0 = I_2$ (chosen as U defined before). Let $X_{-1} = 0$ be an $n \times 2$ matrix. Then

$$\Omega_j = X_{j-1}^T A X_{j-1},$$
$$R_j = A X_{j-1} - X_{j-1}\Omega_j - X_{j-2}\Gamma_{j-1}^T,$$
$$X_j \Gamma_j = R_j$$

The last step is the QR decomposition of R_j such that X_j is $n \times 2$ with $X_j^T X_j = I_2$ and Γ_j is 2×2. The matrix Ω_j is 2×2 and Γ_j is upper triangular.

It may happen that R_j is rank deficient and in that case Γ_j is singular. The solution of this problem is given in [10]. One of the columns of X_j can be chosen arbitrarily. To complete the algorithm, we choose this column to be orthogonal with the previous block vectors X_k. We can for instance choose another vector (randomly) and orthogonalize it against the previous ones.

This algorithm generates a sequence such that

$$X_j^T X_i = \delta_{ij} I_2,$$

where I_2 is the 2×2 identity matrix.

Proposition 4.3

$$X_i = \sum_{k=0}^{i} A^k X_0 C_k^{(i)},$$

where $C_k^{(i)}$ are 2×2 matrices.

Proof: The proof is given by induction.

We define a matrix polynomial $p_i(\lambda)$, a 2×2 matrix, as

$$p_i(\lambda) = \sum_{k=0}^{i} \lambda^k C_k^{(i)}.$$

Thus, we have the following result.

Theorem 4.4

$$X_i^T X_j = \int_a^b p_i(\lambda)^T d\alpha(\lambda) p_j(\lambda) = \delta_{ij} I_2.$$

Proof:

Using the orthogonality of the X_is, we can write

$$\begin{aligned}
\delta_{ij} I_2 = X_i^T X_j &= \left(\sum_{k=0}^{i} (C_k^{(i)})^T X_0^T A^k\right)\left(\sum_{l=0}^{j} A^l X_0 C_l^{(j)}\right) \\
&= \sum_{k,l} (C_k^{(i)})^T X_0^T Q \Lambda^{k+l} Q^T X_0 C_l^{(j)} \\
&= \sum_{k,l} (C_k^{(i)})^T \alpha \Lambda^{k+l} \alpha^T C_l^{(j)} \\
&= \sum_{k,l} (C_k^{(i)})^T \left(\sum_{m=1}^{n} \lambda_m^{k+l} \alpha_m \alpha_m^T\right) C_l^{(j)} \\
&= \sum_{m=1}^{n} \left(\sum_k \lambda_m^k (C_k^{(i)})^T\right) \alpha_m \alpha_m^T \left(\sum_l \lambda_m^l C_l^{(j)}\right).
\end{aligned}$$

The p_js can be considered as matrix orthogonal polynomials for the (matrix) measure α. To compute the polynomials, we need to show that the following recurrence relation holds.

Theorem 4.5 *The matrix valued polynomials p_j satisfy*

$$p_j(\lambda)\Gamma_j = \lambda p_{j-1}(\lambda) - p_{j-1}(\lambda)\Omega_j - p_{j-2}(\lambda)\Gamma_{j-1}^T,$$

$$p_{-1}(\lambda) \equiv 0, \quad p_0(\lambda) \equiv I_2,$$

where λ is a scalar.

Proof: From the previous definition, it is easily shown by induction that p_j can be generated by the given (matrix) recursion.

As we have seen before this can be written as

$$\lambda[p_0(\lambda), \ldots, p_{N-1}(\lambda)] = [p_0(\lambda), \ldots, p_{N-1}(\lambda)] J_N + [0, \ldots, 0, p_N(\lambda)\Gamma_N],$$

and as $P(\lambda) = [p_0(\lambda), \ldots, p_{N-1}(\lambda)]^T$,

$$J_N P(\lambda) = \lambda P(\lambda) - [0, \ldots, 0, p_N(\lambda)\Gamma_N]^T,$$

with J_N defined by 3.8.

Most of the following results on the properties of the matrix polynomials are derived from [1].

Proposition 4.4 *The eigenvalues of J_N are the zeroes of* $\det[p_N(\lambda)]$.

Proof: Let μ be a zero of $\det[p_N(\lambda)]$. As the rows of $p_N(\mu)$ are linearly dependent, there exists a vector v with two components such that

$$v^T p_N(\mu) = 0.$$

This implies that

$$\mu[v^T p_0(\mu), \ldots, v^T p_{N-1}(\mu)] = [v^T p_0(\mu), \ldots, v^T p_{N-1}(\mu)] J_N.$$

Therefore μ is an eigenvalue of J_N. $\det[p_N(\lambda)]$ is a polynomial of degree $2N$ in λ. Hence, there exists $2N$ zeroes of the determinant and therefore all eigenvalues are zeroes of $\det[p_N(\lambda)]$.

Proposition 4.5 *For λ and μ real, we have the analog of the Christoffel–Darboux identity (see [21]) :*

$$p_{N-1}(\mu)\Gamma_N^T p_N^T(\lambda) - p_N(\mu)\Gamma_N p_{N-1}^T(\lambda) = (\lambda - \mu) \sum_{m=0}^{N-1} p_m(\mu) p_m^T(\lambda). \qquad (4.3)$$

Proof: From the previous results, we have

$$\Gamma_{j+1}^T p_{j+1}^T(\lambda) = \lambda p_j^T(\lambda) - \Omega_{j+1} p_j^T(\lambda) - \Gamma_j p_{j-1}^T(\lambda),$$

$$p_{j+1}(\mu)\Gamma_{j+1} = \mu p_j(\mu) - p_j(\mu)\Omega_{j+1} - p_{j-1}(\mu)\Gamma_j^T.$$

Multiplying the first relation by $p_j(\mu)$ on the left and the second one by $p_j^T(\lambda)$ on the right gives

$$p_j(\mu)\Gamma_{j+1}^T p_{j+1}^T(\lambda) - p_{j+1}(\mu)\Gamma_{j+1} p_j^T(\lambda) =$$
$$= (\lambda - \mu) p_j(\mu) p_j^T(\lambda) - p_j(\mu)\Gamma_j p_{j-1}^T(\lambda) + p_{j-1}(\mu)\Gamma_j^T p_j^T(\lambda).$$

Summing these equalities, some terms cancel and we get the desired result.

In particular, if we choose $\lambda = \mu$ in 4.3, we have that $p_N(\lambda)\Gamma_N p_{N-1}^T(\lambda)$ is symmetric.

Proposition 4.6

$$\sum_{m=0}^{N-1} u_s^T p_m(\lambda_s) p_m^T(\lambda_r) u_r = \delta_{rs}.$$

Proof: If we set $\lambda = \lambda_s$ and $\mu = \lambda_r$ and multiply the Christoffel–Darboux identity 4.3 on the left by u_s^T and on the right by u_r, we have $p_N^T(\lambda_r)u_r = 0$ and $p_N^T(\lambda_s)u_s = 0$, and we get the result if $\lambda_s \neq \lambda_r$. Let

$$K_{N-1}(\mu, \lambda) = \sum_{m=0}^{N-1} p_m(\mu) p_m^T(\lambda).$$

As $p_0(\lambda) = I_2$, $K_{N-1}(\lambda, \lambda)$ is a symmetric positive definite matrix and therefore defines a scalar product. If λ_r is a multiple eigenvalue, there exist linearly independent eigenvectors that we could orthonormalize. If λ_r is an eigenvalue of multiplicity q_r, there exist q_r linearly independent vectors $v_r^1, \ldots, v_r^{q_r}$ with two components such that the vectors

$P(\lambda_r)v_r^j, j = 1, \ldots, q_r$ are the eigenvectors associated with λ_r. We can certainly find a set of vectors $(w_r^1, \ldots, w_r^{q_r})$ spanning the same subspace as $(v_r^1, \ldots, v_r^{q_r})$ and such that

$$(w_r^k)^T K_{N-1}(\lambda_r, \lambda_r) w_r^l = \delta_{kl}.$$

This property is nothing else than the orthogonality relation of the eigenvectors.

Proposition 4.7

$$\sum_{r=1}^{2N} p_i^T(\lambda_r) u_r u_r^T p_l(\lambda_r) = \delta_{il} I_2, \quad i = 0, \ldots, N-1, \qquad l = 0, \ldots, N-1.$$

Proof: Note that the eigenvectors of J_N are linearly independent. We take a set of N vectors with two elements : $\{y_0, \ldots, y_{N-1}\}$, and write

$$[y_0^T, \ldots, y_{N-1}^T]^T = \sum_{r=1}^{2N} \alpha_r [u_r^T p_0(\lambda_r), \ldots, u_r^T p_{N-1}(\lambda_r)]^T,$$

or

$$y_l = \sum_{r=1}^{2N} p_l^T(\lambda_r) u_r \alpha_r, \quad l = 0, \ldots, N-1.$$

Multiplying on the left by $u_s^T p_l(\lambda_s)$ and summing :

$$\sum_{l=0}^{N-1} u_s^T p_l(\lambda_s) y_l = \alpha_s, \quad s = 1, \ldots, 2N.$$

Therefore,

$$y_i = \sum_{r=1}^{2N} \sum_{l=0}^{N-1} p_i^T(\lambda_r) u_r u_r^T p_l(\lambda_r) y_l.$$

This gives the desired result.

To prove that the block quadrature rule is exact for polynomials of degree up to $2N-1$, we cannot use the same method as for the scalar case where the given polynomial is factored because of commutativity problems. Therefore, we take another approach that has been used in a different setting in [2]. The following results are taken from [2].

We will consider all the monomials $\lambda^k, k = 1, \ldots, 2N-1$. Let M_k the moment matrix, be defined as

$$M_k = \int_a^b \lambda^k \, d\alpha(\lambda).$$

We write the (matrix) orthonormal polynomials p_j as

$$p_j(\lambda) = \sum_{k=0}^{j} p_k^{(j)} \lambda^k,$$

$p_k^{(j)}$ being a matrix of order 2. Then, we have

$$\int_a^b p_j^T(\lambda) \, d\alpha(\lambda) = \sum_{k=0}^{j} (p_k^{(j)})^T \int_a^b \lambda^k \, d\alpha(\lambda) = \sum_{k=0}^{j} (p_k^{(j)})^T M_k,$$

and more generally

$$\int_a^b p_j^T(\lambda)\lambda^q \, d\alpha(\lambda) = \sum_{k=0}^{j} (p_k^{(j)})^T M_{k+q}.$$

We write these equations for $j = N-1$. Note that because of the orthogonality of the polynomials, we have

$$\int_a^b p_{N-1}^T(\lambda)\lambda^q \, d\alpha(\lambda) = 0, \quad q = 0, \ldots, N-2.$$

Let H_N be the block Hankel matrix of order $2N$, defined as

$$H_N = \begin{pmatrix} M_0 & \cdots & M_{N-1} \\ \vdots & & \vdots \\ M_{N-1} & \cdots & M_{2N-2} \end{pmatrix}.$$

Then

$$H_N \begin{pmatrix} p_0^{(N-1)} \\ \vdots \\ p_{N-2}^{(N-1)} \\ p_{N-1}^{(N-1)} \end{pmatrix} = \begin{pmatrix} 0 \\ \vdots \\ 0 \\ \int_a^b p_{N-1}^T(\lambda)\lambda^{N-1} \, d\alpha(\lambda) \end{pmatrix}.$$

We introduce some additional notation. Let L_N be a block upper triangular matrix of order $2N$,

$$L_N = \begin{pmatrix} p_0^{(0)} & p_0^{(1)} & \cdots & p_0^{(N-1)} \\ 0 & p_1^{(1)} & \cdots & p_1^{(N-1)} \\ & & \ddots & \vdots \\ & & & p_{N-1}^{(N-1)} \end{pmatrix}.$$

Let V_N be a $4N \times 2N$ matrix defined in block form as

$$V_N = \begin{pmatrix} B_1 \\ B_2 \\ \vdots \\ B_l \end{pmatrix},$$

where B_j is a $q_j \times 2N$ matrix,

$$B_j = \begin{pmatrix} I_2 & \mu_j I_2 & \cdots & \mu_j^{N-1} I_2 \\ \vdots & \vdots & \vdots & \vdots \\ I_2 & \mu_j I_2 & \cdots & \mu_j^{N-1} I_2 \end{pmatrix},$$

and the μ_j are the eigenvalues of A.

Let K_i^j be a $2q_i \times 2q_j$ matrix

$$K_i^j = \begin{pmatrix} K_{N-1}(\mu_i,\mu_j) & \cdots & K_{N-1}(\mu_i,\mu_j) \\ \vdots & \vdots & \vdots \\ K_{N-1}(\mu_i,\mu_j) & \cdots & K_{N-1}(\mu_i,\mu_j) \end{pmatrix},$$

and

$$K = \begin{pmatrix} K_1^1 & K_1^2 & \dots & K_1^l \\ K_2^1 & \dots & \dots & K_2^l \\ \vdots & & & \vdots \\ K_l^1 & \dots & \dots & K_l^l \end{pmatrix}.$$

Proposition 4.8

$$V_N L_N = \begin{pmatrix} C_1 \\ C_2 \\ \vdots \\ C_l \end{pmatrix},$$

where C_j *is a* $2q_j \times 2N$ *matrix,*

$$C_j = \begin{pmatrix} p_0(\mu_j) & p_1(\mu_j) & \cdots & p_{N-1}(\mu_j) \\ \vdots & \vdots & \vdots & \vdots \\ p_0(\mu_j) & p_1(\mu_j) & \cdots & p_{N-1}(\mu_j) \end{pmatrix}.$$

Proof: This is straightforward by the definition of the polynomials $p_j(\lambda)$.

Proposition 4.9

$$L_N^T H_N L_N = I.$$

Proof: the generic term of $H_N L_N$ is

$$(H_N L_N)_{ij} = \sum_{s=1}^{j} M_{s+i-2}\, p_{s-1}^{(j-1)}.$$

Therefore the generic term of $L_N^T H_N L_N$ is

$$(L_N^T H_N L_N)_{ij} = \sum_{r=1}^{i} \sum_{s=1}^{j} \int_a^b (p_{r-1}^{(i-1)})^T \lambda^{s+r-2}\, d\alpha(\lambda) p_{s-1}^{(j-1)}.$$

Splitting the power of λ we can easily see that this is

$$(L_N^T H_N L_N)_{ij} = \int_a^b p_{i-1}^T(\lambda)\, d\alpha(\lambda)\, p_{j-1}(\lambda).$$

Therefore because of the orthonormality properties, we have

$$(L_N^T H_N L_N)_{ij} = \begin{cases} I_2 & \text{if } i = j, \\ 0 & \text{otherwise.} \end{cases}$$

This result implies that

$$H_N^{-1} = L_N L_N^T.$$

Proposition 4.10

$$V_N L_N (V_N L_N)^T = K.$$

Proof: this is just using the definition of K_i^j.

Now, we define a $2N \times 4N$ matrix W_N^T whose only non zero components in row i are in position $(i, 2i-1)$ and $(i, 2i)$ and are successively the two components of

$$(w_1^1)^T, \ldots, (w_1^{q_1})^T, (w_2^1)^T, \ldots, (w_l^{q_l})^T.$$

Then because of the way the w_i^j are constructed we have

Proposition 4.11

$$W_N^T K W_N = I.$$

Proposition 4.12 *$W_N^T V_N$ is a non singular $2N \times 2N$ matrix.*

Proof:

$$W_N^T V_N H_N^{-1} V_N^T W_N = W_N^T V_N L_N L_N^T V_N^T W_N = W_N^T K W_N = I.$$

This shows that $W_N^T V_N$ is non singular.

Then, we have the main result

Theorem 4.6 *The quadrature rule 3.9 or 3.10 is exact for polynomials of order less than or equal to $2N-1$.*

Proof: From Proposition 4.12, we have

$$H_N^{-1} = (W_N^T V_N)^{-1} (V_N^T W_N)^{-1}.$$

Therefore,

$$H_N = (V_N^T W_N)(W_N^T V_N).$$

By identification of the entries of the two matrices we have,

$$M_k = \sum_{i=1}^{l} \left(\sum_{j=1}^{q_i} (w_i^j)(w_i^j)^T \right) \mu_i^k, \quad k = 0, \ldots, 2N-2.$$

It remains to prove that the quadrature rule is exact for $k = 2N-1$. As we have,

$$H_{N+1} \begin{pmatrix} p_0^{(N)} \\ \vdots \\ p_{N-1}^{(N)} \\ p_N^{(N)} \end{pmatrix} = \begin{pmatrix} 0 \\ \vdots \\ 0 \\ \int_a^b p_N^T(\lambda) \lambda^N \, d\alpha(\lambda) \end{pmatrix}.$$

Writing the $(N-1)$th block row of this equality, we get

$$M_{2N-1} \, p_N^{(N)} = - \sum_{r=0}^{N-1} M_{N+r-1} \, p_r^{(N)}.$$

We have proved before that

$$M_{N+r-1} = \sum_{i=1}^{l} \left(\sum_{j=1}^{q_i} w_i^j (w_i^j)^T \right) \mu_i^{N+r-1}.$$

By substitution, we get

$$M_{2N-1}\, p_N^{(N)} = - \sum_{r=0}^{N-1} \sum_{i=1}^{l} \sum_{j=1}^{q_i} w_i^j (w_i^j)^T \mu_i^{N+r-1} p_r^{(N)}.$$

We use the fact that

$$(w_i^j)^T \sum_{r=0}^{N-1} \mu_i^r p_r^{(N)} = (w_i^j)^T p_N(\mu_i) - (w_i^j)^T \mu_i^N p_N^{(N)},$$

and

$$(w_i^j)^T p_N(\mu_i) = 0.$$

This shows that

$$M_{2N-1} p_N^{(N)} = \sum_{i=1}^{l} \sum_{j=1}^{q_i} (w_i^j)(w_i^j)^T \mu_i^{2N-1} p_N^{(N)}.$$

As $p_N^{(N)}$ is non singular, we get the result.

To obtain expressions for the remainder, we would like to use a similar approach as for the scalar case. However there are some differences, as the quadrature rule is exact for polynomials of order $2N-1$ and we have $2N$ nodes, we cannot interpolate with an Hermite polynomial and we have to use a Lagrange polynomial. By Theorems 2.1.1.1 and 2.1.4.1 of [18], there exists a polynomial q of degree $2N-1$ such that

$$q(\lambda_j) = f(\lambda_j), \quad j = 1, \ldots, 2N$$

and

$$f(x) - q(x) = \frac{s(x) f^{(2N)}(\xi(x))}{(2N)!},$$

where

$$s(x) = (x - \lambda_1) \cdots (x - \lambda_{2N}).$$

If we can apply the mean value theorem, the remainder $R(f)$ which is a 2×2 matrix can be written as

$$R(f) = \frac{f^{(2N)}(\eta)}{(2N)!} \int_a^b s(\lambda)\, d\alpha(\lambda).$$

Unfortunately s does not have a constant sign over the interval $[a, b]$. Therefore this representation formula for the remainder is of little practical use for obtaining bounds with the knowledge of the sign of the entries of the remainder.

It is easy to understand why we cannot directly obtain bounds with this block approach. We must use $W = (e_i\ e_j)$ and the block Lanczos algorithm with $X_0 = W$. For the block Lanczos algorithm we multiply successively A with the Lanczos vectors. If A

is sparse, most of the components of these products are 0 for the first few iterates of the algorithm. Therefore, it is likely that at the beginning, the estimates that we will get for the non diagonal entries will be 0. This explains why we cannot directly obtain upper or lower bounds with Gauss, Gauss–Radau or Gauss–Lobatto. A way to avoiding this difficulty is to use $W = (e_i + e_j \; e_j)$ but this cannot be done since $X_0^T X_0 \neq I_2$.

However, we will see in the numerical experiments that the estimates we get are often quite good.

5 Application to the inverse of a matrix

In this Section we consider obtaining analytical bounds for the entries of the inverse of a given matrix and simplifying the algorithms to compute numerical bounds and approximations.

We consider

$$f(\lambda) = \frac{1}{\lambda}, \quad 0 < a < b,$$

and hence

$$f^{(2n+1)}(\lambda) = -(2n+1)! \; \lambda^{-(2n+2)},$$

and

$$f^{(2n)}(\lambda) = (2n)! \; \lambda^{-(2n+1)}.$$

Therefore, the even derivatives are positive on $[a, b]$ and the odd derivatives are negative which implies that we can apply Theorems 3.1, 3.2 and 3.3.

Consider a dense non singular matrix $A = (a_{ij})_{i,j=1,\ldots,m}$. We choose $u = x_0 = e_i$ and we apply the Lanczos algorithm. From results of the first iteration we can obtain analytical results. The first step of the Lanczos algorithm gives us

$$\omega_1 = e_i^T A e_i = a_{ii},$$

$$\gamma_1 x_1 = r_1 = (A - \omega_1 I) e_i.$$

Let s_i be defined by

$$s_i^2 = \sum_{j \neq i} a_{ji}^2,$$

and let

$$d_i^T = (a_{1,i}, \ldots, a_{i-1,i}, 0, a_{i+1,i}, \ldots, a_{m,i}).$$

Then

$$\gamma_1 = s_i, \quad x_1 = \frac{1}{s_i} d_i.$$

From this, we have

$$\omega_2 = \frac{1}{s_i^2} \sum_{k \neq i} \sum_{l \neq i} a_{k,i} a_{k,l} a_{l,i}.$$

From this data, we compute the Gauss rule and get a lower bound on the diagonal element:

$$J_2 = \begin{pmatrix} \omega_1 & \gamma_1 \\ \gamma_1 & \omega_2 \end{pmatrix},$$

$$J_2^{-1} = \frac{1}{\omega_1\omega_2 - \gamma_1^2} \begin{pmatrix} \omega_2 & -\gamma_1 \\ -\gamma_1 & \omega_1 \end{pmatrix}.$$

The lower bound is given by

$$\frac{\omega_2}{\omega_1\omega_2 - \gamma_1^2} = \frac{\sum_{k\neq i}\sum_{l\neq i} a_{k,i}a_{k,l}a_{l,i}}{a_{i,i}\sum_{k\neq i}\sum_{l\neq i} a_{k,i}a_{k,l}a_{l,i} - \left(\sum_{k\neq i} a_{k,i}^2\right)^2}.$$

Note that this bound does not depend on the eigenvalues a and b.

Now, we consider the Gauss–Radau rule. Then,

$$\tilde{J}_2 = \begin{pmatrix} \omega_1 & \gamma_1 \\ \gamma_1 & x \end{pmatrix},$$

the eigenvalues λ are the roots of $(\omega_1 - \lambda)(x - \lambda) - \gamma_1^2 = 0$, which gives the relation

$$x = \lambda + \frac{\gamma_1^2}{\omega_1 - \lambda}.$$

To obtain an upper bound we set $\lambda = a$. The solution is

$$x = x_a = a + \frac{\gamma_1^2}{\omega_1 - a}.$$

For the Gauss–Lobatto rule, we have the same problem except that we want $\tilde{J}_2$ to have a and b as eigenvalues. This leads to solving the following linear system,

$$\begin{pmatrix} \omega_1 - a & -1 \\ \omega_1 - b & -1 \end{pmatrix} \begin{pmatrix} x \\ \gamma_1^2 \end{pmatrix} = \begin{pmatrix} a\omega_1 - a^2 \\ b\omega_1 - b^2 \end{pmatrix}.$$

Solving this system and computing the $(1,1)$ element of the inverse gives

$$\frac{a + b - \omega_1}{ab}.$$

Hence we have the following result.

Theorem 5.1 *We have the following bounds*

$$\frac{\sum_{k\neq i}\sum_{l\neq i} a_{k,i}a_{k,l}a_{l,i}}{a_{i,i}\sum_{k\neq i}\sum_{l\neq i} a_{k,i}a_{k,l}a_{l,i} - \left(\sum_{k\neq i} a_{k,i}^2\right)^2} \leq (A^{-1})_{i,i}$$

$$\frac{a_{i,i} - b + \frac{s_i^2}{b}}{a_{i,i}^2 - a_{i,i}b + s_i^2} \leq (A^{-1})_{i,i} \leq \frac{a_{i,i} - a + \frac{s_i^2}{a}}{a_{i,i}^2 - a_{i,i}a + s_i^2}$$

$$(A^{-1})_{i,i} \leq \frac{a + b - a_{ii}}{ab}.$$

It is not too easy to derive analytical bounds from the block Lanczos algorithm as we have to compute repeated inverses of 2×2 matrices.

It is much easier to use the non-symmetric Lanczos method with the Gauss-Radau rule. We are looking at the sum of the (i,i) and (i,j) elements of the inverse. Let

$$t_i = \gamma_1 \beta_1 = \sum_{k \neq i} a_{k,i}(a_{k,i} + a_{k,j}) - a_{i,j}(a_{i,j} + a_{i,i}).$$

Then, the computations are essentially the same as for the diagonal case.

Theorem 5.2 *For $(A^{-1})_{i,j} + (A^{-1})_{i,i}$ we have the two following estimates*

$$\frac{a_{i,i} + a_{i,j} - a + \frac{t_i}{a}}{(a_{i,i} + a_{i,j})^2 - a(a_{i,i} + a_{i,j}) + t_i}, \quad \frac{a_{i,i} + a_{i,j} - b + \frac{t_i}{b}}{(a_{i,i} + a_{i,j})^2 - b(a_{i,i} + a_{i,j}) + t_i}.$$

If $t_i \geq 0$, the first expression with a gives an upper bound and the second one with b a lower bound. Then, we have to subtract the bounds for the diagonal term to get bounds on $(A^{-1})_{i,j}$.

The previous results can be compared with those obtained by other methods in [17]. Results can also be obtained for sparse matrices taking into account the sparsity structure.

In the computations using the Lanczos algorithm for the Gauss, Gauss-Radau and Gauss-Lobatto rules, we need to compute the $(1,1)$ element of the inverse of a tridiagonal matrix. This may be done in many different ways, see for instance [13]. Here, we will show that we can compute this element of the inverse incrementally as we go through the Lanczos algorithm and we obtain the estimates for very few additional operations. This is stated in the following theorem where b_j stands for the bounds for Lanczos iteration j and the ω_js and the γ_js are generated by Lanczos.

Theorem 5.3 *The following algorithm yields a lower bound b_j of A_{ii}^{-1} by the Gauss quadrature rule, a lower bound $\bar{b}_j$ and an upper bound $\hat{b}_j$ through the Gauss-Radau quadrature rule and an upper bound $\breve{b}_j$ through the Gauss-Lobatto rule.*

Let $x_{-1} = 0$ and $x_0 = e_i$, $\omega_1 = a_{ii}$, $\gamma_1 = \|(A - \omega_1 I)e_i\|$, $b_1 = \omega_1^{-1}$, $d_1 = \omega_1$, $c_1 = 1$, $\hat{d}_1 = \omega_1 - a$, $\bar{d}_1 = \omega_1 - b$, $x_1 = (A - \omega_1 I)e_i/\gamma_1$.

Then for $j = 2, \ldots$ we compute

$$r_j = (A - \omega_j I)x_{j-1} - \gamma_{j-1}x_{j-2},$$

$$\omega_j = x_{j-1}^T A x_{j-1},$$

$$\gamma_j = \|r_j\|,$$

$$x_j = \frac{r_j}{\gamma_j},$$

$$b_j = b_{j-1} + \frac{\gamma_{j-1}^2 c_{j-1}^2}{d_{j-1}(\omega_j d_{j-1} - \gamma_{j-1}^2)},$$

$$d_j = \omega_j - \frac{\gamma_{j-1}^2}{d_{j-1}},$$

$$c_j = c_{j-1}\frac{\gamma_{j-1}}{d_{j-1}},$$

$$\hat{d}_j = \omega_j - a - \frac{\gamma_{j-1}^2}{\hat{d}_{j-1}},$$

$$\bar{d}_j = \omega_j - b - \frac{\gamma_{j-1}^2}{\bar{d}_{j-1}},$$

$$\hat{\omega}_j = a + \frac{\gamma_j^2}{\hat{d}_j},$$

$$\bar{\omega}_j = b + \frac{\gamma_j^2}{\bar{d}_j},$$

$$\hat{b}_j = b_j + \frac{\gamma_j^2 c_j^2}{d_j(\hat{\omega}_j d_j - \gamma_j^2)},$$

$$\bar{b}_j = b_j + \frac{\gamma_j^2 c_j^2}{d_j(\bar{\omega}_j d_j - \gamma_j^2)},$$

$$\breve{\omega}_j = \frac{\hat{d}_j \bar{d}_j}{\bar{d}_j - \hat{d}_j}\left(\frac{b}{\hat{d}_j} - \frac{a}{\bar{d}_j}\right),$$

$$\breve{\gamma}_j^2 = \frac{\hat{d}_j \bar{d}_j}{\bar{d}_j - \hat{d}_j}(b - a),$$

$$\breve{b}_j = b_j + \frac{\breve{\gamma}_j^2 c_j^2}{d_j(\breve{\omega}_j d_j - \breve{\gamma}_j{}^2)}.$$

Proof: We have from 3.4

$$J_N = \begin{pmatrix} \omega_1 & \gamma_1 & & & \\ \gamma_1 & \omega_2 & \gamma_2 & & \\ & \ddots & \ddots & \ddots & \\ & & \gamma_{N-2} & \omega_{N-1} & \gamma_{N-1} \\ & & & \gamma_{N-1} & \omega_N \end{pmatrix}.$$

Let $x_N^T = (0 \ \ldots \ 0 \ \gamma_N)$, so that

$$J_{N+1} = \begin{pmatrix} J_N & x_N \\ x_N^T & \omega_{N+1} \end{pmatrix}.$$

Letting

$$\tilde{J} = J_N - \frac{x_N x_N^T}{\omega_{N+1}},$$

the upper left block of J_{N+1}^{-1} is $\tilde{J}^{-1}$. This can be obtained through the use of the Sherman–Morrison formula (see [11]),

$$\tilde{J}^{-1} = J_N^{-1} + \frac{(J_N^{-1}x_N)(x_N^T J_N^{-1})}{\omega_{N+1} - x_N^T J_N^{-1} x_N}.$$

Let $j_N = J_N^{-1} e_N$ be the last column of the inverse of J_N. With this notation, we have

$$\tilde{J}^{-1} = J_N^{-1} + \frac{\gamma_N^2 j_N j_N^T}{\omega_{N+1} - \gamma_N^2 (j_N)_N}.$$

Therefore, it is clear that we only need the first and last elements of the last column of the inverse of J_N. This can be obtained using the Cholesky decomposition of J_N. It is easy to check that if we define

$$d_1 = \omega_1, \quad d_i = \omega_i - \frac{\gamma_{i-1}^2}{d_{i-1}}, \quad i = 2, \ldots, N$$

then

$$(j_N)_1 = (-1)^{N-1} \frac{\gamma_1 \cdots \gamma_{N-1}}{d_1 \cdots d_N}, \quad (j_N)_N = \frac{1}{d_N}.$$

When we put all these results together we get the proof of the Theorem.

The algorithm is essentially the same for the non-symmetric case. The modified algorithm is the following, using $f_1 = 1$:

$$r_j = (A - \omega_j I)x_{j-1} - \beta_{j-1} x_{j-2},$$

$$\hat{r}_j = (A - \omega_j I)\hat{x}_{j-1} - \gamma_{j-1} \hat{x}_{j-2},$$

$$\omega_j = \hat{x}_{j-1}^T A x_{j-1},$$

$$\gamma_j \beta_j = \hat{r}_j^T r_j.,$$

$$x_j = \frac{r_j}{\gamma_j},$$

$$\hat{x}_j = \frac{\hat{r}_j}{\beta_j},$$

$$b_j = b_{j-1} + \frac{\gamma_{j-1} \beta_{j-1} c_{j-1} f_{j-1}}{d_{j-1}(\omega_j d_{j-1} - \gamma_{j-1} \beta_{j-1})},$$

$$d_j = \omega_j - \frac{\gamma_{j-1}^2}{d_{j-1}},$$

$$c_j = c_{j-1} \frac{\gamma_{j-1}}{d_{j-1}},$$

$$f_j = f_{j-1} \frac{\beta_{j-1}}{d_{j-1}},$$

$$\hat{d}_j = \omega_j - a - \frac{\gamma_{j-1}\beta_{j-1}}{\hat{d}_{j-1}},$$

$$\bar{d}_j = \omega_j - b - \frac{\gamma_{j-1}\beta_{j-1}}{\bar{d}_{j-1}},$$

$$\hat{\omega}_j = a + \frac{\gamma_j\beta_j}{\hat{d}_j},$$

$$\bar{\omega}_j = b + \frac{\gamma_j\beta_j}{\bar{d}_j},$$

$$\hat{b}_j = b_j + \frac{\gamma_j\beta_j c_j f_j}{d_j(\hat{\omega}_j d_j - \gamma_j\beta_j)},$$

$$\bar{b}_j = b_j + \frac{\gamma_j\beta_j c_j f_j}{d_j(\bar{\omega}_j d_j - \gamma_j\beta_j)},$$

$$\breve{\omega}_j = \frac{\hat{d}_j\bar{d}_j}{\bar{d}_j - \hat{d}_j}\left(\frac{b}{\hat{d}_j} - \frac{a}{\bar{d}_j}\right),$$

$$\breve{\gamma}_j^2 = \frac{\hat{d}_j\bar{d}_j}{\bar{d}_j - \hat{d}_j}(b - a),$$

$$\breve{b}_j = b_j + \frac{\breve{\gamma}_j\beta_j c_j f_j}{d_j(\breve{\omega}_j d_j - \breve{\gamma}_j\beta_j)}.$$

We have the analog for the block case. For simplicity we only consider the Gauss rule. Then, in 3.8

$$J_N = \begin{pmatrix} \Omega_0 & \Gamma_0^T & & & \\ \Gamma_0 & \Omega_1 & \Gamma_1^T & & \\ & \ddots & \ddots & \ddots & \\ & & \Gamma_{N-3} & \Omega_{N-2} & \Gamma_{N-2}^T \\ & & & \Gamma_{N-2} & \Omega_{N-1} \end{pmatrix},$$

and

$$J_{N+1} = \begin{pmatrix} J_N & x_N \\ x_N^T & \omega_{N+1} \end{pmatrix},$$

with

$$x_N^T = (0 \ \ldots \ 0 \ \Gamma_{N-1}),$$

and

$$\tilde{J} = J_n - x_N\Omega_N^{-1}x_N^T.$$

Here, we use the Sherman–Morrison–Woodbury formula (see [11]) which is a generalization of the formula we used before. Then,

$$\tilde{J}^{-1} = J_N^{-1} + J_N^{-1}x_N(\Omega_N - x_N^T J_N^{-1}x_N)^{-1}x_N^T J_N^{-1}.$$

In order to compute all the elements, we need a block Cholesky decomposition of J_N. We obtain the following algorithm which gives a 2×2 matrix B_i, the block element of the inverse that we need.

Theorem 5.4 *Let* $B_0 = \Omega_0^{-1}$, $D_0 = \Omega_0$, $C_0 = I$, *for* $i = 1, \ldots$ *we compute*

$$B_i = B_{i-1} + C_{i-1} D_{i-1}^{-1} \Gamma_{i-1}^T (\Omega_i - \Gamma_{i-1} D_{i-1}^{-1} \Gamma_{i-1}^T)^{-1} \Gamma_{i-1} D_{i-1}^{-1} C_{i-1}^T,$$

$$D_i = \Omega_i - \Gamma_{i-1} D_{i-1}^{-1} \Gamma_{i-1}^T,$$

$$C_i = C_{i-1} \Gamma_{i-1}^T D_{i-1}^{-1}.$$

These recurrences for 2×2 matrices can be easily computed. Hence, the approximations can be computed as we apply the block Lanczos algorithm.

Given these algorithms to compute the estimates, we see that almost all of the operations are a result of the Lanczos algorithm. Computing the estimate has a complexity independent of the problem size.

To compute a diagonal entry, the Lanczos algorithm needs per iteration the following operations: 1 matrix–vector product, $4N$ multiplies and $4N$ adds. To compute two diagonal entries and a non diagonal one, the block Lanczos algorithm needs per iteration: 2 matrix–vector products, $9N$ multiplies, $7N$ adds plus the QR decomposition which is $8N$ flops (see [11]). The non–symmetric Lanczos algorithm requires per iteration: 1 matrix–vector product, $6N$ multiplies and $6N$ adds.

Therefore, if we only want to estimate diagonal elements it is best to use the Lanczos algorithm. If we want to estimate a non diagonal element, it is best to use the block Lanczos algorithm since we get three estimates in one run while for the non–symmetric Lanczos method we need also to have an estimate of a diagonal element. The number of flops is the same but for the block Lanczos we have three estimates instead of two with the non–symmetric Lanczos. On the other hand, the non–symmetric Lanczos gives bounds but the block Lanczos yields only estimates.

As we notice before, we can compute bounds for the non diagonal elements by considering $\frac{1}{2}(e_i + e_j)^T A^{-1}(e_i + e_j)$. For this, we need to run three Lanczos algorithms that is per iteration: 3 matrix–vector products, $12N$ multiplies and $12N$ adds to get 3 estimates. This no more operations than in the block Lanczos case but here, we can get bounds. With the non–symmetric Lanczos, we have 2 bounds with 2 matrix–vector products, $10N$ multiplies and $10N$ adds.

One can ask why in the case of the inverse are we not solving the linear system

$$Au = e_i$$

to obtain the ith column of the inverse at once. To our knowledge, it is not possible then to tell if the estimates are upper or lower bounds. Moreover this can be easily added to the algorithm of Theorem 5.3.

Let $Q_N = [x_0, \ldots, x_{N-1}]$ be the matrix of the Lanczos vectors. Then we have the approximate solution

$$u_N = Q_N y_N,$$

where y_N is the solution of

$$J_N y_N = Q_N^T e_i = e_1.$$

This tridiagonal linear system can be easily solved incrementally from the LDL^T decomposition, see [11]. This yields a variant of the conjugate gradient algorithm. We give numerical examples in Section 6 and show that our methods give better bounds.

6 Numerical examples

In this Section, we first describe the examples we use and then we give numerical results for some specific functions f.

6.1 Description of the examples

First we look at examples of small dimension for which the inverses are known. Then, we will turn to larger examples arising from the discretization of partial differential equations. Most of the numerical computations have been done with Matlab 3.5 on an Apple Macintosh Powerbook 170 and a few ones on a Sun workstation.

Example 1.

First, we consider

$$A = I + uu^T, \quad u^T = (1, 1, \ldots, 1)$$

This matrix has two distinct eigenvalues 1 and $n+1$. Therefore, the minimal polynomial is of degree 2 and the inverse can be written as

$$A^{-1} = \frac{2+n}{1+n} I - \frac{1}{1+n} A.$$

Example 2.

The entries of the Hilbert matrix are given by $\frac{1}{i+j-1}$. We consider a matrix of dimension 5 which is

$$A(\alpha) = \alpha I_5 + \begin{pmatrix} 1 & 1/2 & 1/3 & 1/4 & 1/5 \\ 1/2 & 1/3 & 1/4 & 1/5 & 1/6 \\ 1/3 & 1/4 & 1/5 & 1/6 & 1/7 \\ 1/4 & 1/5 & 1/6 & 1/7 & 1/8 \\ 1/5 & 1/6 & 1/7 & 1/8 & 1/9 \end{pmatrix}.$$

The inverse of $A(0)$ is

$$A^{-1} = \begin{pmatrix} 25 & -300 & 1050 & -1400 & 630 \\ -300 & 4800 & -18900 & 26880 & -1260 \\ 1050 & -18900 & 79380 & -117600 & 56700 \\ -1400 & 26880 & -117600 & 179200 & -88200 \\ 630 & -1260 & 56700 & -88200 & 44100 \end{pmatrix},$$

and the eigenvalues of $A(0)$ are

$$(3.288\ 10^{-6},\ 3.059\ 10^{-4},\ 1.141\ 10^{-2},\ 0.209,\ 1.567)$$

Example 3.

We take an example of dimension 10,

$$A = \frac{1}{11}\begin{pmatrix} 10 & 9 & 8 & 7 & 6 & 5 & 4 & 3 & 2 & 1 \\ 9 & 18 & 16 & 14 & 12 & 10 & 8 & 6 & 4 & 2 \\ 8 & 16 & 24 & 21 & 18 & 15 & 12 & 9 & 6 & 3 \\ 7 & 14 & 21 & 28 & 24 & 20 & 16 & 12 & 8 & 4 \\ 6 & 12 & 18 & 24 & 30 & 25 & 20 & 15 & 10 & 5 \\ 5 & 10 & 15 & 20 & 25 & 30 & 24 & 18 & 12 & 6 \\ 4 & 8 & 12 & 16 & 20 & 24 & 28 & 21 & 14 & 7 \\ 3 & 6 & 9 & 12 & 15 & 18 & 21 & 24 & 16 & 8 \\ 2 & 4 & 6 & 8 & 10 & 12 & 14 & 16 & 18 & 9 \\ 1 & 2 & 3 & 4 & 5 & 6 & 7 & 8 & 9 & 10 \end{pmatrix}.$$

It is easily seen (cf. [13]) that the inverse is a tridiagonal matrix

$$A^{-1} = \begin{pmatrix} 2 & -1 & & & \\ -1 & 2 & -1 & & \\ & \ddots & \ddots & \ddots & \\ & & -1 & 2 & -1 \\ & & & -1 & 2 \end{pmatrix}.$$

The eigenvalues of A are therefore distinct and given by

$$(0.2552,\ 0.2716,\ 0.3021,\ 0.3533,\ 0.4377,\ 0.5830,\ 0.8553,\ 1.4487,\ 3.1497,\ 12.3435).$$

Example 4.

We have

$$A = \begin{pmatrix} 3 & -1 & 0 & 0 & 0 \\ -1 & 2 & -1 & 0 & 0 \\ 0 & -1 & 2 & -1 & 0 \\ 0 & 0 & -1 & 2 & -1 \\ 0 & 0 & 0 & -1 & 1 \end{pmatrix},$$

whose inverse is

$$A^{-1} = \frac{1}{2}\begin{pmatrix} 1 & 1 & 1 & 1 & 1 \\ 1 & 3 & 3 & 3 & 3 \\ 1 & 3 & 5 & 5 & 5 \\ 1 & 3 & 5 & 7 & 7 \\ 1 & 3 & 5 & 7 & 9 \end{pmatrix},$$

whose eigenvalues are

$$(0.0979,\ 0.8244,\ 2.0000,\ 3.1756,\ 3.9021).$$

Example 5.

We use a matrix of dimension 10 constructed with the TOEPLITZ function of Matlab,

$$A = 21I_{10} + \text{TOEPLITZ}(1:10).$$

This matrix has distinct eigenvalues but most of them are very close together:

(0.5683, 14.7435, 18.5741, 19.5048, 20.0000, 20.2292, 20.3702, 20.4462, 20.4875, 65.0763).

Example 6.

This example is the matrix arising from the 5–point finite difference of the Poisson equation in a unit square. This gives a linear system

$$Ax = b$$

of order m^2, where

$$A = \begin{pmatrix} T & -I & & & \\ -I & T & -I & & \\ & \ddots & \ddots & \ddots & \\ & & -I & T & -I \\ & & & -I & T \end{pmatrix}$$

each block being of order m and

$$T = \begin{pmatrix} 4 & -1 & & & \\ -1 & 4 & -1 & & \\ & \ddots & \ddots & \ddots & \\ & & -1 & 4 & -1 \\ & & & -1 & 4 \end{pmatrix}.$$

For $m = 6$, the minimum and maximum eigenvalues are 0.3961 and 7.6039.

Example 7.

This example arises from the 5–point finite difference approximation of the following equation in a unit square,

$$-\text{div}(a\nabla u)) = f,$$

with Dirichlet boundary conditions. $a(x, y)$ is a diagonal matrix with equal diagonal elements. This element is equal to 1000 in a square $]1/4, 3/4[\times]1/4, 3/4[$, 1 otherwise.

For $m = 6$, the minimum and maximum eigenvalues are 0.4354 and 6828.7.

6.2 Results for a polynomial function

To numerically check some of the previous theorems, f was chosen as a polynomial of degree q,

$$f(\lambda) = \prod_{i=1}^{q}(\lambda - i).$$

We chose Example 6 with $m = 6$, that is a matrix of order 36.

1) We compute the $(2, 2)$ element of $f(A)$ and we vary the order of the polynomial. In the next table, we give, as a function of the degree q of the polynomial, the value of N to have an "exact" result (4 digits in Matlab) for the Gauss rule.

q	2	3	4	5	6
N	2	2	3	3	4

>From these results, we can conclude that the maximum degree for which the results are exact is $q = 2N - 1$, as predicted by the theory.

For the Gauss–Radau rule, we get

q	2	3	4	5	6	7	8	9
N	2	2	2	3	3	4	4	5

>From this we deduce $q = 2N$ as predicted.

For the Gauss–Lobatto rule, we have

q	2	3	4	5	6	7	8	9
N	1	1	2	2	3	3	4	4

This shows that $q = 2N + 1$ which is what we expect.

2) If we consider the block case to compute the $(3, 1)$ element of the polynomial, we get the same results, therefore the block Gauss rule is exact for the degree $2N - 1$, the block Gauss–Radau rule is exact for degree $2N$ and the block Gauss–Lobatto is exact for degree $2N + 1$.

3) The same is also true for the non–symmetric Lanczos algorithm if we want to compute the sum of the $(3, 1)$ and $(3, 3)$ elements.

6.3 Bounds for the inverse

6.3.1 Diagonal elements

Now, we turn to some numerical experiments using Matlab on the examples described above. Usually the results will be given using the "short" format of Matlab. In the following results, Nit denotes the number of iterations of the Lanczos algorithm. This corresponds to N for the Gauss and Gauss–Radau rules and $N - 1$ for the Gauss–Lobatto rule.

Example 1.

Because of the properties of the matrix we should get the answer in two steps. We have $\omega_0 = 2$ and $\gamma_0 = n - 1$, therefore, the lower bound from Gauss–Radau is $\frac{n}{n+1}$ and the upper bound is $\frac{n}{n+1}$, the exact result. If we look at the lower bound from the Gauss rule, we find the same value. This is also true for the numerical experiments as well as for Gauss–Lobatto.

Example 2.

Let us consider $(A(0)^{-1})_{33}$ whose exact value is 79380. The Gauss rule, as a function of the degree of the quadrature, gives

Results from Gauss rule

Nit=1	2	3	4	5
5	26.6	1808.3	3666.8	79380

The Gauss–Radau rule gives upper and lower bounds. For a and b, we use the computed values from the EIG function of Matlab.

Results from the Gauss–Radau rule

	Nit=1	2	3	4	5
lw bnd	23.74	1801.77	3666.58	3559.92	79380
up bnd	257674	216812	202814	79380	79380

Results from Gauss–Lobatto rule

Nit=1	2	3	4	5
265330	216870	202860	79268	79380

The results are not as good as expected. The exact results should have been obtained for $Nit = 3$. The discrepancy comes from round off errors, particularly for the lower bound, because of the eigenvalue distribution of A and a poor convergence rate of the Lanczos algorithm in this case. To see how this is related to the conditioning of A, let us vary α. For simplicity we consider only the Gauss rule. The following tables give results for different values of α and the $(3, 3)$ element of the inverse. The exact values are $(\alpha = 0.01,\ 70.3949)$, $(\alpha = 0.1,\ 7.7793)$, $(\alpha = 1,\ 0.9054)$.

Lower bound from Gauss rule for $\alpha = 0.01$

Nit=2	3	4	5
20.5123	69.7571	70.3914	70.3949

Lower bound from Gauss rule for $\alpha = 0.1$

Nit=2	3	4	5
6.7270	7.7787	7.7793	7.7793

Lower bound from Gauss rule for $\alpha = 1$

Nit=2	3	4	5
0.9040	0.9054	0.9054	0.9054

We see that when A is well conditioned, the numerical results follow the theory. The discrepancies probably arise from the poor convergence of the smallest eigenvalues of J_N towards those of A.

Example 3.

We are looking for bounds for $(A^{-1})_{55}$ whose exact value is, of course, 2.

Lower bounds for $(A^{-1})_{55}$ from the Gauss rule

Nit=1	2	3	4	5	6	7
0.3667	1.3896	1.7875	1.9404	1.9929	1.9993	2

Results for $(A^{-1})_{55}$ from the Gauss–Radau rule

	Nit=1	2	3	4	5	6	7
b_1	1.3430	1.7627	1.9376	1.9926	1.9993	2.0117	2
b_2	3.0330	2.2931	2.1264	2.0171	2.0020	2.0010	2

Upper bounds for $(A^{-1})_{55}$ from the Gauss–Lobatto rule

Nit=1	2	3	4	5	6	7
3.1341	2.3211	2.1356	2.0178	2.0021	2.0001	2

In this example 5 or 6 iterations should be sufficient, so we are a little off the theory.

Example 4.

We look at bounds for $(A^{-1})_{55}$ whose exact value is 4.5

Lower bounds for $(A^{-1})_{55}$ from the Gauss rule

Nit=1	2	3	4	5
1	2	3	4	4.5

Lower and upper bounds for $(A^{-1})_{55}$ from the Gauss–Radau rule

	Nit=1	2	3	4	5
lw bnd	1.3910	2.4425	3.4743	4.5	4.5
up bnd	5.8450	4.7936	4.5257	4.5	4.5

Upper bounds for $(A^{-1})_{55}$ from the Gauss–Lobatto rule

Nit=1	2	3	4	5
7.8541	5.2361	4.6180	4.5	4.5

Example 5.

We get for $(A^{-1})_{55}$, whose value is 0.0595,

Lower bounds for $(A^{-1})_{55}$ from the Gauss rule

Nit=1	2	3	4	5
0.0455	0.0511	0.0523	0.0585	0.0595

Lower and upper bounds for $(A^{-1})_{55}$ from the Gauss–Radau rule

	Nit=1	2	3	4	5
lw bnd	0.0508	0.0522	0.0582	0.0595	0.0595
up bnd	0.4465	0.0721	0.0595	0.0595	0.0595

Upper bounds for $(A^{-1})_{55}$ from the Gauss–Lobatto rule

Nit=1	2	3	4	5
1.1802	0.0762	0.0596	0.0595	0.0595

Because some eigenvalues are very close together, we get the exact answers a little sooner than it is predicted by theory.

Example 6.

Consider $m = 6$. Then we have a system of order 36 and we look for bounds on $(A^{-1})_{18,18}$ whose value is 0.3515. There are 19 distinct eigenvalues, therefore we should get the exact answer in about 10 iterations for Gauss and Gauss–Radau and 9 iterations for Gauss–Lobatto.

Lower bounds for $(A^{-1})_{18,18}$ from the Gauss rule

Nit=1	2	3	4	8	9
0.25	0.3077	0.3304	0.3411	0.3512	0.3515

Lower and upper bounds for $(A^{-1})_{18,18}$ from the Gauss–Radau rule

	Nit=1	2	3	4	8	9
lw bnd	0.2811	0.3203	0.3366	0.3443	0.3514	0.3515
up bnd	0.6418	0.4178	0.3703	0.3572	0.3515	0.3515

Upper bounds for $(A^{-1})_{18,18}$ from the Gauss–Lobatto rule

Nit=1	2	3	4	8
1.3280	0.4990	0.3874	0.3619	0.3515

Now, we consider $m = 16$ which gives a matrix of order 256. We want to compute bounds for the $(125, 125)$ element whose value is 0.5604. In this case there are 129 distinct eigenvalues, so we should find the exact answer in about 65 iterations at worst. These computations for a larger problem have been done on a Sun Sparcstation 1+. We find the following results.

Lower bounds for $(A^{-1})_{125,125}$ from the Gauss rule

Nit=2	3	4	5	6	7	8	9	10	20
0.3333	0.3929	0.4337	0.4675	0.4920	0.5084	0.5201	0.5301	0.5378	0.5600

Lower and upper bounds for $(A^{-1})_{125,125}$ from the Gauss–Radau rule

	Nit=2	3	4	5	6	7	8	10	20
lw bnd	0.3639	0.4140	0.4514	0.4804	0.5006	0.5146	0.5255	0.5414	0.5601
up bnd	1.5208	1.0221	0.8154	0.7130	0.6518	0.6139	0.5925	0.5730	0.5604

Upper bounds for $(A^{-1})_{125,125}$ from the Gauss–Lobatto rule

Nit=2	3	4	5	6	7	8	9	10	18
2.1011	1.2311	0.8983	0.7585	0.6803	0.6310	0.6012	0.5856	0.5760	0.5604

We have very good estimates much sooner than predicted. This is because there are distinct eigenvalues which are very close together.

We also ran two other examples with $m = 10$ and $m = 20$ that show that the number of iterations to reach a "correct" value with four exact digits grows like m. This can be expected from the Lanczos method.

Example 7.

We took $m = 6$ as in the previous example. So we have a matrix of dimension 36. The $(2, 2)$ element of the inverse has an "exact" value of 0.3088 and there are 23 distinct eigenvalues so that the exact answer should be obtained after 12 iterations but the matrix is ill conditioned. We get the following results:

Lower bounds for $(A^{-1})_{2,2}$ from the Gauss rule

Nit=1	2	3	4	5	6	8	10	12	15
0.25	0.2503	0.2510	0.2525	0.2553	0.2609	0.2837	0.2889	0.3036	0.3088

Lower and upper bounds for $(A^{-1})_{2,2}$ from the Gauss–Radau rule

	Nit=2	3	4	5	6	7	8	10	12	15
bnd	0.2504	0.2516	0.2538	0.2583	0.2699	0.2821	0.2879	0.2968	0.3044	0.3088
bnd	0.5375	0.5202	0.5121	0.5080	0.5060	0.5039	0.5013	0.3237	0.3098	0.3088

Upper bounds for $(A^{-1})_{2,2}$ from the Gauss–Lobatto rule

Nit=1	2	3	4	5	6	8	10	12	15
2.2955	0.5765	0.5289	0.5156	0.5093	0.5065	0.5020	0.3237	0.3098	0.3088

6.3.2 Non diagonal elements with non–symmetric Lanczos

Here, we use the non–symmetric Lanczos algorithm to get estimates on non diagonal elements.

Example 1.

The matrix is of dimension $n = 5$. All the non diagonal elements are $-1/6 = -0.1667$ and the diagonal elements are equal to 0.8333.

We compute the sum of the $(2, 2)$ and $(2, 1)$ elements (e.g. $\delta = 1$), that is 0.6667. With the Gauss rule, after 1 iteration we get 0.3333 and after 2 iterations 0.6667. With Gauss–Radau, we obtain the result in one iteration as well as with Gauss–Lobatto.

We note that for $\delta = 1$, the measure is not positive but for $\delta = 2$ the measure is positive and increasing.

Example 2.

Let us consider first the Hilbert matrix and $(A(0)^{-1})_{2,1}$ whose exact value is 79380. We compute the sum of the $(2, 2)$ and $(2, 1)$ elements, (i.e. $\delta = 1$) that is 4500. The Gauss rule, as a function of the number of iterations, gives

Bounds from the non–symmetric Gauss rule

Nit=1	2	3	4	5
1.2	-21.9394	73.3549	667.1347	4500

Consider now the non–symmetric Gauss–Radau rule.

Bounds from the non–symmetric Gauss–Radau rule

	Nit=1	2	3	4	5
b1	-17.5899	73.1917	667.1277	667.0093	4500
b2	144710	155040	51854	4500	4500

Bounds from the non–symmetric Gauss–Lobatto rule

Nit=1	2	3	4	5
142410	155570	51863	3789.2	4500

Note that the measure is positive and increasing therefore, we obtain a lower bound with the Gauss rule, b_1 is a lower bound and b_2 an upper bound with Gauss–Radau and Gauss–Lobatto gives an upper bound.

Again the results are not so good. Consider now the results of the non–symmetric Gauss rule with a better conditioned problem by looking at $A(0.1)$. The sum of the elements we compute is 5.1389. In the last line we indicate if the product of the non diagonal coefficients is positive (p) or negative (n). If it is positive we should have a lower bound, an upper bound otherwise. Note that in this case, the measure is positive but not increasing.

Estimates from the non–symmetric Gauss rule for $\alpha = 0.1$

Nit=1	2	3	4	5
1.0714	6.1735	5.1341	5.1389	5.1389
p	n	p	p	p

We see that the algorithm is able to determine if it is computing a lower or an upper bound.

Estimates from the non–symmetric Gauss–Radau rule

	Nit=1	2	3	4
b1	6.5225	5.1338	5.1389	5.1389
b2	5.0679	5.2917	5.1390	5.1389

We remark that b_1 and b_2 are alternatively upper and lower bounds.

Estimates from the non–symmetric Gauss–Lobatto rule

Nit=1	2	3	4
5.0010	5.3002	5.1390	5.1389

Again, we do not have an upper bound with Gauss–Lobatto but the results oscillate around the exact value. In this case, this can be fixed by using a value $\delta = 3$ that gives a positive increasing measure.

Example 3.

We are looking for estimates for the sum of the $(2,2)$ and $(2,1)$ elements whose exact value is 1. First, we use $\delta = 1$ for which the measure is positive but not increasing.

Estimates from the non–symmetric Gauss rule

Nit=1	2	3	4	5	6	7
0.4074	0.6494	0.8341	0.9512	0.9998	1.0004	1
p	p	p	p	p	n	p

Estimates from the non–symmetric Gauss–Radau rule

	Nit=1	2	3	4	5	6	7
b1	0.6181	0.8268	0.9488	0.9998	1.0004	1.0001	1
b2	2.6483	1.4324	1.0488	1.0035	1.0012	0.9994	1

Estimates from the non–symmetric Gauss–Lobatto rule

Nit=1	2	3	4	5	6	7	8
3.2207	1.4932	1.0529	1.0036	1.0012	0.9993	0.9994	1

Here we have a small problem at the end near convergence, but the estimates are quite good. Note that for $\delta = 4$ the measure is positive and increasing.

Example 4.

This example illustrates some of the problems that can happen with the non–symmetric Lanczos algorithm. We would like to compute the sum of the $(2,2)$ and $(2,1)$ elements that is 2. After 2 iterations we have a breakdown of the Lanczos algorithm as $\gamma\beta = 0$. The same happens at the first iteration for the Gauss–Radau rule and at the second one for the Gauss–Lobatto rule. Choosing a value of δ different from 1 cures the breakdown problem. We can obtain bounds with a value $\delta = 10$ (with a positive and increasing measure). Then the value we are looking for is 1.55 and the results follow.

Bounds from the non–symmetric Gauss rule

Nit=1	2	3	4	5
0.5263	0.8585	1.0333	1.4533	1.55

Bounds from the non–symmetric Gauss–Radau rule

	Nit=2	3	4
b1	1.0011	1.2771	1.55
b2	1.9949	1.5539	1.55

Bounds from the non–symmetric Gauss–Lobatto rule

Nit=2	3	4
2.2432	1.5696	1.55

Example 5.

The sum of the $(2,2)$ and $(2,1)$ elements is 0.6158.

Bounds from the non-symmetric Gauss rule

Nit=1	2	3	4	5
0.0417	0.0974	0.4764	0.6155	0.6158
p	p	p	p	p

Bounds from the non-symmetric Gauss-Radau rule

	Nit=1	2	3	4
b1	0.0847	0.4462	0.6154	0.6158
b2	0.9370	0.6230	0.6158	0.6158

Bounds from the non-symmetric Gauss-Lobatto rule

Nit=1	2	3	4
1.1261	0.6254	0.6159	0.6158

Example 6.

We consider $m = 6$, then, we have a system of order 36 and we look for estimates of the sum of the $(2,2)$ and $(2,1)$ elements which is 0.4471. Remember there are 19 distinct eigenvalues.

Bounds from the non-symmetric Gauss rule

Nit=1	2	3	4	5	6	7	8	9
0.3333	0.4000	0.4262	0.4369	0.4419	0.4446	0.4461	0.4468	0.4471
p	p	p	p	p	p	p	p	p

Bounds from the non-symmetric Gauss-Radau rule

	Nit=1	2	3	4	5	6	7	8	9
b1	0.3675	0.4156	0.4320	0.4390	0.4436	0.4456	0.4466	0.4470	0.4471
b2	0.7800	0.5319	0.4690	0.4537	0.4490	0.4476	0.4472	0.4472	0.4471

Bounds from the non-symmetric Gauss-Lobatto rule

Nit=1	2	3	4	5	6	7	8	9	10
1.6660	0.6238	0.4923	0.4596	0.4505	0.4480	0.4473	0.4472	0.4472	0.4471

Example 7.

We took $m = 6$ as in the previous example. So we have a matrix of dimension 36. The sum of the $(2,2)$ and $(2,1)$ elements of the inverse is 0.3962 and there are 23 distinct eigenvalues. We get the following results:

Bounds from the non-symmetric Gauss rule

Nit=1	2	3	4	5	6	8	10	12	15
0.3333	0.3336	0.3340	0.3348	0.3363	0.3396	0.3607	0.3689	0.3899	0.3962
p	p	p	p	p	p	p	p	p	p

Bounds from the non-symmetric Gauss–Radau rule

	Nit=2	3	4	5	6	8	10	12	15
b1	0.3337	0.3343	0.3355	0.3380	0.3460	0.3672	0.3803	0.3912	0.3962
b2	0.6230	0.5930	0.5793	0.5725	0.5698	0.5660	0.4078	0.3970	0.3962

Bounds from the non-symmetric Gauss–Lobatto rule

Nit=1	2	3	4	5	6	8	10	12	15
2.2959	0.6898	0.6081	0.5850	0.5746	0.5703	0.5664	0.4078	0.3970	0.3962

Finally, one can ask why we do not store the Lanczos vectors x_j and compute an approximation to the solution of $Au = e_i$. This can be done doing the following. Let

$$Q_N = [x_0, \ldots, x_{N-1}].$$

If we solve

$$J_N y_N = e_1,$$

then the approximate solution is given by $Q_N y_N$.

Unfortunately this does not give bounds and even the approximations are not as good as with our algorithms. Consider Example 5 and computing the fifth column of the inverse. For the element (1,5) whose "exact" value is 0.460, we find

Estimates from solving the linear system

Nit= 2	3	4	5	6
-0.0043	-0.0046	0.0382	0.0461	0.0460

By computing bounds for the sum of the (5,5) and (1,5) elements and subtracting the bounds for the (5,5) element, we obtain

Bounds from the Gauss–Radau quadrature rules

	Nit= 2	3	4
lw bnd	0.0048	0.0451	0.0460
up bnd	0.0551	0.0473	0.0460

We see that we get good bounds quite fast.

6.3.3 Non diagonal elements with block Lanczos

Here, we use the block Lanczos algorithm to get estimates on non diagonal elements. Unfortunately, most of the examples are too small to be of interest as for matrices of dimension 5 we cannot go further than 2 block iterations. Nevertheless, let us look at the results.

Example 1.

Consider a matrix of dimension $n = 5$. Then all the non diagonal elements are $-1/6 = -0.1667$.

We compute the $(2,1)$ element. With the block Gauss rule, after 2 iterations we get -0.1667. With block Gauss–Radau, we get the exact answer in 1 iteration as well as with Gauss–Lobatto.

Example 2.

The $(2,1)$ element of the inverse of the Hilbert matrix $A(0)$ of dimension 5 is -300.

With the block Gauss rule, after 2 iterations we find -90.968. Note that this is an upper bound. With block Gauss–Radau, 2 iterations give -300.2 as a lower bound and -300 as an upper bound. Block Gauss–Lobatto gives -5797 as a lower bound.

Now we consider $A(0.1)$ for which the $(2,1)$ element of the inverse is -1.9358. After 2 iterations, block Gauss gives -2.2059 a lower bound and block Gauss–Radau and Gauss–Lobatto give the exact answer.

Example 3.

The $(2,1)$ element is -1, the $(3,1)$ element is 0. After 2 iterations we get the exact answers with Gauss as well as with Gauss–Radau. Gauss–Lobatto gives -0.0609, a lower bound. Three iterations give the exact answer.

Example 4.

The $(2,1)$ element is $1/2$. After 2 iterations, we have 0.3182 which is a lower bound. Gauss–Radau gives 0.2525 and 0.6807. Gauss–Lobatto gives 0.7236 which is an upper bound.

Example 5.

The $(2,1)$ element is 0.2980. Two iterations give 0.2329, a lower bound and 3 iterations give the exact answer. Gauss–Radau and Gauss–Lobatto also give the exact answer in 3 iterations.

Example 6.

This example uses $n = 36$. The $(2,1)$ element is 0.1040. Remember that we should do about 10 iterations. We get the following figures.

Estimates from the block Gauss rule

Nit=2	3	4	5	6	7	8
0.0894	0.0974	0.1008	0.1024	0.1033	0.1037	0.1040

Estimates from the block Gauss–Radau rule

Nit=2	3	4	5	6	7	8
0.0931	0.0931	0.1017	0.1029	0.1035	0.1038	0.1040
0.1257	0.1103	0.1059	0.1046	0.1042	0.1041	0.1040

Estimates from the block Gauss–Lobatto rule

Nit=2	3	4	5	6	7	8
0.1600	0.1180	0.1079	0.1051	0.1041	0.1043	0.1041

Note that here everything works. Gauss gives a lower bound, Gauss–Radau a lower and an upper bound and Gauss–Lobatto an upper bound.

Example 7.

We would like to obtain estimates of the $(2,1)$ whose value is 0.0874. We get the following results.

Estimates from the block Gauss rule

Nit=2	4	6	8	10	12	14	15
0.0715	0.0716	0.0722	0.0761	0.0789	0.0857	0.0873	0.0874

Estimates from the block Gauss–Radau rule

Nit=2	4	6	8	10	12	14	15
0.0715	0.0717	0.0731	0.0782	0.0831	0.0861	0.0873	0.0874
0.1375	0.1216	0.1184	0.1170	0.0894	0.0876	0.0874	0.0874

Estimates from the block Gauss–Lobatto rule

Nit=2	4	6	8	10	12	14
0.1549	0.1237	0.1185	0.1176	0.0894	0.0876	0.0874

Note that in this example we obtain bounds. Now, to illustrate what we said before about the estimates being 0 for some iterations, we would like to estimate the $(36,1)$ element of the inverse which is 0.005.

Estimates from the block Gauss rule

Nit=2	4	6	8	10	11
0.	0.	0.0023	0.0037	0.0049	0.0050

Estimates from the block Gauss–Radau rule

Nit=2	4	6	8	10	11
0.	0.	0.0023	0.0037	0.0049	0.0050
0.	0.	0.0024	0.0050	0.0050	0.0050

Estimates from the block Gauss–Lobatto rule

Nit=2	4	6	8	10
0.	0.	0.0022	0.0043	0.0050

6.3.4 Dependence on the eigenvalue estimates

In this sub–Section, we numerically investigate how the bounds and estimates of the Gauss–Radau rules depend on the accuracy of the estimates of the eigenvalues of A. We take Example 6 with $m = 6$ and look at the results given by the Gauss–Radau rule as a function of a and b. Remember that in the previous experiments we took for a and b the values returned by the EIG function of Matlab.

It turns out that the estimates are only weakly dependent of the values of a and b (for this example). We look at the number of iterations needed to obtain an upper for the element $(18, 18)$ with four exact digits and with an "exact" value of b. The results are given in the following table.

a=10^{-4}	10^{-2}	0.1	0.3	0.4	1	6
15	13	11	11	8	8	9

We have the same properties when b is varied.

Therefore, we see that the estimation of the extreme eigenvalues does not seem to matter very much and can be obtained with a few iterations of Lanczos or with the Gerschgorin circles.

6.4 Bounds for the exponential

In this Section we are looking for bounds of diagonal elements of the exponential of the matrices of some of the examples.

6.4.1 diagonal elements

Example 1

We consider the $(2, 2)$ element whose value is 82.8604. With Gauss, Gauss–Radau and Gauss–Lobatto we obtain the exact value in 2 iterations.

Example 2

We would like to compute the $(3, 3)$ element whose value is 1.4344. Gauss gives the answer in 3 iterations, Gauss–Radau and Gauss–Lobatto in 2 iterations.

Example 3

The $(5, 5)$ entry is $4.0879\ 10^4$. Gauss obtains the exact value in 4 iterations, Gauss–Radau and Gauss–Lobatto in 3 iterations.

Example 6

We consider the $(18, 18)$ element whose value is 197.8311. We obtain the following results.

Lower bounds from the Gauss rule

Nit=2	3	4	5	6	7
159.1305	193.4021	197.5633	197.8208	197.8308	197.8311

Lower and upper bounds from the Gauss–Radau rule

	Nit=2	3	4	5	6
lw bnd	182.2094	196.6343	197.7779	197.8296	197.8311
up bnd	217.4084	199.0836	197.8821	197.8325	197.8311

Upper bounds from the Gauss–Lobatto rule

Nit=2	3	4	5	6	7
273.8301	203.4148	198.0978	197.8392	197.8313	197.8311

We remark that to compute diagonal elements of the exponential the convergence rate is quite fast.

6.4.2 Non diagonal elements

Here we consider only Example 6 and we would like to compute the element $(2,1)$ whose value is -119.6646. First, we use the block Lanczos algorithm which give the following results.

Results from the block Gauss rule

Nit=2	3	4	5	6
-111.2179	-119.0085	-119.6333	-119.6336	-119.6646

Results from the block Gauss–Radau rule

	Nit=2	3	4	5	6
b1	-115.9316	-119.4565	-119.6571	-119.6644	-119.6646
b2	-122.2213	-119.7928	-119.6687	-119.6647	-119.6646

Results from the block Gauss–Lobatto rule

Nit=2	3	4	5	6
-137.7050	-120.6801	-119.7008	-119.6655	-119.6646

Now, we use the non–symmetric Lanczos algorithm. The sum of the $(2,2)$ and $(2,1)$ elements of the exponential is 73.9023.

Results from the non–symmetric Gauss rule

Nit=2	3	4	5	6	7
54.3971	71.6576	73.7637	73.8962	73.9021	73.9023

Results from the non–symmetric Gauss–Radau rule

	Nit=2	3	4	5	6
b1	65.1847	73.2896	73.8718	73.9014	73.9023
b2	84.0323	74.6772	73.9323	73.9014	73.9023

Results from the non–symmetric Gauss–Lobatto rule

Nit=2	3	4	5	6	7
113.5085	77.2717	74.0711	73.9070	73.9024	73.9023

6.5 Bounds for other functions

When one uses domain decomposition methods for matrices arising from the finite difference approximation of partial differential equations in a rectangle, it is known that the matrix

$$A = \sqrt{T + \frac{1}{4}T^2},$$

where T is the matrix of the one dimensional Laplacian, is a good preconditioner for the Schur complement matrix. It is interesting to see if we can estimate some elements of the matrix A to generate a Toeplitz tridiagonal approximation to A.

We have

$$T = \begin{pmatrix} 2 & -1 & & & \\ -1 & 2 & -1 & & \\ & \ddots & \ddots & \ddots & \\ & & -1 & 2 & -1 \\ & & & -1 & 2 \end{pmatrix},$$

and we choose an example of dimension 100. We estimate the $(50, 50)$ element whose exact value is 1.6367 with the Gauss–Radau rule. We obtain the following results.

Estimates from the Gauss–Radau rule

	Nit=2	3	4	5	10	15	20
lw	1.6014	1.6196	1.6269	1.6305	1.6355	1.6363	1.6365
up	1.6569	1.6471	1.6430	1.6409	1.6378	1.6371	1.6369

We estimate the non diagonal elements by using the block Gauss rule. We choose the $(49, 50)$ element.

Estimates from the block Gauss rule

Nit=2	3	4	5	10	15	20
-0.6165	-0.6261	-0.6302	-0.6323	-0.6354	-0.6361	-0.6363

Now, we construct a Toeplitz tridiagonal matrix C whose elements are chosen from the estimates given at the fifth iteration. We took the average of the Gauss–Radau values for the diagonal (1.6357) and -0.6323 for the non diagonal elements. We look at the spectrum of $C^{-1}A$. The condition number is 13.35 the minimum eigenvalue being 0.0837 and the maximum one being 1.1174, but there are 86 eigenvalues between 0.9 and the maximum eigenvalue. Therefore, the matrix C (requiring only 5 iterations of some Lanczos algorithms) seems to be a good preconditioner for A which is itself a good preconditioner for the Schur complement.

7 Conclusions

We have shown how to obtain bounds (or in certain cases estimates) of the entries of a function of a symmetric positive definite matrix. The proposed algorithms use the Lanczos algorithm to estimate diagonal entries and either the non–symmetric Lanczos or block Lanczos algorithms for the non diagonal entries.

The algorithms are particularly simple for the inverse of a matrix. Analytical bounds are derived by considering one or two iterations of these algorithms. We have seen in the numerical experiments that very good approximations are obtained in a few iterations.

References

[1] F.V. Atkinson, "Discrete and continuous boundary problems", (1964) Academic Press

[2] S. Basu, N.K. Bose, "Matrix Stieltjes series and network models", SIAM J. Math. Anal. v14 n2 (1983) pp 209–222

[3] G. Dahlquist, S.C. Eisenstat and G.H. Golub, "Bounds for the error of linear systems of equations using the theory of moments", J. Math. Anal. Appl. 37 (1972) pp 151–166

[4] P. Davis, P. Rabinowitz, "Methods of numerical integration", Second Edition (1984) Academic Press

[5] W. Gautschi, "Construction of Gauss–Christoffel quadrature formulas", Math. Comp. 22 (1968) pp 251–270

[6] W. Gautschi, "Orthogonal polynomials– constructive theory and applications", J. of Comp. and Appl. Math. 12 & 13 (1985) pp 61–76

[7] G.H. Golub, J.H. Welsch, "Calculation of Gauss quadrature rule" Math. Comp. 23 (1969) pp 221–230

[8] G.H. Golub, "Some modified matrix eigenvalue problems", SIAM Review v15 n2 (1973) pp 318–334

[9] G.H. Golub, "Bounds for matrix moments", Rocky Mnt. J. of Math., v4 n2 (1974) pp 207–211

[10] G.H. Golub, R. Underwood, "The block Lanczos method for computing eigenvalues", in Mathematical Software III, Ed. J. Rice (1977) pp 361–377

[11] G.H. Golub, C. van Loan, "Matrix Computations", Second Edition (1989) Johns Hopkins University Press

[12] R. Haydock, "Accuracy of the recursion method and basis non–orthogonality", Computer Physics Communications 53 (1989) pp 133–139

[13] G. Meurant, "A review of the inverse of tridiagonal and block tridiagonal matrices", SIAM J. Matrix Anal. Appl. v13 n3 (1992) pp 707–728

[14] C.M. Nex, "Estimation of integrals with respect to a density of states", J. Phys. A, v11 n4 (1978) pp 653–663

[15] C.M. Nex, "The block Lanczos algorithm and the calculation of matrix resolvents", Computer Physics Communications 53 (1989) pp 141–146

[16] D.P. O'Leary, "The block conjugate gradient algorithm and related methods", Linear Alg. and its Appl. v29 (1980) pp 293–322

[17] P.D. Robinson, A. Wathen, "Variational bounds on the entries of the inverse of a matrix", IMA J. of Numer. Anal. v12 (1992) pp 463–486

[18] J. Stoer, R. Bulirsch, "Introduction to numerical analysis", Second Edition (1983) Springer Verlag

[19] G.W. Struble, "Orthogonal polynomials: variable–signed weight functions", Numer. Math v5 (1963) pp 88–94

[20] G. Szegö, "Orthogonal polynomials", Third Edition (1974) American Mathematical Society

[21] H.S. Wilf, "Mathematics for the physical sciences", (1962) Wiley

Acknowledgement

This paper was presented as the first A.R. Mitchell lecture in Dundee, Scotland, July 1991. We dedicate this paper to Ron Mitchell who has given intellectual leadership and generous support to all.
The work of the first author was supported in part by the National Science Foundation under Grant NSF CCR-8821078

Gene H. Golub
Computer Science Department, Stanford University
Stanford CA 94305, USA
golub@sccm.stanford.edu

Gérard Meurant
CEA, Centre d'Etudes de Limeil-Valenton,
94195 Villeneuve St Georges cedex, France
meurant@etca.fr

J GROENEWEG AND M N SPIJKER

On the error due to the stopping of the Newton iteration in implicit linear multistep methods

Abstract This paper concerns the numerical solution of initial value problems for non-linear ordinary differential equations by implicit linear multistep methods.

For non-stiff problems, fixed point iteration (Picard iteration), is a classical approach to the solution of the system of algebraic equations occurring in each time step of these methods. The order of the error due to the stopping of this process after a fixed number of iterations is well understood.

For stiff problems, Picard iteration is not appropriate and some variant of the Newton method is usually used instead. This paper addresses the problem of estimating the stopping error of Newton-like iterations. We aim for an understanding of this error comparable to what is known about Picard iterations. Because of stiffness, the theory is more delicate than for Picard iterations.

1 Introduction

In this paper we deal with the numerical solution of the initial value problem

$$U'(t) = f(U(t)) \quad \text{for} \quad 0 \le t \le T, \quad U(0) = u_0. \tag{1}$$

Here u_0 is a given vector in the s-dimensional real vectorspace $\mathbb{R}^s$, and $U(t) \in \mathbb{R}^s$ is unknown. Further, f is a given mapping from an open, convex set $\mathcal{D} \subset \mathbb{R}^s$ into $\mathbb{R}^s$. We assume that the partial derivatives of f up to the 2nd order exist and are continuous on $\mathcal{D}$.

Let $h > 0$ be a given *stepsize*, and consider *gridpoints* $t_n = nh$ in $[0, T]$ for integer values of n. Numerical approximations u_n to $U(t_n)$ can be obtained by using the *linear multistep method*

$$\begin{aligned} &u_n = \alpha_1 u_{n-1} + \cdots + \alpha_k u_{n-k} + h[\beta_0 f_n + \beta_1 f_{n-1} + \cdots + \beta_k f_{n-k}] \\ &\text{for } n = k, k+1, k+2, \ldots . \end{aligned}$$

Here $k \ge 1$, and α_i, β_i are judiciously chosen constants with $\alpha_1 + \alpha_2 + \cdots + \alpha_k = 1$. The approximations $u_k, u_{k+1}, u_{k+2}, \ldots \in \mathbb{R}^s$ are computed sequentially by putting $f_{n-i} = f(u_{n-i})$ and applying the linear multistep formula with starting vectors $u_0, u_1, \ldots, u_{k-1} \in \mathbb{R}^s$ (see e.g. Henrici (1962), Hairer et al. (1987), Hairer & Wanner (1991), Lambert (1991)).

In the following the coefficient β_0 is assumed to satisfy

$$\beta_0 > 0,$$

in which case the multistep method is called *implicit*. For a given $n \geq k$ the application of the multistep formula now amounts to solving the (nonlinear) equation

$$F(x) = 0, \tag{2}$$

where

$$F(x) \equiv -x + h\beta_0 f(x) + \sum_{i=1}^{k}(\alpha_i u_{n-i} + h\beta_i f_{n-i}).$$

The solution x^* to the equation (2) may be found numerically by computing first an initial guess $x_0 \simeq x^*$, and then applying the so-called *Picard iteration*. Here x^* is approximated by one of the iterates x_j, with $j \geq 1$, where

$$x_j = h\beta_0 f(x_{j-1}) + \sum_{i=1}^{k}(\alpha_i u_{n-i} + h\beta_i f_{n-i}) \quad \text{for} \quad j = 1, 2, 3, \ldots.$$

This classical iteration process was studied by many authors, and is well understood for an important class of problems. Let $|x|$ denote an arbitrary norm for vectors $x \in \mathbb{R}^s$. Assuming that f satisfies a Lipschitz condition

$$|f(y) - f(x)| \leq L|y - x|,$$

the error committed by stopping the iteration after j steps can easily be estimated. Since $|x^* - x_j| = h\beta_0|f(x^*) - f(x_{j-1})| \leq (h\beta_0 L)|x^* - x_{j-1}|$, one has $|x^* - x_j| = O(h^j)|x^* - x_0|$. Assume the initial guess x_0 satisfies

$$|x^* - x_0| = O(h^q) \tag{3}$$

(with O-constant of moderate size, and $q > 0$). Then the stopping error can be estimated by

$$|x^* - x_j| = O(h^{q+j}) \quad \text{for} \quad j \geq 1. \tag{4}$$

This relation gives a useful insight in the error due to stopping the Picard iteration.

But, in the following we focus on *stiff problems* (1), which may be characterized by

$$hL >> 1$$

(see e.g. Gear (1971), Dekker & Verwer (1984), Hairer & Wanner (1991), Lambert (1991)). Here h stands for a 'natural' stepsize, chosen according to the smoothness of the solution $U(t)$. It is well known that for stiff problems the above error analysis is misleading, as now the factor $(h\beta_0 L)$ is, in general, greater than 1. Accordingly, for such problems one has to abandon Picard iteration in favour of *Newton's method* or a variant thereof.

In the following we study, for $j \geq 1$, the order of the error committed by stopping a Newton-like iteration after j steps. We shall measure this error using an arbitrary norm $|x|$ (for $x \in \mathbb{R}^s$) and relate it to the stepsize h. We aim in the stiff situation for an understanding of the Newton stopping error comparable to what is known in the nonstiff situation about Picard iterations.

2 An analysis of the Newton stopping error while neglecting stiffness

We consider the so-called *modified Newton process*

$$F'(x_0)(x_j - x_{j-1}) = -F(x_{j-1}) \quad \text{for} \quad j = 1, 2, 3, \ldots. \tag{5}$$

Here $F'(x)$ denotes the Jacobian matrix of F at x.

We assume again (3). Using Taylor series expansions in a straightforward way the corresponding errors $x^* - x_j$ can be estimated. In this manner one can derive

$$|x^* - x_j| = O(h^{R(j)}) \quad \text{with} \quad R(j) = (j+1)q + j \quad \text{for} \quad j \geq 1. \tag{6}$$

The estimate (6) was already obtained in the important pioneering work by Liniger (1971).

However, there is a weak point in the above derivation. The relation (6) is obtained, similarly as (4), by putting

$$hL = O(h)$$

and by replacing also 2nd order derivatives of $h \cdot f(x)$ simply by $O(h)$. Clearly, for stiff problems such quantitites $O(h)$ cannot be interpreted, in the standard fashion, as the product of a moderate O-constant and a small stepsize h. Therefore, in the stiff case, (6) might be misleading.

We thus arrive at the question of whether the estimate (6) is reliable and the O-constant in (6) is of moderate size.

3 A numerical experiment

In order to check the relevance of (6) to stiff problems (1) we consider

Example 1

$$\begin{aligned}
&U_1'(t) = 1 + 10^8[U_2(t) - (U_1(t) - 2)^3], && U_1(0) = 0, \\
&U_2'(t) = 3(U_1(t) - 2)^2 - 10^8[U_2(t) - (U_1(t) - 2)^3], && U_2(0) = -8, \\
&0 \leq t \leq 1/2
\end{aligned}$$

The true solution equals

$$U_1(t) = t, \quad U_2(t) = (t-2)^3,$$

so that any 'natural' stepsize for the numerical solution of this problem need not be small. The large factor 10^8 causes the problem to be stiff. This stiffness also reflects itself in the eigenvalues μ_1, μ_2 of the corresponding Jacobian matrix $f'(x)$ at $x = u_0 = (0, -8)^T$. We have $\mu_1 \simeq -10^9$, $\mu_2 \simeq -1$.

We consider the backward Euler method, i.e.

$$k = 1, \quad \alpha_1 = 1, \quad \beta_0 = 1, \quad \beta_1 = 0.$$

Further, we consider the corresponding equation (2) with

$$t_n = 0.1, \quad u_{n-1} = U(t_{n-1}),$$

and we choose the natural initial guess

$$x_0 = u_{n-1}.$$

We consider the ℓ_1-norm $|x| = |x|_1$ for $x \in \mathbb{R}^2$. It can be proved that the estimate (3) holds with an O-constant of moderate size and

$$q = 1.$$

For $j = 1$ the estimate (6) thus reduces to

$$|x^* - x_1| = O(h^3).$$

In order to check this estimate we have listed some actual ratios $|x^* - x_1|/h^3$.

h	10^{-1}	10^{-2}	10^{-3}	10^{-4}	10^{-5}	10^{-6}
$\lvert x^* - x_1\rvert/h^3$	0.10×10^2	0.97×10^2	0.96×10^3	0.96×10^4	0.96×10^5	0.96×10^6

From the table it is evident that the estimate (6) is not reliable; it provides no insight regarding the actual order of the error $x^* - x_1$.

4 A class of nonlinear stiff problems

In the following we present a framework in which reliable estimates for $|x^* - x_j|$ can be derived.

We assume, in addition to the assumptions concerning f made in Section 1, that

(A1) $f'(y) = f'(x)\{I + e[x, y]\}$ with a matrix $e[x, y]$ satisfying $\|e[x, y]\| \leq \lambda|x - y|$ for all $x, y \in \mathcal{D}$;

(A2) $|W(\tau_1) - V(\tau_1)| \leq |W(\tau_0) - V(\tau_0)|$ whenever V, W are any two solutions of the differential equation on $[\tau_0, \tau_1] \subset \mathbb{R}$.

In the above assumptions $|x|$ denotes an arbitrary norm for the vectors $x \in \mathbb{R}^s$, and $\|e[x, y]\|$ stands for the corresponding induced matrix norm of $e[x, y]$. Further, the parameter λ is assumed to be of moderate size. It provides an upperbound for the (relative) variation of the Jacobian matrix $f'(y)$ when y varies through $\mathcal{D}$.

We stress that our framework is rather general due to the fact that we allow of an *arbitrary norm* in $\mathbb{R}^s$. By chosing the ℓ_2-norm $|x| = |x|_2$ our assumption (A2) encompasses a setting used by Dahlquist (1978), Crouzeix (1979), Burrage & Butcher (1979) and others. Further, choosing the ℓ_∞-norm $|x| = |x|_\infty$ our framework encompasses a setting related to one used by Alexander (1991).

There exist arbitrarily stiff, nonlinear problems satisfying (A1), (A2) with λ of moderate size.

As an illustration we consider Example 1. Here the function

$$f(x) = \begin{pmatrix} 1 + 10^8[\xi_2 - (\xi_1 - 2)^3] \\ 3(\xi_1 - 2)^2 - 10^8[\xi_2 - (\xi_1 - 2)^3] \end{pmatrix}, \quad \text{with} \quad x = \begin{pmatrix} \xi_1 \\ \xi_2 \end{pmatrix},$$

satisfies (A1), (A2) with

$$\begin{aligned} \mathcal{D} &= \{(\xi_1, \xi_2)^T : \ -1/4 < \xi_1 < 1, \quad -\infty < \xi_2 < \infty\}, \\ |x| &= |x|_1 = |\xi_1| + |\xi_2| \quad \text{for} \quad x = (\xi_1, \xi_2)^T \in \mathbb{R}^2, \\ \lambda &= 12. \end{aligned}$$

It is worth noting that with this $\mathcal{D}$, $|x|$ and λ the assumptions (A1), (A2) would still be satisfied by f in case the factor 10^8 would be replaced by *any* factor $\sigma \geq 1$.

As a further illustration we consider

Example 2

$$\begin{aligned} &U_1'(t) = -10^8 U_1(t) + 10^8 U_2(t)[1 - U_1(t) - U_2(t)], && U_1(0) = 1/6, \\ &U_2'(t) = 10^8 U_1(t) - 10^8 U_2(t)[1 - U_1(t) - U_2(t)] - [U_2(t)]^2, && U_2(0) = 1/3, \\ &0 \leq t \leq 1 \end{aligned}$$

This example stems from the following chemical reaction

$$\begin{aligned} A &\xrightarrow{k_1} B \\ 2B &\xrightarrow{k_2} B + C \\ B + C &\xrightarrow{k_3} A + C, \end{aligned}$$

which is also considered in Hairer & Wanner (1991, p3). Suppose the initial concentrations of the chemical compounds A, B, C are equal to 1/6, 1/3, 1/2, respectively, and assume the rate constants are given by

$$k_1 = 10^8, \quad k_2 = 1, \quad k_3 = 10^8.$$

Denoting the concentrations of A, B at time $t \geq 0$ by $U_1(t)$, $U_2(t)$ one easily arrives at the above initial value problem.

In Example 2 the graphs of the true $U_1(t)$, $U_2(t)$, for $0 < t \leq 1$, are quite smooth, and have a negative slope. The large factor 10^8 makes the problem stiff. The eigenvalues

μ_1, μ_2 of the corresponding Jacobian matrix $f'(x)$ at $x = u_0 = (1/6, 1/3)^T$ satisfy $\mu_1 \simeq -2 \times 10^8$, $\mu_2 \simeq -0.6$.

A straightforward calculation shows that the function f corresponding to Example 2 satisfies (A1), (A2) with

$$\begin{aligned}
\mathcal{D} &= \{(\xi_1, \xi_2)^T : \ 0 < \xi_1 < 1 - 2\xi_2 \quad \text{and} \quad \xi_2 > 1/4\}, \\
|x| &= |x|_1 = |\xi_1| + |\xi_2| \quad \text{for} \quad x = (\xi_1, \xi_2)^T \in \mathbb{R}^2, \\
\lambda &= 8
\end{aligned}$$

5 A rigorous error analysis relevant to nonlinear stiff problems

We make the same assumptions regarding f as in Section 1, and we assume U satisfies (1). We denote the distance of the set $\{U(t) : \ 0 \le t \le T\}$ to the boundary of $\mathcal{D}$ by δ; in case $\mathcal{D} = \mathbb{R}^s$ we put $\delta = \infty$.

Let $n \ge k$, $t_n \in (0, T]$ and $u_{n-i} \in \mathcal{D}$, $f_{n-i} \in \mathbb{R}^s$ (for $1 \le i \le k$) be given. In the following theorems we use the assumption

$$\begin{aligned}
&|u_{n-i} - U(t_{n-i})| \le \alpha \cdot \delta, \quad h|f_{n-i} - U'(t_{n-i})| \le \alpha \cdot \delta \quad (\text{for} \quad 1 \le i \le k), \\
&h|U'(t)| \le \alpha \cdot \delta \quad (\text{for} \quad 0 \le t \le T).
\end{aligned} \tag{7}$$

Theorem 1 *Assume (A2). Then there is a factor $\alpha > 0$ such that (2) has a unique solution $x^* \in \mathcal{D}$ whenever u_{n-i}, f_{n-i}, h satisfy (7). Here α only depends on the coefficients α_i, β_i of the linear multistep method.*

Let $K > 0$, $q > 0$ be given. The next theorem relates the assumption

$$x_0 \in \mathcal{D}, \quad |x_0 - x^*| \le Kh^q \le \tfrac{1}{4}\min[\delta, 1/\lambda] \tag{8}$$

to the following property of the corresponding modified Newton process:

$$\begin{aligned}
&\text{There are unique } x_j \in \mathcal{D} \text{ generated by process (5) with initial guess } x_0, \text{ and} \\
&|x^* - x_j| \le (3\lambda)^j (Kh^q)^{j+1} \quad \text{for } j \ge 1.
\end{aligned} \tag{9}$$

Theorem 2 *Assume (A1), (A2), and let α be as in Theorem 1. Then (9) holds whenever u_{n-i}, f_{n-i}, h, x_0, K and q satisfy (7), (8).*

The Theorems 1, 2 easily follow from the material in Dorsselaer & Spijker (1992). The last theorem shows that, under the assumptions (A1), (A2), the relation (3) (with $\mathcal{O}$-constant K of moderate size) implies

$$|x^* - x_j| = \mathcal{O}(h^{Q(j)}) \text{ with } Q(j) = (j+1)q \text{ for } j \ge 1. \tag{10}$$

In (10) we have an $\mathcal{O}$-constant

$$K_j = (3\lambda)^j K^{j+1},$$

which is not affected by the stiffness of (1). The constant K_j is of moderate size when j is limited in a realistic way.

Note that in the situation of Section 3 our relation (10) reduces to the error estimate

$$|x^* - x_1| = \mathcal{O}(h^2).$$

This estimate is nicely in agreement with the numerical experiments displayed in the table in Section 3.

6 Various related questions

6.1 References to related work

The discussion in Section 2 about the relevance in stiff problems of estimates involving terms $\mathcal{O}(h^R)$ is analogous to considerations regarding B-consistency in Frank et al. (1985), Dekker & Verwer (1984), Hairer & Wanner (1991). The difference $Q(j) - R(j)$ (see (6), (10)) can be viewed as an order reduction due to stiffness. Similar order reductions regarding consistency can be found in the works just mentioned.

Under the conditions of Theorem 2 the iterates x_j converge linearly towards x^*. This conclusion is related to interesting investigations by Alexander (1991), who studied the rate of convergence of similar iterates. The class of nonlinear stiff problems considered by Alexander has a nonempty intersection with our class specified by (A1), (A2); but neither class is contained in the other one.

We note that Theorem 1 is related to conclusions obtained in Desoer & Haneda (1972), Söderlind (1992), Alexander (1991), Hairer & Wanner (1991).

6.2 Related iterative processes

Orders $R(j)$ and $Q(j)$ analogous to those in (6), (10) can be derived for the *exact Newton process*

$$F'(x_{j-1})(x_j - x_{j-1}) = -F(x_{j-1}) \quad \text{for} \quad j = 1, 2, 3, \ldots \tag{5'}$$

Estimating the errors $x^* - x_j$ by Taylor series expansions while neglecting stiffness one arrives at

$$|x^* - x_j| = \mathcal{O}(h^{R(j)}) \quad \text{with} \quad R(j) = 2^j q + 2^j - 1 \quad \text{for} \quad j \geq 1. \tag{6'}$$

This estimate was first given in Liniger (1971). Taking stiffness into account, and assuming (A1), (A2) as before, it can be seen from Dorsselaer & Spijker (1992) that

$$|x^* - x_j| = \mathcal{O}(h^{Q(j)}) \quad \text{with} \quad Q(j) = 2^j q \quad \text{for} \quad j \geq 1, \tag{10'}$$

with an $\mathcal{O}$-constant that is not affected by stiffness.

Also orders for the simplified Newton process

$$(5'') \qquad F'(y_0)(x_j - x_{j-1}) = -F(x_{j-1}) \quad \text{for} \quad j = 1, 2, 3, \ldots$$

were derived. Assume, in addition to (3), that

$$|x^* - y_0| = \mathcal{O}(h^r)$$

(with $\mathcal{O}$-constant of moderate size, and $0 < r \leq q$). In this case the analogues of (6), (10) are as follows:

$$(6'') \qquad |x^* - x_j| = \mathcal{O}(h^{R(j)}) \quad \text{with} \quad R(j) = jr + q + j \quad \text{for} \quad j \geq 1,$$

and

$$(10'') \qquad |x^* - x_j| = \mathcal{O}(h^{Q(j)}) \quad \text{with} \quad Q(j) = jr + q \quad \text{for} \quad j \geq 1$$

(see Dorsselaer & Spijker (1992)).

6.3 Global stopping errors

Let v_n denote the approximations to $U(t_n)$ generated by the theoretical linear multistep method starting with $v_i = u_i$ (for $0 \leq i \leq k-1$). By u_n we denote the approximations obtained when, at each gridpoint t_n, the solution to the equation (2) is approximated by performing j iteration steps with a Newton-like method. Clearly, the *global stopping errors* defined by

$$v_n - u_n$$

are related to stopping errors of the type $x^* - x_j$ as discussed above. The above estimates for $x^* - x_j$ can be used to obtain insight into the order of $v_n - u_n$.

For the ease of presentation we assume in the following that

$$\beta_1 = \beta_2 = \ldots = \beta_k = 0,$$

and we confine ourselves to the modified Newton process (5) with initial guess

$$x_0 = u_{n-1}.$$

We define functions Ψ and Φ, in an obvious way, so that we can write

$$v_n = \Psi(\underline{v}_{n-1}), \quad u_n = \Phi(\underline{u}_{n-1}),$$

with

$$\underline{v}_{n-1} = (v_{n-1}, \ldots, v_{n-k}), \quad \underline{u}_{n-1} = (u_{n-1}, \ldots, u_{n-k}).$$

We have

$$v_n - u_n = \left[\Psi(\underline{v}_{n-1}) - \Psi(\underline{U}(t_{n-1}))\right] + \left[\Phi(\underline{U}(t_{n-1})) - \Phi(\underline{u}_{n-1})\right] + d_n$$

with a *local stopping error* d_n defined by

$$d_n = \Psi(\underline{U}(t_{n-1})) - \Phi(\underline{U}(t_{n-1})) = x^* - x_j.$$

Using (10) (with $x_0 = U(t_{n-1}),\ q = 1$) we may thus arrive at

$$v_n - u_n \simeq \Psi(\underline{v}_{n-1}) - \Psi(\underline{u}_{n-1}) + \mathcal{O}(h^{j+1}) \tag{11}$$

with an $\mathcal{O}$-constant that is not affected by stiffness (see also Dorsselaer & Spijker (1992)).

In view of (11) we expect, for stable linear multistep methods, that the global stopping error satisfies

$$v_n - u_n = \mathcal{O}(h^j) \quad (\text{for} \quad t_n \in (0, T]), \tag{12}$$

with an $\mathcal{O}$-constant that is again not affected by stiffness.

We have performed numerical experiments with the Examples 1, 2 (see the Sections 3, 4) in order to check (12). The conclusion of these experiments is that (12) is unnecessarily pessimistic. In fact, the actual global stopping errors in our experiments all satisfy

$$v_n - u_n = \mathcal{O}(h^{j+1}) \quad (\text{for} \quad t_n \in (0, T]) \tag{13}$$

for $j = 1, 2, 3, 4$ and for the so-called backward differentiation formulae of orders $k = 1, 2, 3$ (see e.g. Henrici (1962), Hairer & Wanner (1991), Lambert (1991), Gear (1971)).

We believe that the 'superconvergence' result (13) is in agreement with (10), (11) and that it can be explained by means of a device used by Hundsdorfer & Steiniger (1991) in their analysis of global discretization errors.

References

Alexander, R. (1991): *The modified Newton method in the solution of stiff ordinary differential equations.* Math. Comp. **57**, 673–701.

Burrage, K. & J.C. Butcher (1979): *Stability criteria for implicit Runge–Kutta processes.* SIAM J. Numer. Anal. **16**, 46–57.

Crouzeix, M. (1979): *Sur la B-stabilité des méthodes de Runge–Kutta.* Numer. Math. **32**, 75–82.

Dahlquist, G. (1978): *G-stability is equivalent to A-stability.* Bit **18**, 384–401

Dekker, K. & J.G. Verwer (1984): *Stability of Runge–Kutta methods for stiff nonlinear differential equations.* North Holland (Amsterdam).

Desoer, C.A. & H. Haneda (1972): *The measure of a matrix as a tool to analyse computer algorithms for circuit analysis.* IEEE Trans. Circuit Theory **19**, 480–486.

Dorsselaer, J.L.M. van & M.N. Spijker (1992): *The error committed by stopping the Newton iteration in the numerical solution of stiff initial value problems.* To appear in IMANA Journ. Numer. Anal.

Frank, R., J. Schneid & C.W. Ueberhuber (1985): *Order results for implicit Runge–Kutta methods.* SIAM Journ. Numer. Anal. **22**, 515–534.

Gear, C.W. (1971): *Numerical initial value problems in ordinary differential equations.* Prentice–Hall (Englewood Cliffs).

Hairer, E. & G. Wanner (1991): *Solving ordinary differential equations*, Vol. II, Springer (Berlin).

Hairer, E., S.P. Nørstett & G. Wanner (1987): *Solving ordinary differential equations I.* Springer (Berlin).

Henrici (1962): *Discrete variable methods in ordinary differential equations*, Wiley (New York).

Hundsdorfer, W.H. & B.I. Steiniger (1991): *Convergence of linear multistep and one-leg methods for stiff nonlinear initial value problems.* BIT **31**, 124–143.

Lambert, J.D. (1991): *Numerical methods for ordinary differential equations.* John Wiley & Sons (Chichester).

Liniger, W. (1971): *A stopping criterion for the Newton–Raphson method in implicit multistep integration algorithms for nonlinear systems of ordinary differential equations.* Comm. ACM **14**, 600–601.

Söderlind, G. (1992): *The Lipschitz algebra and its extensions.* Manuscript.

Prof.Dr. M.N. Spijker & Drs. J. Groeneweg
University of Leiden
Department of Mathematics and Computer Science
Niels Bohrweg 1
2333 CA Leiden
The Netherlands

D J HIGHAM

The dynamics of a discretised nonlinear delay differential equation

Abstract We consider a delay-logistic differential equation that has important applications in population dynamics. We set up a constant-stepsize discretisation of the problem and examine the long term behaviour of the solution. Key differences between this problem and the more familiar non-delay version are that
(i) introducing a small delay improves the stability of the iteration
(ii) for a certain parameter range, the stable steady state loses stability through a complex eigenvalue, and
(iii) spurious solutions exist even when the stepsize is stable in a linear sense.
We also consider a variable-stepsize version of the discrete method, based on a standard local error estimate. Tests reveal that the error control suppresses spurious behaviour, and we give some theoretical arguments to support these observations.

1 Introduction

Standard numerical methods for initial value problems replace a continuous problem by a discrete map. When long-time calculations are performed on nonlinear problems it is imperative that the discrete map mimics the asymptotic behaviour of the true solution. Recently, several authors have addressed this issue using ideas from dynamical systems and bifurcation theory; [2, 8, 9, 15] are key references in the area.

A simple example of a nonlinear ordinary differential equation (ODE) is the logistic equation, $y'(t) = y(t)(\alpha - y(t))$, $y(0) = y_0 \in \mathbb{R}$, $\alpha > 0$. The behaviour of $y(t)$ can be summarised very simply: $y(t) \to \alpha$ for positive initial values, and $y(t)$ blows up in finite time for negative initial values. It is well-known, however, that the corresponding map arising from Euler's method possesses an extremely rich variety of dynamics. This was pointed out by May [12], and the issue has since been examined by many other authors; see, for example, the references in [1]. The familiar bifurcation diagram for this map appears in many textbooks and papers, including [2].

Griffiths and Mitchell [1] generalised the logistic ODE by adding diffusion in one space-dimension, and they examined the discrete map arising from central differences in space combined with Euler's method in time. In this work, we alter the logistic ODE by introducing a *delay*. This delay-logistic ODE, and various discrete analogues, have often been proposed as population models in mathematical biology. In the next section we describe the continuous equation and set up our discretisation. In section 3 we analyse the fixed points of the discrete map and illustrate our results numerically. The behaviour is contrasted with that for the non-delay case. Section 4 gives an analysis of a more sophisticated, variable stepsize version of the numerical method, where a local error estimate is controlled on each step.

2 Continuous Problem and Discrete Algorithm

We consider the scalar delay-logistic equation

$$\begin{aligned} y'(t) &= y(t)(\alpha - y(t-1)), \quad \text{for } t \geq 0, \\ y(t) &= \Phi(t), \quad \text{for } t \in [-1, 0], \end{aligned} \tag{2.1}$$

where the initial function $\Phi(t)$ is continuous and $\alpha > 0$. The problem (2.1) arises in the study of population dynamics [13] and has been analysed extensively. Here, we confine our attention to the linear stability of the fixed points $y(t) \equiv 0$ and $y(t) \equiv \alpha$; for a detailed analysis of (2.1), see, for example, [4, Section 11.4]. Linearising about the zero fixed point gives the unstable equation $y'(t) = \alpha y(t)$. Setting $z(t) = \alpha + \epsilon(t)$ and linearising leads to $\epsilon'(t) = -\alpha\epsilon(t-1)$. Inserting $\epsilon(t) = e^{\gamma t}$ then yields the characteristic equation $\gamma + \alpha e^{-\gamma} = 0$. It can be shown that this equation has purely real, negative roots for $\alpha \leq 1/e$ and has complex roots with negative real parts for $1/e < \alpha < \pi/2$. In summary,

- $y(t) \equiv 0$ is never a linearly stable fixed point.
- $y(t) \equiv \alpha$ is a linearly stable fixed point for $\alpha < \pi/2$.

The standard numerical approach for solving delay differential equations such as (2.1) is to apply a formula for ordinary differential equations, using an interpolant $q(t)$ to approximate delayed values of the solution; see, for example, [3]. In this paper we analyse the case where Euler's method is combined with piecewise linear Lagrange interpolation. This may be thought of as a direct analogue of the Euler-based schemes studied in [1] and [12]. The ODE method and interpolant are 'compatible' in the sense that they have matching orders of local accuracy [6, 14], and the use of the linear Lagrange interpolant with a general class of methods that includes Euler's method has been studied in [11].

We suppose first that the stepsize $h > 0$ is constant. (A variable stepsize scheme is examined in section 4.) Setting $t_n = nh$, we let y_n denote the approximation to $y(t_n)$. Euler's method applied to (2.1) gives

$$y_{n+1} = y_n + hy_n(\alpha - q(t_n - 1)),$$

where $q(t_n - 1) \approx y(t_n - 1)$. If $t_n - 1 \leq 0$ then we may take $q(t_n - 1) = \Phi(t_n - 1)$. Otherwise, let m be the smallest integer such that $mh \geq 1$, so that $t_n - 1$ lies in $[t_{n-m}, t_{n-m+1})$. The linear Lagrange interpolant based on the values y_{n-m} and y_{n-m+1} is then given by $q(t_n - 1) = (1 - \sigma)y_{n-m} + \sigma y_{n-m+1}$, where $\sigma = m - 1/h$. Hence, on a general step with $t_n > 1$, we have the recurrence

$$y_{n+1} = y_n + hy_n(\alpha - \sigma y_{n-m+1} - (1 - \sigma)y_{n-m}). \tag{2.2}$$

Note that this is an $(m+1)$-step recurrence, and the value of m depends upon the stepsize, h. Our aim is to study the long term behaviour of y_n.

To conclude this section, we mention that there is no loss of generality in specifying a unit delay in (2.1). The equation

$$Y'(t) = Y(t)(\frac{\alpha}{\tau} - Y(t - \tau)), \tag{2.3}$$

reduces to $y'(t) = y(t)(\alpha - y(t-1))$ under the transformation $y(t) = \tau Y(t\tau)$. Similarly, applying the numerical method to (2.3) with stepsize $h\tau$ produces the numerical solution $\{y_n/\tau\}$.

3 Fixed Points

We begin the analysis with the case where $m = 1$ in (2.2). Here, we have $h \geq 1$, so the stepsize is no smaller than the size of the delay. Using $\sigma h = h - 1$, the recurrence reduces to

$$y_{n+1} = y_n(1 + h\alpha - (h-1)y_n - y_{n-1}). \tag{3.1}$$

This two-step scalar recurrence may be written in vector form as

$$\begin{bmatrix} y_{n+1} \\ y_n \end{bmatrix} = \begin{bmatrix} y_n(1 + h\alpha - (h-1)y_n - y_{n-1}) \\ y_n \end{bmatrix} =: F\left(\begin{bmatrix} y_n \\ y_{n-1} \end{bmatrix}\right). \tag{3.2}$$

The map (3.2) has fixed points that correspond to those of the underlying continuous problem, $y_n \equiv 0$ and $y_n \equiv \alpha$. Based on a standard linearisation argument (see, for example, [1] or [9]), we say that a fixed point is linearly stable if the Jacobian of F at that point has spectral radius less than unity. For the zero fixed point,

$$\frac{\partial F}{\partial y}\left(\begin{bmatrix} 0 \\ 0 \end{bmatrix}\right) = \begin{bmatrix} 1 + h\alpha & 0 \\ 1 & 0 \end{bmatrix}.$$

Hence $y_n \equiv 0$ is never linearly stable. For $y_n \equiv \alpha$,

$$\frac{\partial F}{\partial y}\left(\begin{bmatrix} \alpha \\ \alpha \end{bmatrix}\right) = \begin{bmatrix} 1 - h\alpha + \alpha & -\alpha \\ 1 & 0 \end{bmatrix}.$$

The eigenvalues of this matrix are the roots of the polynomial $p(\lambda) = \lambda^2 - \lambda(1 - h\alpha + \alpha) + \alpha$. Using the Routh-Hurwitz criterion [10], we find that necessary and sufficient conditions for $p(\lambda)$ to be *Schur* (that is, to have roots satisfying $|\lambda| < 1$) are that α and h lie in the region $\mathrm{R}_{m=1}$, defined by

$$\mathrm{R}_{m=1} := \{h, \alpha \ : \ \alpha < 1, \ 2\alpha > h\alpha - 2\}.$$

This region of the $h\alpha, \alpha$ plane corresponds to the quadrilateral shaded area in Figure 3.1. The region should be interpreted as follows. Given a value of α and a stepsize $h \geq 1$, the fixed point $y_n \equiv \alpha$ is linearly stable if $(h\alpha, \alpha)$ lies in $\mathrm{R}_{m=1}$. Note that since we are assuming $h \geq 1$, only the points in $\mathrm{R}_{m=1}$ below the line $\alpha = h\alpha$ are relevant. (The $\alpha > h\alpha$ portion corresponds to the case where extrapolation is used; that is, the interpolant based on y_{n-1} and y_n is used for an approximation at a point to the left of t_{n-1}.) We also deduce from Figure 3.1 that, given $\alpha < 1$, if we perform the iteration a number of times, each time slightly increasing the stepsize, then as h passes through $2(1 + 1/\alpha)$ the fixed point becomes unstable. When $h\alpha = 2(\alpha+1)$, the quadratic $p(\lambda)$ has roots $\lambda = -\alpha$ and $\lambda = -1$. Hence instability is caused by a root passing through -1. This suggests that period two solutions may bifurcate from $h = 2(1 + 1/\alpha)$.

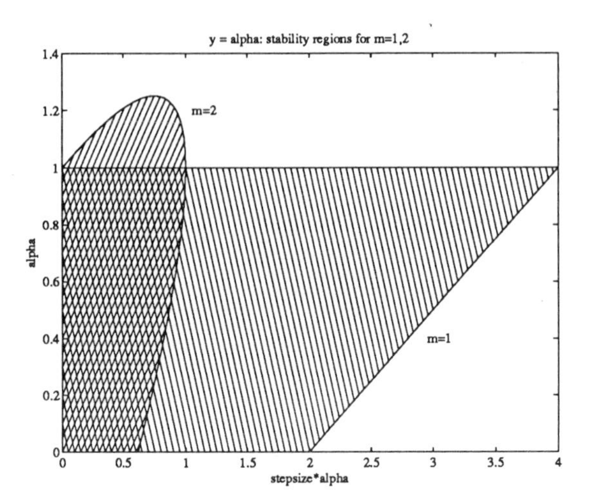

Figure 3.1: Regions of Stability with $m = 1, 2$ for $y_n \equiv \alpha$.

To search for period two solutions, we look for u and v with $u \neq v$ such that $y_{2n} \equiv u$ and $y_{2n+1} \equiv v$. We must therefore have

$$\begin{aligned} v &= u(1 + h\alpha - (h-1)u - v) && (3.3) \\ u &= v(1 + h\alpha - (h-1)v - u) && (3.4) \end{aligned}$$

Each equation is linear in one variable. Substituting for v from (3.3) into (3.4) leaves a quartic polynomial in u. Eliminating the roots $u = 0$ and $u = \alpha$ (which correspond to the period one solutions) leaves the equation

$$(h-1)(2-h)u^2 + (h-2)(h\alpha + 2)u + (-2 - h\alpha) = 0.$$

The discriminant of this quadratic polynomial is $h(h\alpha+2)(h-2)(\alpha(h-2)-2)$. It follows that real, period two solutions exist when either

1. $h - 2 < 0$ and $\alpha(h-2) - 2 < 0$, which reduces to $h < 2$.

2. $h - 2 > 0$ and $\alpha(h-2) - 2 > 0$, which reduces to $h > 2(1 + 1/\alpha)$.

Hence, we see that in addition to a period two solution bifurcating out of $h = 2(1 + 1/\alpha)$, a period two solution also exists for $h < 2$. The linear stability of the period two solution is determined by the spectral radius of the product

$$\frac{\partial F}{\partial y}\left(\begin{bmatrix} u \\ v \end{bmatrix}\right) \frac{\partial F}{\partial y}\left(\begin{bmatrix} v \\ u \end{bmatrix}\right). \qquad (3.5)$$

Numerical tests indicate that for any $1 < h < 2$ the period two solution is unstable. Further, it is straightforward to show that $[u, v]^T$ becomes unbounded as h approaches the values 1 and 2. To check the stability of the period two solution bifurcating out of the period one solution, we let $h = 2(1 + 1/\alpha) + \epsilon^2$. (We use ϵ^2 rather than ϵ so that valid Taylor expansions exist.) Using a symbolic algebra package, we found that for $\alpha < 1$, the dominant eigenvalue of (3.5) has an expansion that begins with $1 - 4\alpha\epsilon^2/(1-\alpha)$. Hence, we see that for $\alpha < 1$, if h is sufficiently close to $2(1 + 1/\alpha)$ then the period two solution is stable.

In Figure 3.2 we present a bifurcation diagram for the case $\alpha = .9$. Here the iteration (2.2) was computed for a range of stepsizes between .5 and 5. For each stepsize, we performed 6 different runs with constant initial functions $\Phi(t) \equiv .7, .8, .85, .95, 1.0$ and 1.1. Two hundred steps were taken in each case, and the figure plots the last twenty values, $\{y_n\}_{n=181}^{200}$. (The iteration was aborted if $|y_j| > 10^6$ for any j.) For $\alpha = .9$, we know that the fixed point $y_n \equiv \alpha$ becomes unstable (with $m = 1$) when $h = 2(1+1/\alpha) \approx 4.22$, and a stable period two solution arises as h increases beyond this value. This can be seen in the bifurcation diagram. It is also apparent that with $3 < h < 3.5$ and for some initial data, the iterates are not attracted to the fixed point $y_n \equiv \alpha$. More detailed numerical tests revealed the existence of stable period three solutions and extremely complex behaviour.

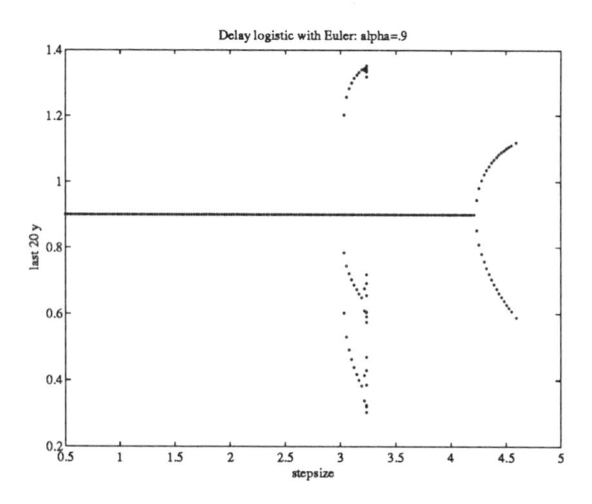

Figure 3.2: Bifurcation diagram with $\alpha = .9$.

When $h < 1 \leq 2h$, the recurrence (2.2) has $m = 2$. Using $\sigma = 2 - 1/h$, we have

$$y_{n+1} = y_n(1 + h\alpha - (2h-1)y_{n-1} - (1-h)y_{n-2}). \tag{3.6}$$

The three-step recurrence (3.6) has fixed points $y_n \equiv 0$ and $y_n \equiv \alpha$, and their linear stability can be examined in the manner described above for the $m = 1$ case. We find that the zero fixed point is always unstable. For the fixed point α, the eigenvalues of the relevant Jacobian matrix are the roots of

$$\lambda^3 - \lambda^2 + \lambda\alpha(2h-1) + \alpha(1-h). \tag{3.7}$$

Applying the Routh-Hurwitz criterion, we find that necessary and sufficient conditions for a Schur polynomial are

1. $\alpha < 1 + h\alpha/2$.

2. $\alpha > 5h\alpha/4 - 1$.

3. $\alpha < 3h\alpha/2 + 1$.

4. $\alpha^2(1-h)^2 < 1 - h\alpha$.

It is easily verified that condition 4 is dominant, and the region of stability, $\mathrm{R}_{m=2}$, is given by

$$\mathrm{R}_{m=2} := \{h, \alpha \ : \ h\alpha - \sqrt{1 - h\alpha} < \alpha < h\alpha + \sqrt{1 - h\alpha}\}.$$

The region $\mathrm{R}_{m=2}$ in the $h\alpha, \alpha$ plane is displayed in Figure 3.1. This result applies to the case where $h < 1 \leq 2h$, and hence only the subregion of $\mathrm{R}_{m=2}$ between the lines $\alpha = h\alpha$ and $\alpha = 2h\alpha$ is relevant for our algorithm.

The nontrivial boundary of $\mathrm{R}_{m=2}$ is given by the curves $\alpha = h\alpha \pm \sqrt{1 - h\alpha}$ for $0 < h\alpha \leq 1$. Along the upper curve, the polynomial (3.7) becomes

$$\lambda^3 - \lambda^2 + \lambda(h\alpha - \sqrt{1 - h\alpha}) + \sqrt{1 - h\alpha}.$$

The roots of this polynomial are

$$-\sqrt{1 - h\alpha}, \qquad \frac{1 + \sqrt{1 - h\alpha} \pm i\sqrt{2 + h\alpha - 2\sqrt{1 - h\alpha}}}{2},$$

which shows that linear instability is caused by nonreal roots leaving the unit circle.

A bifurcation diagram for the case $\alpha = 1.025$ is given in Figure 3.3. In this case, as h increases the fixed point $y_n \equiv \alpha$ becomes unstable when $\alpha = h\alpha + \sqrt{1 - h\alpha}$; that is, $h \approx .975$. The relevant eigenvalues are nonreal, and hence we do not see period two solutions evolving.

We now turn to the case of general $m > 1$ in the recurrence (2.2). In vector form, we have

$$\begin{bmatrix} y_{n+1} \\ y_n \\ \vdots \\ \vdots \\ y_{n-m+1} \end{bmatrix} = \begin{bmatrix} y_n(1 + h\alpha - h\sigma y_{n-m+1} - h(1-\sigma)y_{n-m}) \\ y_n \\ \vdots \\ \vdots \\ y_{n-m+1} \end{bmatrix} =: F\left(\begin{bmatrix} y_n \\ y_{n-1} \\ \vdots \\ \vdots \\ y_{n-m} \end{bmatrix}\right), \tag{3.8}$$

where $\sigma \in [0, 1)$ is given by $\sigma = m - 1/h$. There are two fixed points, $y_n \equiv 0$ and $y_n \equiv \alpha$. At $y_n \equiv 0$, $\partial F/\partial y$ has an eigenvalue $1 + h\alpha$, and hence this fixed point is always linearly unstable. At $y_n \equiv \alpha$, the eigenvalues of the Jacobian are the roots of the polynomial

$$\lambda^{m+1} - \lambda^m + h\alpha\sigma\lambda + (1 - \sigma)h\alpha. \tag{3.9}$$

Figure 3.4 shows the results of a numerical computation using a grid search in the $h\alpha, h$ plane. The shaded region shows where (3.9) is Schur. (The relevant parts of the $m = 1$ and $m = 2$ regions in Figure 3.1 can be clearly seen.)

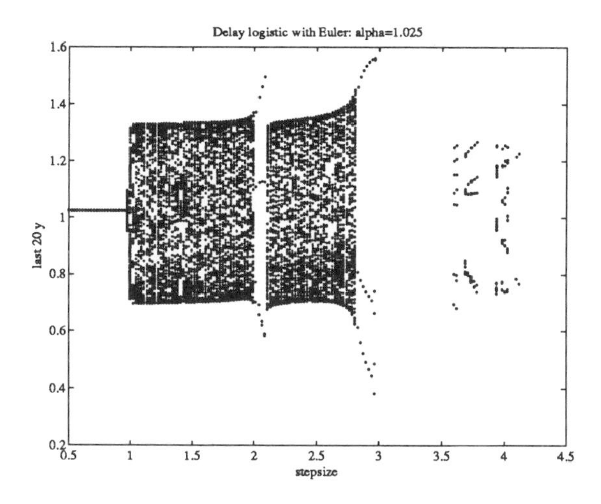

Figure 3.3: Bifurcation diagram with $\alpha = 1.025$.

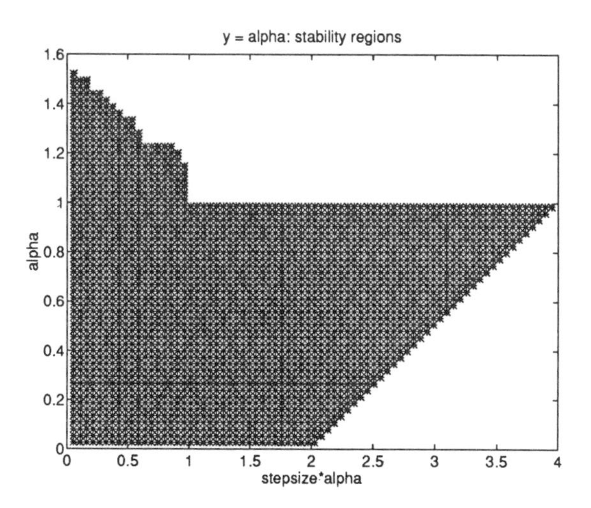

Figure 3.4: Region of Stability for $y_n \equiv \alpha$.

Recall from section 2 that the continuous problem has $y(t) \equiv \alpha$ linearly stable for $\alpha < \pi/2$. From Figure 3.4 it seems plausible that the stability region for $y_n \equiv \alpha$ intersects the α axis at $\pi/2$. In other words, it appears that whenever $y(t) \equiv \alpha$ is linearly stable there exists a stepsize such that $y_n \equiv \alpha$ is linearly stable. This can be verified as follows. Note that (3.9) may be written

$$\lambda^{m+1} - \lambda^m = -\alpha \left(\frac{\lambda\sigma + 1 - \sigma}{m - \sigma} \right). \tag{3.10}$$

If m is large (so that h is small) then the right-hand side of (3.10) is negligible and the roots of the polynomial will be close to 0 and 1. We may assume that the root close to 1 has the expansion

$$\lambda = 1 + c/m + O(1/m^2), \quad \text{for some constant } c \in \mathbb{C}. \tag{3.11}$$

Substituting the expansion (3.11) into (3.10) and letting $m \to \infty$ leads to the relation

$$e^c c = -\alpha. \tag{3.12}$$

Now, from (3.11), if $c = a + ib$ with $a, b \in \mathbb{R}$ then $|\lambda| = 1 + a/m + O(1/m^2)$, so the polynomial is Schur for sufficiently large m if $a < 0$. Finally, note that (3.12) is precisely the characteristic equation that arose for the continuous problem, so $a < 0$ is also the condition for $y(t) \equiv \alpha$ to be linearly stable.

Based on the results of this section, we find that the introduction of a delay has three main effects on the Euler-based discretisation.

First, *introducing a small delay improves the stability of the iteration*, in the sense that the nonzero period one solution is stable for a larger range of stepsizes. To see this, note that for the untransformed problem (2.3) if we use a stepsize H with $m = 1$, and if $\alpha/\tau < 1/\tau$, then the fixed point $Y_n \equiv \alpha/\tau$ becomes unstable at $H = 2\tau(1 + 1/\alpha)$. By contrast, for the standard ODE $Y'(t) = Y(t)(\alpha/\tau - Y(t))$ the fixed point becomes unstable at $H = 2\tau/\alpha$ (see, for example, [2]).

Second, we saw that with $m = 2$ ($h < 1 \leq 2h$) the loss of stability of the nonzero fixed point is caused by *nonreal eigenvalues*. This, of course, cannot happen for the non-delay equation, since in that case the Jacobian is a 1×1 matrix. Similarly, it cannot happen when diffusion, discretised via central differences, is added to the non-delay ODE version, since the discretised diffusion operator has real eigenvalues [1].

The third, and perhaps most important, observation is that the iteration studied in this section *admits spurious behaviour* (including attractive period three solutions) *for stepsizes that are stable in a linear sense.* This implies that choosing the stepsize based on a simple linearisation can lead to qualitatively incorrect results. Such behaviour has not been observed for the non-delay logistic ODE, although similar effects have been seen on other nonlinear ODEs (see, for example, Figure 2 of [2]).

4 Variable Stepsize Algorithm

The analysis in the previous section applies when the stepsize h is constant. In modern software, however, the stepsize is allowed to vary from step to step, according to an error control mechanism. In this section we add a standard local error estimate and stepsize selection formula to the numerical method. Our aim is to investigate the long-term behaviour of the adaptive algorithm. A similar investigation appears in [7, section 5]—the problem that we consider here is more specific, and hence we are able to make more precise statements. Our overall approach follows that of Hall [5].

The usual technique for local error estimation is to compare two formulas of different orders. We suppose here that the first order Euler method is combined with the second order Improved Euler method [10, page 155]. When applied to the scalar ODE $y'(t) = f(t, y(t))$, one step of length h_n with the overall algorithm begins by computing

$$k_1 = f(t_n, y_n), \tag{4.1}$$
$$k_2 = f(t_n + h_n, y_n + h_n k_1), \tag{4.2}$$
$$\mathrm{err}_{n+1} = \frac{h_n}{2}|k_2 - k_1|, \tag{4.3}$$
$$h_{\mathrm{new}} = \left(\frac{\theta\,\mathrm{TOL}}{\mathrm{err}_{n+1}}\right)^{1/2} h_n. \tag{4.4}$$

If $\mathrm{err}_{n+1} \leq \mathrm{TOL}$, where TOL is a parameter supplied by the user, then the step is accepted and we update the current values to

$$\begin{aligned} y_{n+1} &= y_n + h_n k_1, \\ t_{n+1} &= t_n + h_n, \\ h_{n+1} &= h_{\mathrm{new}}. \end{aligned}$$

Otherwise we reject the step, set $h_n = h_{\mathrm{new}}$ and repeat the process until $\mathrm{err}_{n+1} \leq \mathrm{TOL}$. The new stepsize, h_{new}, can be justified by an asymptotic ($h_n \to 0$) expansion, since err_{n+1} behaves as $O(h_n^2)$. The fixed safety factor $\theta \in (0, 1)$ is used to reduce the likelihood of step rejections. (We used $\theta = .81$ in our tests.)

For the delayed ODE (2.1) the algorithm above can be adapted so that the evaluations in (4.1) and (4.2) make use of interpolated values $q(t_n - 1)$ and $q(t_n + h_n - 1)$. We use the linear Lagrange interpolant defined in section 2 for $q(t)$. Also, since we are interested in the behaviour around the linearly stable fixed point α of (2.1) we consider the linearised problem $y'(t) = -\alpha y(t - 1)$.

We can regard the variable stepsize algorithm as a discrete iteration that updates y_n and h_n on every step. If $t_n - 1$ lies in the interval $[t_{n-m}, t_{n-m+1})$ then the iteration has the form

$$\begin{bmatrix} y_{n+1} \\ h_{n+1} \\ \vdots \\ \vdots \\ y_{n-m+1} \\ h_{n-m+1} \end{bmatrix} = G\left(\begin{bmatrix} y_n \\ h_n \\ \vdots \\ \vdots \\ y_{n-m} \\ h_{n-m} \end{bmatrix}\right). \tag{4.5}$$

However, it should be noted that the value of m may vary from step to step.

Now, following the analysis in [7, section 5] we look for a period two fixed point of the iteration (4.5) with the stepsize constant, say $h_n \equiv h_D$, and the numerical solution oscillating, say $y_{n+k} = (-1)^k y_D$. Inserting these values into the algorithm we find that a solution of this form is always possible for $m = 1$ with

$$h_D = 2(1 + 1/\alpha) \quad \text{and} \quad y_D = \frac{\theta \mathrm{TOL}}{2}. \tag{4.6}$$

Note that the value $h_D = 2(1 + 1/\alpha)$ arose in the constant stepsize analysis of section 3. When $\alpha < 1$, h_D is the largest stepsize that we could recommend, based on a linearisation of the problem, and hence it is reasonable to expect a good error control mechanism to choose stepsizes at this level.

Having identified a period two fixed point of the iteration, our next task is to establish whether the fixed point is linearly stable. Hence, we must take the $m = 1$ version of the iteration (4.5) and evaluate the corresponding Jacobian at the two points. Writing

$$\sigma_{n-1} = 1 - \frac{1}{h_{n-1}}, \qquad \sigma_n = 1 - \frac{1}{h_n}, \tag{4.7}$$

the iteration may be written

$$\begin{aligned}
k_1 &= -\alpha((1 - \sigma_{n-1})y_{n-1} + \sigma_{n-1}y_n), && (4.8)\\
y_{n+1} &= y_n + h_n k_1, && (4.9)\\
k_2 &= -\alpha((1 - \sigma_n)y_n + \sigma_n y_{n+1}), && (4.10)\\
\mathrm{err}_{n+1} &= \frac{h_n}{2}|k_2 - k_1|, && (4.11)\\
h_{n+1} &= \left(\frac{\theta\,\mathrm{TOL}}{\mathrm{err}_{n+1}}\right)^{1/2} h_n. && (4.12)
\end{aligned}$$

Since y_D is positive in (4.6) we have $k_2 - k_1 > 0$, so that the modulus signs can be ignored in (4.11). Some straightforward analysis using (4.7)–(4.12) shows that the Jacobian of G in (4.5) at the point $[y_D, h_D, -y_D, h_D]^T$ is given by

$$\begin{bmatrix}
\frac{\alpha+1}{-1} & \frac{\alpha\theta\mathrm{TOL}}{-2(\alpha+1)} & -\alpha & \frac{\alpha^2\theta\mathrm{TOL}}{-2(\alpha+1)}\\
\frac{-(\alpha+1)(\alpha^2+3\alpha+4)}{2\alpha\theta\mathrm{TOL}} & \frac{\alpha}{-2} & \frac{(\alpha+3)(\alpha+1)}{-2\theta\mathrm{TOL}} & \frac{\alpha(\alpha+3)}{4}\\
1 & 0 & 0 & 0\\
0 & 1 & 0 & 0
\end{bmatrix} =: J. \tag{4.13}$$

Similarly, the Jacobian at the point $[-y_D, h_D, y_D, h_D]^T$ is

$$DJD, \quad \text{where} \quad D = \mathrm{diag}(-1, 1, -1, 1). \tag{4.14}$$

Hence, the condition for linear stability of the period two point becomes

$$\rho(JDJD) < 1, \quad \text{or, equivalently,} \quad \rho(JD) < 1, \tag{4.15}$$

where $\rho(\cdot)$ denotes the spectral radius. It is easily seen that the similarity transformation $JD \to \widehat{D}(JD)\widehat{D}^{-1}$, with $\widehat{D} = \mathrm{diag}(1, \theta\,\mathrm{TOL}, 1, \theta\,\mathrm{TOL})$ produces a matrix that is independent of θ and TOL, and hence the condition (4.15) is independent of θ and TOL.

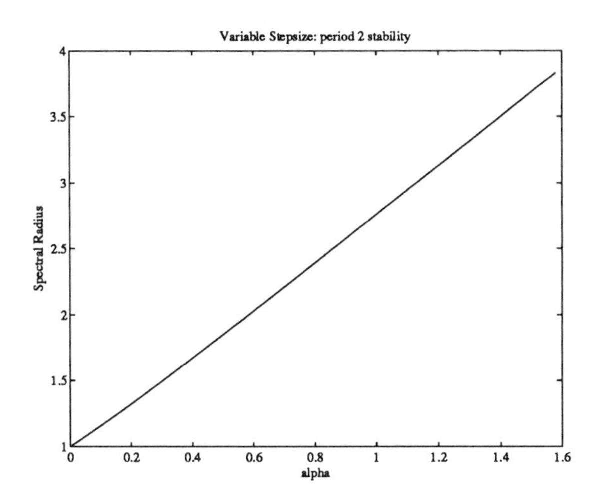

Figure 4.1: Stability of period two solution.

(This property is proved to hold under more general circumstances in [7].) In summary, the condition (4.15) expresses the linear stability of the fixed point defined in (4.6) as a function of the parameter α. Figure 4.1 plots $\rho(JD)$ against α for $0 < \alpha < \pi/2$, and we see that the fixed point is always unstable, with the spectral radius increasing monotonely with α.

Figure 4.2 illustrates the behaviour of the algorithm. Here, with the initial function $\Phi(t) \equiv 2$, we ran the variable stepsize iteration for 5,000 steps over a range of α between 0 and $\pi/2$. The figure plots the last twenty values of $\{h_n\alpha\}$; scaling by α on the y-axis allows us to see the results more clearly. The average of $h_n\alpha$ over the last twenty steps is marked with the symbol 'o'. The straight line $h\alpha = 2(1 + \alpha)$ corresponding to the fixed point (4.6) is also superimposed in dashed type. We see that for small α the average stepsize is very close to this value. Generally, there are oscillations about this fixed point, and the oscillations increase as α increases towards 1. This is to be expected, since from Figure 4.1 we know that the fixed point becomes more unstable as α increases. Typically, stepsizes that are too large lead to rejected steps, causing the stepsize to be drawn back towards the unstable period two level. A dramatic difference in behaviour takes place as α crosses 1. This can be explained by referring to the constant stepsize analysis of the previous section. From Figure 3.1 we see that for $\alpha > 1$ the value $h\alpha = 2(1+\alpha)$ no longer represents the boundary of linearised stability for the constant stepsize iteration, and hence it is extremely unlikely that the error control test would be passed with stepsizes at that level. Overall, for $\alpha < 1$, the iterates remain close to the unstable period two solution defined by (4.6), but for $\alpha > 1$ this period two solution is no longer relevant and the error control mechanism chooses smaller, more appropriate stepsizes.

In all our tests, we found that the standard technique of varying the stepsize according to a local error estimate completely suppressed the potential for spurious behaviour that is inherent in the constant stepsize algorithm.

References

[1] D.F. Griffiths and A.R. Mitchell, *Stable periodic bifurcations of an explicit discretization of a nonlinear partial differential equation in reaction diffusion*, IMA J. Numer. Anal. (1988) 8, 435–454.

[2] D.F. Griffiths, P.K. Sweby and H.C. Yee, *On spurious asymptotic numerical solutions of explicit Runge-Kutta methods*, IMA J. Numer. Anal (1992) 12, 319–338.

[3] E. Hairer, S.P. Nørsett and G. Wanner, *Solving Ordinary Differential Equations I*, (1987), Springer-Verlag, Berlin.

[4] J.K. Hale, *Theory of Functional Differential Equations*, (1977), Springer-Verlag, Berlin.

[5] G. Hall, *Equilibrium states of Runge Kutta schemes*, ACM Trans. Math. Soft. (1985) 11, 289-301.

[6] D.J. Higham, *Error control for initial value problems with discontinuities and delays*, Applied Numerical Mathematics (1993) 12, 315–330.

[7] D.J. Higham and I.Th. Famelis, *Stability of adaptive algorithms for delay differential equations*, Numerical Analysis Report NA/146, University of Dundee, 1992.

[8] A.R. Humphries, *Spurious solutions of numerical methods for initial value problems*, IMA J. Numer. Anal. (1993) 13, 263–290.

[9] A. Iserles, *Stability and dynamics of numerical methods for nonlinear ordinary differential equations*, IMA J. Numer. Anal (1990) 10, 1–30.

[10] J.D. Lambert, *Numerical Methods for Ordinary Differential Systems*, (1991) Wiley.

[11] M.Z. Liu and M.N. Spijker *The stability of the θ-methods in the numerical solution of delay differential equations*, IMA J. Numer. Anal. (1990) 10, 31–48.

[12] R.M. May, *Simple mathematical models with very complicated dynamics*, Nature (1976) 261, 459–467.

[13] R.M. May, *Mathematical aspects of the dynamics of animal populations*, in Studies in Mathematical Biology, part II; S.A. Levin, Ed., The Mathematical Association of America (1978), 317–366.

[14] H.J. Oberle and H.J. Pesch, *Numerical treatment of delay differential equations by Hermite interpolation*, Num. Math. (1981) 37, 235–255.

[15] H.C. Yee, P.K. Sweby and D.F. Griffiths, *Dynamical approach study of spurious steady-state numerical solutions of nonlinear differential equations. 1. The dynamics of time discretization and its implications for algorithm development in computational fluid dynamics* J. Comp. Phys. (1991) 97, 249–310.

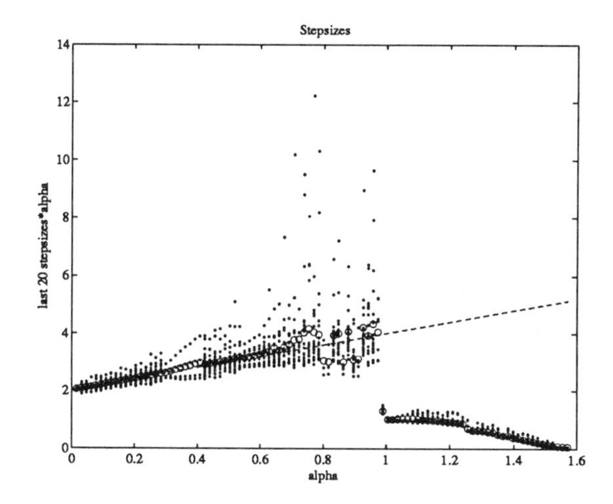

Figure 4.2: Stepsizes chosen by the error control mechanism.

Acknowledgement

I have benefited from many useful discussions with Dave Griffiths and Ron Mitchell about long term dynamics. I thank Nick Higham and Tasneem Sardar for suggesting improvements to this manuscript. This research was supported by the Science and Engineering Research Council.

Dr. Desmond J Higham
Dept. of Mathematics and Computer Science
University of Dundee
Dundee, DD1 4HN
UK

A R HUMPHRIES, D A JONES AND A M STUART

Approximation of dissipative partial differential equations over long time intervals

Abstract In this article the numerical analysis of dissipative semilinear evolution equations with sectorial linear part is reviewed. In particular the approximation theory for such equations over long time intervals is discussed. Emphasis is placed on studying the effect of approximation on certain invariant objects which play an important role in understanding long time dynamics. Specifically the existence of absorbing sets, the upper and lower semicontinuity of global attractors and the existence and convergence of attractive invariant manifolds, such as the inertial manifold and unstable manifolds of equilibrium points, is studied.

1 Introduction

In this paper we consider initial value problems of the form

$$u_t = f(u), \quad u(0) = u_0. \tag{1.1}$$

In particular, our interest is in the approximation of the equation as $t \to \infty$. Recall that standard error estimates typically grow exponentially with the time interval under consideration and are hence of no direct value in this context. Our study will be focussed on partial differential equations but, to introduce the main ideas, we consider several illustrative examples in ordinary differential equations.

Examples

(i) *Dissipativity.* Consider the equation (1.1) with vector field $f : \mathbb{R} \to \mathbb{R}$ defined by

$$f(u) = u - u^3. \tag{1.2}$$

It is straightforward to show that

$$\frac{1}{2}\frac{d}{dt}|u(t)|^2 \leq 1 - |u|^2$$

and hence that

$$|u(t)|^2 \leq 1 + e^{-2t}[|u(0)|^2 - 1].$$

Thus there exists $T = T(|u(0)|, \epsilon)$ such that

$$|u(t)|^2 \leq 1 + \epsilon, \quad \forall t \geq T.$$

This shows that the solution satisfies an asymptotic bound which is independent of initial data. Many physical systems exhibit such a property and it is often a mathematical manifestation of some form of energy dissipation. In some cases it may be important to preserve the property under approximation.

The backward Euler method applied to (1.1), (1.2) yields the map

$$U^{n+1} = U^n + \Delta t[U^{n+1} - (U^{n+1})^3].$$

A little calculation shows that

$$|U^{n+1}|^2 \leq |U^n|^2 + 2\Delta t[1 - |U^{n+1}|^2]$$

and hence that

$$|U^{n+1}|^2 \leq (1 + 2\Delta t)^{-1}[|U^n|^2 + 2\Delta t].$$

Induction yields

$$|U^n|^2 \leq 1 + (1 + 2\Delta t)^{-n}[|U^0|^2 - 1].$$

Thus there exists $N = N(|U^0|, \epsilon)$ such that

$$|U^n|^2 \leq 1 + \epsilon, \quad \forall n \geq N.$$

This shows that the dissipativity of the equation is preserved under discretization for any $\Delta t > 0$.

In contrast, the explicit Euler scheme applied to (1.1),(1.2) is not dissipative: we obtain the map

$$U^{n+1} = U^n + \Delta t[U^n - (U^n)^3].$$

If we let

$$V = [1 + \frac{2}{\Delta t}]^{\frac{1}{2}}$$

and set $U^0 = V$, then the map admits the solution

$$U^n = (-1)^n V.$$

Since $|U^n| = |U^0|$, it is clear that the amplitude of the solution is not bounded independently of initial data for any n, however large; thus dissipativity does not occur for all fixed $\Delta t > 0$. Indeed, if $|U^0| > 1$, to obtain dissipativity for the explicit Euler scheme requires the restriction

$$\Delta t \leq \frac{2}{|U^0|^2 - 1}.$$

These two examples illustrate a general principle – only certain methods (such as backward Euler) will dissipate on the whole phase space for any $\Delta t > 0$. Most methods (such as forward Euler) will dissipate on any bounded set, but this requires a time-step restriction dependent upon the size of that set. In section 3 we shall discuss the dissipativity of numerical schemes. □

(ii) *The Global Attractor.* The notion of dissipativity observed in the previous example can be abstracted thus: a system is dissipative if there is a bounded set $\mathcal{B}$, independent of initial data, which all trajectories starting within any bounded set E enter and remain inside after a finite time $T = T(E, \mathcal{B})$. An analogous definition can be made for maps.

The set $\mathcal{B}$ is known as an *absorbing set.* For the equation (1.1), (1.2), and its backward Euler approximation, $\mathcal{B} = [-1-\epsilon, 1+\epsilon]$.

The global attractor $\mathcal{A}$ is found by mapping the set $\mathcal{B}$ forward under the equation (1.1) and seeing what remains as $t \to \infty$. A precise definition is given in section 4. For equation (1.1), (1.2) the global attractor is simply the interval $[-1, 1]$. This may be understood by noting that, since $f : \mathbb{R} \mapsto \mathbb{R}$, the flow is in gradient form as (1.2) may be written as

$$f(u) = -F'(u), \quad F(u) = \frac{1}{4}(1-u^2)^2.$$

Hence

$$\frac{d}{dt}\{F(u(t)\} = -u_t^2.$$

Thus $F(u)$ acts as a potential well for the equation (1.1), (1.2); since F has minima at $u = \pm 1$ and a maximum at $u = 0$ it follows that any bounded set in $\mathbb{R}$ is mapped into an ϵ neighbourhood of the interval $[-1, 1]$ in a finite time $T = T(\epsilon)$. Figure 1.1 shows the potential well F governing the flow of (1.1), (1.2).

In many situations it is of interest to understand the effect of approximation on the global attractor for (1.2). However, the global attractor may be very sensitive to perturbation and can undergo discontinuous shrinking under arbitrarily small perturbations. Consider (1.2) with vector field given by $f_\epsilon(u)$:

$$f_\epsilon(u) = \left\{ \begin{array}{cc} -(u+1)^3 + \epsilon, & u \leq -1 \\ \epsilon(u^3/2 - 3u/2), & -1 < u < 1 \\ -(u-1)^3 - \epsilon, & u \geq 1 \end{array} \right\} \tag{1.3}$$

This vector field is $C^1(\mathbb{R}, \mathbb{R})$ for every $\epsilon \geq 0$. Furthermore, using the gradient structure of the equation it is straightforward to show that the problem is dissipative with absorbing set $\mathcal{B} = [-1-\delta, 1+\delta]$ for any $\delta > 0$ and hence has an attractor $\mathcal{A}_\epsilon$. The gradient flow has potential $F(u)$ shown in Figure 1.2 for $\epsilon > 0$ and in Figure 1.3 for $\epsilon = 0$.

The important point to observe from these figures is that for every $\epsilon > 0$ the attractor

$$\mathcal{A}_\epsilon = \{0\}, \quad \epsilon > 0,$$

a single point whilst for $\epsilon = 0$

$$\mathcal{A}_0 = [-1, 1],$$

an entire interval. This shows that the attractor $\mathcal{A}_0$ is *upper-semicontinuous* with respect to $\epsilon > 0$ but it is not *lower-semicontinuous.* Although the perturbation induced by ϵ in this example is not directly analogous to a numerical approximation, it nonetheless indicates an important point – without strong assumptions it may be difficult to prove lower-semicontinuity of attractors with respect to perturbations of any kind, including those induced by numerical approximation. This will be clearly illustrated in section 4. Roughly speaking, the difficulty associated with the attractor $\mathcal{A}_0$ is its lack of hyperbolicity, or viewed another way the fact that it is not exponentially attracting. □

(iii) Attractive Invariant Manifolds. Since attractors may vary discontinuously under approximation, it is sometimes of interest to study objects which are more robust under

perturbation. An important example is an exponentially attractive invariant manifold. For the purposes of this article it is sufficient to think of an invariant manifold as a graph relating one subset (or projection) of the solution variables to another subset (or projection); in this context the invariance means that solutions starting on the graph remain on the graph for all time.

Consider the equations

$$\begin{aligned} p_t &= p - p^3, \quad p(0) = p_0, \\ q_t &= -q + 3p^2 - 2p^4, \quad q(0) = q_0. \end{aligned} \tag{1.4}$$

For (1.4), setting

$$w = q - p^2$$

yields

$$w_t = -w.$$

Thus the graph $w = 0$, that is $q = p^2$, is invariant for the equation and, furthermore, the set of points $q = p^2$ is exponentially attracting since $w(t) = \exp(-t)w(0)$. Thus $w = 0$ is an exponentially attracting invariant manifold.

In the context in which we are interested, attractive invariant manifolds are important since they are either contained within the attractor (*unstable manifolds*) or contain the attractor (*inertial manifolds*) – see section 5. In general, the exponential attraction of the manifolds in question ensures good stability properties under perturbation. □

In this paper we review the analysis of the effect of numerical approximation on dissipativity, attractors and certain attractive invariant manifolds. Most of the results appear elsewhere in the literature but some are presented in the context of partial differential equations for the first time. Furthermore, an overview of the subject is given which relates a variety of different topics concerned with numerical approximation over long time intervals.

In section 2 we describe the mathematical setting for the partial differential equations studied and for their approximations. In section 3 we discuss discretization to preserve dissipativity. Section 4 contains a study of the upper and lower semicontinuity of the global attractor under numerical approximation. Section 5 is concerned with the upper and lower semicontinuity of attractive invariant manifolds; in particular unstable manifolds and inertial manifolds will be considered.

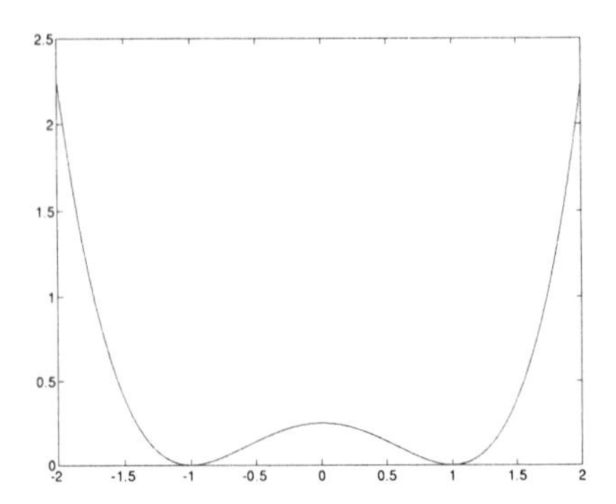

Figure 1.1: Potential Well for Equations (1.1),(1.2)

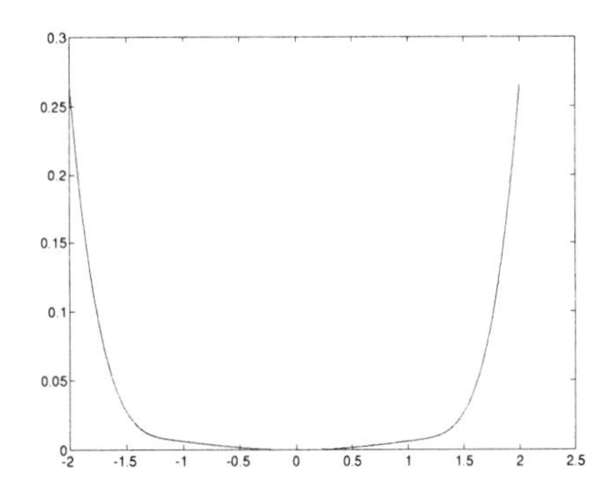

Figure 1.2: Potential Well for Equations (1.1), (1.3) with $\epsilon > 0$

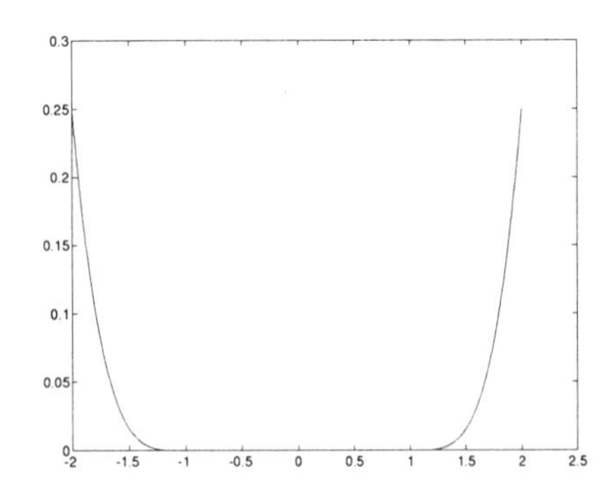

Figure 1.3: Potential Well for Equations (1.1), (1.3) with $\epsilon = 0$

FOR ALL FIGURES $F(u)$ IS ON THE VERTICAL AXIS AND u ON THE HORIZONTAL AXIS

2 Mathematical Setting

In the remainder of this paper we study the behaviour of the abstract evolution equation

$$\frac{du}{dt} + Au = F(u), \quad u(0) = u_0. \tag{2.1}$$

We consider (2.1) as an ordinary differential equation in a separable Hilbert space X with inner product $\langle \bullet, \bullet \rangle$ and norm $| \bullet |^2 = \langle \bullet, \bullet \rangle$. We assume that A is a densely defined sectorial operator with compact inverse, eigenvalues $\{\lambda_i\}$ and associated eigenfunctions $\{\varphi_i\}$. We also assume, without loss of generality, that A is positive definite and that the eigenvalues λ_i are ordered so that

$$0 < Re\{\lambda_1\} \leq Re\{\lambda_2\} \leq \ldots.$$

As in [39] we set $X^\alpha = D(A^\alpha)$ where $A^\alpha = (A^{-\alpha})^{-1}$ and for $\alpha > 0$

$$A^{-\alpha} = \frac{1}{\Gamma(\alpha)} \int_0^\infty t^{\alpha-1} e^{-At} dt.$$

For $\alpha = 0$ we define $A^0 = I$. Then X^α is a Hilbert space with norm $|\bullet|_\alpha^2 = \langle A^{\alpha/2}\bullet, A^{\alpha/2}\bullet \rangle$. The operator A generates an analytic semigroup $L(t)$. (In section 5 we encounter the case where A is not positive definite in our study of unstable manifolds; in that case the spaces X^α and their associated norms are defined by considering the operator $\tilde{A} = A + \zeta I$ for $\zeta > 0$ chosen to make $\tilde{A}$ positive definite).

We assume that F satisfies conditions sufficient that (2.1) generates a semigroup $S(t)$ with the properties that

$$\left\{ \begin{array}{c} \exists S(t) : X^\gamma \mapsto X^\gamma, \quad \gamma \in (0,1), \text{ such that } u(t) = S(t)u_0, \\ \\ \exists K = K(t,x,y) > 0 : |S(t)x - S(t)y|_\gamma \leq K|x-y|_\gamma, \\ \\ \{S(t)u_0\}_{t \geq 0} \text{ is relatively compact in } X^\gamma. \end{array} \right\} \tag{2.2}$$

Thus (2.1) has solution $u(t) = S(t)u_0$ for every $u_0 \in X^\gamma$. Conditions on $F(\bullet)$ which yield (2.2) may be found, for example, in [23].

In the following we employ the notation

$$\mathcal{B}_\gamma(v,r) = \{u \in X^\gamma : |u - v|_\gamma < r\}.$$

Three examples, to which the theory considered in the remainder of the paper applies, are now described.

Examples

(i) *Reaction-Diffusion.*

$$\begin{gathered} u_t = \Delta u + \lambda g(u), \quad x \in \Omega, \\ u = 0, \quad x \in \partial\Omega, \\ g(u) = \textstyle\sum_{j=1}^{2p-1} b_j u^j, \quad b_{2p-1} < 0. \end{gathered} \tag{2.3}$$

We take $X = L^2(\Omega)$ in this case. Existence of a Lipschitz continuous semigroup may be proved for $\gamma = \frac{1}{2}$ so that $S(t) : H_0^1(\Omega) \mapsto H_0^1(\Omega)$.

(ii) *Navier-Stokes equation in 2 Dimensions.*

$$u_t + u \cdot \nabla u = \tfrac{1}{R}\Delta u - \nabla p + h(x), \quad x \in \Omega,$$

$$\nabla \cdot u = 0, x \in \Omega, \tag{2.4}$$

$$u = 0, \quad x \in \partial\Omega.$$

Here X is the Hilbert space $\mathcal{H}$ comprising divergence free velocity fields contained in the space $L^2(\Omega)^2$ – see, for example, [44]. Existence of a Lipschitz continuous semigroup maybe proved for $\gamma = \frac{1}{2}$.

(iii) *Lorenz Equations.*

$$\begin{aligned} x_t &= -\sigma(x-y) \\ y_t &= rx - y - xz \\ z_t &= -bz + xy \end{aligned} \tag{2.5}$$

Here $X = \mathbb{R}^3$ and, again, existence of a Lipschitz continuous semigroup may be proved for all $t \geq 0$. □

We shall consider the approximation of (2.1) in space by a finite difference, finite element or spectral method to yield the equation

$$\frac{dU}{dt} + A^h U = F^h(U), \quad U(0) = U_0 \tag{2.6}$$

for $U(t) \in \mathcal{V} \subset X$. Here $\mathcal{V}$ is a finite dimensional subspace of X.

For the temporal discretization we will consider the θ method in the form

$$\frac{U^{n+1} - U^n}{\Delta t} + A^h U^{n+\theta} = F^h(U^{n+\theta}), \quad U(0) = U_0, \tag{2.7}$$

where

$$U^{n+\theta} := \theta U^{n+1} + (1-\theta)U^n, \quad \theta \in [0,1].$$

We assume that A^h defines a sectorial operator on $\mathcal{V}$ and note that we may then define $X^{h,\alpha}$ to be the Hilbert space comprised of elements of $\mathcal{V}$ with norm $|\bullet|^2 = \langle A^{h,\alpha/2}\bullet, A^{h,\alpha/2}\bullet\rangle$. We employ the following notation:

$$\mathcal{B}_{h,\alpha}(v,r) = \{u \in X^{h,\alpha} : |u-v|_{h,\alpha} < r\}.$$

We also assume that (2.6) and (2.7) generate discrete Lipschitz continuous semigroups $S^h(t) : \mathcal{V} \mapsto \mathcal{V}$ and $S_n^{h,\Delta t} : \mathcal{V} \mapsto \mathcal{V}$ respectively.

3 Dissipativity

We start by making a precise definition of dissipativity, motivated by the Example (i) in section 1.

Definition 3.1 *Equation* (2.1) *(resp.* (2.6), (2.7)*) is said to be* dissipative *in* $X^\beta, \beta \in [0,\infty)$, *(resp.* $X^{h,\beta}$*) if there exists* $\rho > 0$ *such that for any* $r > 0$ *there is a* $T = T(\rho, r)$ *(resp.* $N = N(\rho, r)$*) such that*

$$S(t)\mathcal{B}_\gamma(0,r) \subseteq \mathcal{B}_\beta(0,\rho) \ \ \forall t > T$$

(resp.

$$S^h(t)\mathcal{B}_{h,\gamma}(0,r) \subseteq \mathcal{B}_{h,\beta}(0,\rho) \ \ \forall t > T$$
$$S_n^{h,\Delta t}\mathcal{B}_{h,\gamma}(0,r) \subseteq \mathcal{B}_{h,\beta}(0,\rho) \ \ \forall n > N.)$$

Remarks
(i) The notion of dissipativity for ordinary differential equation is discussed, for example, in [40] and [46]. Generalizations to partial differential equations may be found in [20] and [44].
(ii) For the finite dimensional approximations (2.6) and (2.7) it is clear that, for any fixed $h > 0$, all the spaces $X^{h,\alpha}$ are equivalent for all $\alpha > 0$ so that dissipativity in one space implies dissipativity in them all; however, in most cases, interest is in deriving discrete absorbing sets that have radii ρ bounded independently of $h \to 0$, for any given $\alpha > 0$.
(iii) In many applications of the partial differential equation the initial data is taken in $\mathcal{B}_0(0,r)$; however, the smoothing properties induced by e^{-At} often means that it is equivalent to take data in $\mathcal{B}_\gamma(0,r)$ for some $\gamma \in (0,1)$ – see [23], [44].
(iv) Frequently dissipativity is proved initially in X. Use of the uniform Gronwall lemma [44] or, for certain gradient systems, a Lyapunov function [20] enables this to be extended to dissipativity in X^β for some $\beta > 0$. □

Throughout this section we assume that A is self-adjoint. We consider the case where $\gamma = \frac{1}{2}$ and

$$f(u) := F(u) - Au \tag{3.1}$$

satisfies the structural assumption

$$\exists a, b > 0 : \langle f(u), u\rangle \leq a - b|u|^2, \quad \forall u \in X^{\frac{1}{2}} \subset X. \tag{3.2}$$

Note that, since $X^{\frac{1}{2}} = D(A^{\frac{1}{2}})$, it follows that $\langle Au, u\rangle = |A^{\frac{1}{2}}u|^2$ is well-defined for $u \in X^{\frac{1}{2}}$ in the weak sense. There are many examples of equations satisfying (3.2) described in, for example, [44]. In particular the three examples described in section 2 all satisfy (3.2). Under (3.1), (3.2) it follows from (2.1) that

$$\frac{1}{2}\frac{d}{dt}|u|^2 \leq a - b|u|^2$$

and application of the Gronwall lemma shows that (2.1) is dissipative in X with

$$\rho = [(a+\epsilon)/b]^{\frac{1}{2}} \tag{3.3}$$

for any $\epsilon > 0$.

Under approximation in space, equation (2.1) yields (2.6), where now we define

$$f^h(u) := F^h(u) - A^h u. \tag{3.4}$$

In some applications it is of interest to perform the spatial approximation in such a way that $f^h(\bullet)$ satisfies a structural assumption analogous to (3.2), namely that

$$\exists a, b > 0 : \langle f^h(u), u \rangle \leq a - b|u|^2, \quad \forall u \in \mathcal{V} \subset X. \tag{3.5}$$

Note that, without loss of generality, we have assumed that the same constants a and b appear in (3.2) as in (3.5). The question of spatial discretization to retain the dissipativity of the underlying problem has been addressed by several authors. The earliest work in this direction appears to be [42], [31], [16], [15]. In [42] the dissipative properties of a Legendre-Galerkin approximation to a reaction-diffusion equation (2.3) were studied in both $L^2(\Omega)$ and $H_0^1(\Omega)$; however the restrictions on the discretization parameters required in this analysis are *initial data dependent* and it is natural to seek schemes for which this is not required. In [15] a finite difference scheme is constructed for the Kuramoto-Sivishinsky equation which preserves a condition closely related to (3.2) *without* initial data dependent restrictions on the discretization parameters; it is interesting to note that the work of [15] employs the same energy conserving approximation to uu_x analysed by Fornberg in [19]. In addition to proving dissipativity of the scheme, the paper [15] also studies conditions under which numerical approximations will blow up if energy-conserving discretizations of uu_x are not used; related issues for the viscous Burger's equations have been studied in [11]. The dissipative properties of certain nonlinear Galerkin methods are studied in [31] and [16]. In [13] analogous properties to those derived in [15] are proved for finite difference and finite element methods applied to a reaction-diffusion equation and in [37] for the Ginzburg-Landau equation. The dissipativity of certain finite element methods for the Navier-Stokes equation is studied in [2] and spectral methods for a nonlinear convection-diffusion equation in [26].

Henceforth we assume that spatial discretization has been performed in such a way that (3.5) holds. It is then natural to study the effect of time-discretization on dissipativity. The following result is of interest in this context. We employ the notation ρ given by (3.3) and define

$$R = \sup_{|v| \leq \rho} |f^h(v)|.$$

Recall that the true absorbing set has radius ρ. Note that R may be unbounded as $h \to 0$ since $f^h(\bullet) = F^h(\bullet) - A^h \bullet$ and A^h approximates an operator A with domain $D(A) \subset X$.

Theorem 3.1 Dissipativity *Consider application of* (2.7) *to* (2.6) *under the structural assumption* (3.4), (3.5). *Then* (2.7) *is dissipative in* X *for every* $\theta \in [\frac{1}{2}, 1]$; *furthermore, the absorbing set* $\mathcal{B}_0(0, \sigma)$ *has radius* σ *given by*

$$\sigma = \left\{ \begin{array}{c} \rho + \frac{\Delta t}{2} R, \quad \theta = \frac{1}{2}, \\ \rho + \Delta t(1-\theta)R, \quad \theta \in (\frac{1}{2}, 1), \quad \Delta t \in (0, \frac{2\rho}{(2\theta-1)R}] \\ \frac{1}{2\theta - 1}\rho, \quad \theta \in (\frac{1}{2}, 1), \quad \Delta t \in (\frac{2\rho}{(2\theta-1)R}, \infty) \\ \rho, \quad \theta = 1. \end{array} \right\} \tag{3.6}$$

Proof. We have

$$\frac{U^{n+1} - U^n}{\Delta t} = f(U^{n+\theta}), \tag{3.7}$$

where f satisfies (3.5). Since

$$U^{n+\theta} = \frac{U^{n+1} + U^n}{2} + (\theta - \frac{1}{2})(U^{n+1} - U^n) \tag{3.8}$$

the inner product with $U^{n+\theta}$ yields, for $\theta \in [\frac{1}{2}, 1]$,

$$\frac{|U^{n+1}|^2 - |U^n|^2}{2\Delta t} \leq a - b|U^{n+\theta}|^2. \tag{3.9}$$

Let

$$\sigma_1 = \rho + \Delta t(1-\theta)R, \quad \sigma_2 = \frac{\rho}{(2\theta - 1)}. \tag{3.10}$$

We show that $\mathcal{B}_0(0, \sigma_1)$ is an absorbing set for all $\theta \in [\frac{1}{2}, 1]$ and that $\mathcal{B}_0(0, \sigma_2)$ is absorbing for $\theta \in (\frac{1}{2}, 1]$. The result then follows – for $\theta \in (\frac{1}{2}, 1)$ the minimum of σ_1, σ_2 has been chosen for each $\Delta t \in (0, \infty)$.

First consider $\mathcal{B}_0(0, \sigma_1)$. Let $\epsilon > 0$. It is clear that either

$$a - b|U^{n+\theta}|^2 \leq -\epsilon, \tag{3.11}$$

or

$$a - b|U^{n+\theta}|^2 \geq -\epsilon. \tag{3.12}$$

If (3.12) holds then we have

$$|U^{n+\theta}| \leq \rho$$

and, since $U^{n+\theta} = U^{n+1} + (\theta - 1)(U^{n+1} - U^n)$, we have $|U^{n+1}| \leq \sigma_1$. Now, if (3.11) holds and if $|U^n| \leq \sigma_1$ then, by (3.9), we have $|U^{n+1}| \leq |U^n| \leq \sigma_1$. These two observations show that $\mathcal{B}_0(0, \sigma_1)$ is positively invariant since, under either (3.11) or (3.12) we have $|U^n| \leq \sigma_1 \Rightarrow |U^{n+1}| \leq \sigma_1$. Furthermore, it follows from (3.9), (3.11) that iterates starting in any bounded set containing $\mathcal{B}_0(0, \sigma_1)$ enter $\mathcal{B}_0(0, \sigma_1)$ in a finite number of steps. Thus we have exhibited an absorbing set $\mathcal{B}_0(0, \sigma_1)$.

Now consider $\mathcal{B}_0(0, \sigma_2)$. Since $U^{n+\theta} = \theta U^{n+1} + (1-\theta)U^n$, equation (3.9) yields

$$\begin{aligned}
\frac{|U^{n+1}|^2 - |U^n|^2}{2\Delta t} &\leq a - b[\theta^2|U^{n+1}|^2 + 2\theta(1-\theta)\langle U^n, U^{n+1}\rangle + (1-\theta)^2|U^n|^2] \\
&\leq a - b[\theta^2|U^{n+1}|^2 - \theta(1-\theta)|U^{n+1}|^2 - \theta(1-\theta)|U^n|^2 + (1-\theta)^2|U^n|^2] \\
&\leq a - b[\theta(2\theta-1)|U^{n+1}|^2 + (1-\theta)(1-2\theta)|U^n|^2].
\end{aligned}$$

Hence

$$[1 + 2\Delta t b\theta(2\theta - 1)]|U^{n+1}|^2 \leq [1 - 2\Delta t b(1-\theta)(1-2\theta)]|U^n|^2 + 2a\Delta t. \tag{3.13}$$

Algebra shows that, for $\theta \in (\frac{1}{2}, 1]$,

$$\frac{|1 - 2\Delta t b(1-\theta)(1-2\theta)|}{1 + 2\Delta t b\theta(2\theta - 1)|} < 1, \quad \forall \Delta t > 0.$$

Thus applying the Gronwall lemma to (3.13) yields

$$\limsup_{n\to\infty} |U^n|^2 \leq \frac{a}{b(2\theta - 1)^2}$$

and it follows that $\mathcal{B}_0(0, \sigma_2)$ is absorbing. This completes the proof. □.

Remarks

(i) Note that, for $\theta = 1$ the absorbing set has radius identical to that arising in the continuous case. For $\theta = \frac{1}{2}$ the dissipaitivty is rather weak since, because of the dependence of R on h, the absorbing set depends on $\Delta t/h^p$ for some $p > 0$. Thus a form of Courant restriction is required to make the absorbing set mesh independent. For example, if we consider the reaction-diffusion equation (2.3) and a standard finite element approximation based on piecewise linear triangulation of Ω, then $p = 2$.

(ii) In [27] the two-step backward differentiation formula is analysed for problems satisfying (3.2) and shown to be dissipative. For gradient systems it is possible to find second order time-accurate schemes which preserve dissipativity, without requiring a Courant restriction, by exploiting a discrete version of a Lyapunov function; see [13], where a modification of the Crank-Nicolson method constructed in [12] and the two-step backward differentiation formula are analysed in this context.

(iii) Theorem 3.1 is a synthesis of the ideas contained in [13], [28] and [2]. In [13] the dissipativity of the backward Euler scheme is studied for reaction-diffusion equations; in that paper a discrete version of the uniform Gronwall lemma is proved and employed to establish dissipativity in $H_0^1(\Omega)$ as well as $L^2(\Omega)$. In [28] Runge-Kutta methods are studied for ordinary differential equations (1.1) satisfying (3.2); it is proved that irreducible algebraically stable methods preserve dissipativity, making a connection with the exsiting theory of contractive methods derived in [6]. In [2] it was observed that, in the context of the Navier-Stokes equations (2.4), a linearization of the $\theta-$method for $\theta \in (\frac{1}{2}, 1]$ has an absorbing set independent of $\Delta t > 0$. This interesting fact has been generalized slightly here and incorporated into our proof. A similar analysis for convection-diffusion equations may be found in [26].

(iv) The linearisation of the $\theta-$method considered in [2] is of some practical interest since it allows the direct implementation of a dissipative numerical method involving only the inversion of *linear systems* at each step. This approach can be used in cases where $F(\bullet)$ may be represented as the sum of a bilinear form and a forcing function so that

$$F(v) = B(v, v) + g$$

and the numerical approximation $F^h(\bullet)$ may be represented as the sum of a bilinear form and a forcing function

$$F^h(v) = B^h(v, v) + g.$$

It is then possible to consider the time-stepping scheme

$$\frac{U^{n+1} - U^n}{\Delta t} + A^h U^{n+\theta} = B^h(U^n, U^{n+\theta}) + g, \quad U(0) = U_0.$$

If B satisfies the property

$$\langle v, B(w, v)\rangle = 0 \;\; \forall v, w \in X^{\frac{1}{2}}$$

(as in the Navier-Stokes equations, for example) and a spatially discrete analogue of this condition also holds then it is possible to generalise the analysis of Theorem 3.1 to cope with this case – see [2] for details. □

This concludes our analysis of dissipativity. We remark that the dissipativity of linear multistep methods is still an open question, as is the matter of the effect of the nonlinear solver on implicit methods such as (2.7). We also note that we have asked here for dissipativity of numerical schemes on the *whole phase space* $\mathcal{V}$ for fixed $h, \Delta t$; this is an extremely strong condition. In many cases we might expect to obtain dissipativity on any compact set $E \subset \mathcal{V}$ for $\Delta t, h$ sufficiently small in terms of the size of E. Such results have been proved for Runge-Kutta methods applied to the ordinary differential equation (1.1) under (1.3) in [29], [30], for nonlinear Galerkin methods in [10] and [41] and for spectral approximation of reaction diffusion convection equations in [26]; however there are still many open questions in this area for the approximation of partial differential equations.

4 The Global Attractor

In this section we define the global attractor for (2.1), (2.6) and (2.7) and then study the relationship between the true and approximate attractors. We employ the notation

$$\mathrm{dist}_\beta(u, A) = \inf_{v\in A} |u - v|_\beta$$

$$\mathrm{dist}_\beta(B, A) = \sup_{u\in B} \mathrm{dist}_\beta(u, A)$$

$$\mathcal{N}_\beta(A, \epsilon) = \{u \in X^\beta : \mathrm{dist}_\beta(u, A) < \epsilon\}$$

$$\mathcal{N}_{h,\beta}(A, \epsilon) = \{u \in X^{h,\beta} : \mathrm{dist}_{h,\beta}(u, A) < \epsilon\}.$$

Notice that $\mathrm{dist}_\beta(B, A) = 0$ is equivalent to the statement $\bar{B} \subseteq \bar{A}$ so that "dist" defines a semidistance – the asymmetric Hausdorff semidistance. We will also find the concept of an *invariant set* useful. A set B is said to be invariant under $S(t)$ if $S(t)B \equiv B$ for all $t \geq 0$.

Definition 4.1 *A set $\mathcal{A} \in X^\gamma$ (resp. $X^{h,\gamma}$) is said to* attract *a set $U \in X^\gamma$ (resp. $X^{h,\gamma}$) under a semigroup $S(t)$ (resp. $S^h(t), S_n^{h,\Delta t}$) if, for any $\epsilon > 0$ $\exists T = T(\epsilon)$ (resp. $N = N(\epsilon)$) such that*

$$S(t)U \in \mathcal{N}_\gamma(\mathcal{A}, \epsilon) \;\; \forall t \geq T,$$

(resp.

$$S^h(t)U \in \mathcal{N}_{h,\gamma}(\mathcal{A}, \epsilon) \;\; \forall t \geq T,$$
$$S_n^{h,\Delta t}U \in \mathcal{N}_{h,\gamma}(\mathcal{A}, \epsilon) \;\; \forall n \geq N.)$$

An attractor *is a compact invariant set which attracts an open neighbourhood of itself. An attractor $\mathcal{A}$ is said to be a* global attractor *if it attracts every bounded set $U \in X^\gamma$ (resp. $U \in X^{h,\gamma}$.)*

We now show how the existence of an absorbing set, together with some compactness, gives the existence of a global attractor [44]. To formalise this idea we require the following definition:

Definition 4.2 *The $\omega-$limit set of a bounded set $U \in X^\gamma$ (resp. $X^{h,\gamma}$) for $S(t)$ (resp. $S^h(t), S_n^{h,\Delta t}$) is defined by*

$$\omega(U) := \bigcap_{s\geq 0} \overline{\bigcup_{t\geq s} S(t)U},$$

(resp.

$$\omega(U) := \bigcap_{s\geq 0} \overline{\bigcup_{t\geq s} S^h(t)U},$$

$$\omega(U) := \bigcap_{m\geq 0} \overline{\bigcup_{n\geq m} S_n^{h,\Delta t}U}.)$$

The following theorem is proved in [44]:

Theorem 4.1 *If $S(t)$ (resp. $S^h(t), S_n^{h,\Delta t}$) has an absorbing set $\mathcal{B}_\gamma(0,\rho)$ (resp. $\mathcal{B}_{h,\gamma}(0,\rho)$) and if $\exists t_c > 0$ (resp. $n_c > 0$) such that $\{S(t)E\}_{t\geq t_c}$ (resp. $\{S^h(t)E\}_{t\geq t_c}, \{S_n^{h,\Delta t}E\}_{n\geq n_c}$) is relatively compact for any bounded set $E \in X^\gamma$ (resp. $X^{h,\gamma}$), then $\omega(\mathcal{B}_\gamma(0,\rho))$ (resp. $\omega(\mathcal{B}_{h,\gamma}(0,\rho))$ is a global attractor for $S(t)$ (resp. $S^h(t), S_n^{h,\Delta t}$).*

Remarks

(i) In [44] Theorem 4.1 is used to construct attractors for $S(t) : X \mapsto X$ by finding absorbing sets in X and $X^\beta, \beta > 0$. In [20] the global attractor for gradient systems is constructed directly by using a Lyapunov function and the existence of absorbing sets is a consequence of the existence of an attractor or, alternatively, can be deduced from the existence of a Lyapunov function.

(ii) In finite dimensions relative compactness is automatic and the existence of an absorbing set in X implies the existence of an attractor. This has been used in a number of cases to construct global attractors for numerical schemes – see [42], [15], [13], [37] and [28]; in addition, the papers [42], [13] and [37] also prove $h-$ independent estimates on the size of the attractor in $X^{h,\beta}$ for some $\beta > 0$. Furthermore, discrete Lyapunov functions can also be used to construct a global attractor for the numerical approximation of gradient systems; see [13].

(iii) In many cases it is not possible to show that an approximate scheme has a *global* attractor as the example of the forward Euler method applied to (1.1), (1.2) shows; however, the work of [22] shows that *local* attractors can be constructed for the approximation under reasonable hypotheses. □

For the remainder of this section we assume that it has been established that (2.1), (2.6), (2.7) have global attractors $\mathcal{A}, \mathcal{A}^h, \mathcal{A}^{h,\Delta t}$ respectively and that, furthermore, these sets may be constructed as $\omega-$limit sets of absorbing sets:

$$\begin{gathered} \exists \beta \in [\gamma, 1), \ \textit{absorbing sets } \mathcal{B}_\beta(0,\rho), \ \mathcal{B}_{h,\beta}(0,\rho) \ \textit{for } S(t), S^h(t), S_n^{h,\Delta t}: \\ \mathcal{A} = \omega(\mathcal{B}_\beta(0,\rho)), \ \mathcal{A}^h = \omega(\mathcal{B}_{h,\beta}(0,\sigma)), \ \mathcal{A}^{h,\Delta t} = \omega(\mathcal{B}_{h,\beta}(0,\sigma)). \end{gathered} \tag{4.1}$$

We assume also that $\mathcal{B}_{h,\beta}(0,\sigma)$ is bounded in X^γ so that,

$$\exists r = r(\sigma) > 0 : U \in \mathcal{B}_{h,\beta}(0,\sigma) \Rightarrow U \in \mathcal{B}_\gamma(0,r). \tag{4.2}$$

In the light of our preceding remarks we know that these assumptions may be quite strong as only local attractors may exist for the approximate semigroup; however, since our aim is to convey the essential ideas, we proceed in this framework. These ideas can be modified to allow for the case where $\mathcal{A}, \mathcal{A}^h$ or $\mathcal{A}^{h,\Delta t}$ are only local attractors as in [22].

Our aim is to study the relationship between $\mathcal{A}$ and $\mathcal{A}^h$ or $\mathcal{A}$ and $\mathcal{A}^{h,\Delta t}$. ¿From the example in section 1 we expect that it may be possible to show that $\text{dist}_\gamma(\mathcal{A}^h, \mathcal{A})$ is small (upper-semicontinuity) but that showing that $\text{dist}_\gamma(\mathcal{A}, \mathcal{A}^h)$ is small (lower-semicontinuity) will not, in general, be possible. The following theorem concerns upper-semicontinuity.

Theorem 4.2 Upper-Semicontinuity of Attractors *Assume that:*
(I) equations (2.1), (2.6) *have global attractors $\mathcal{A}$ and $\mathcal{A}^h$ given by* (4.1) *and that* (4.2) *holds;*
(II) for any $T, \delta, \sigma > 0$ $\exists H = H(T,\delta,\sigma) > 0$ such that

$$|S(t)U - S^h(t)U|_\gamma \leq \delta, \quad T \leq t \leq 2T$$

for all $U \in \mathcal{B}_{h,\beta}(0,\sigma)$ and $h \in (0,H)$.
Then $\text{dist}_\gamma(\mathcal{A}^h, \mathcal{A}) \to 0$ as $h \to 0$.

Proof. Recall that, since $\mathcal{A}^h = \omega(\mathcal{B}_{h,\beta}(0,\sigma))$, it is sufficient to show that for any $\epsilon > 0$ $\exists T > 0$ and $H > 0$ such that, if $h \in (0,H)$, then $S^h(t)\mathcal{B}_{h,\beta}(0,\sigma) \subset \mathcal{N}_\gamma(\mathcal{A},\epsilon)$ $\forall t \geq T$.

Now, by (4.2) we have that $\mathcal{B}_{h,\beta}(0,\sigma) \subseteq \mathcal{B}_\gamma(0,r)$. Thus, for any $\epsilon > 0$ $\exists T > 0$:

$$S(t)\mathcal{B}_{h,\beta}(0,\sigma) \subset \mathcal{N}_\gamma(\mathcal{A},\epsilon/2) \ \forall t \geq T, \tag{4.3}$$

since $\mathcal{A}$ is the global attractor for $S(t)$ in X^γ. Furthermore, by assumption II of the theorem, for any $\epsilon > 0$ $\exists H > 0$:

$$\text{dist}_\gamma(S^h(t)\mathcal{B}_{h,\beta}(0,\sigma), S(t)\mathcal{B}_{h,\beta}(0,\sigma)) \leq \epsilon/2, \quad T \leq t \leq 2T, \tag{4.4}$$

for $h \in (0,H)$. Equations (4.3) and (4.4) show that, provided $h \in (0,H)$,

$$S^h(t)\mathcal{B}_{h,\beta}(0,\sigma) \subset \mathcal{N}_\gamma(\mathcal{A},\epsilon), \quad T \leq t \leq 2T.$$

For the purposes of induction assume that, provided $h \in (0,H)$

$$S^h(t)\mathcal{B}_{h,\beta}(0,\sigma) \subset \mathcal{N}_\gamma(\mathcal{A},\epsilon), \quad kT \leq t \leq (k+1)T \tag{4.5}$$

and note that we have proved this for $k = 1$. Since $S^h(kT)\mathcal{B}_{h,\beta}(0,\sigma) \subseteq \mathcal{B}_{h,\beta}(0,\sigma)$ (by definition of an absorbing set) we may repeat the arguments above (with the role of $t = 0$ taken by $t = kT$) to deduce that

$$\text{dist}_\gamma(S^h(t)\mathcal{B}_{h,\beta}(0,\sigma), S(t)\mathcal{B}_{h,\beta}(0,\sigma)) \leq \epsilon/2, \quad (k+1)T \leq t \leq (k+2)T$$

for $h \in (0, H)$. Hence by (4.3) we deduce that, provided that $h \in (0, H)$ we have

$$S^h(t)\mathcal{B}_{h,\beta}(0,\sigma) \subset \mathcal{N}_\gamma(\mathcal{A},\epsilon), \quad (k+1)T \leq t \leq (k+2)T. \tag{4.6}$$

Since (4.5) implies (4.6), induction on k shows that, for any $\epsilon > 0 \; \exists T > 0, H > 0$ such that

$$S^h(t)\mathcal{B}_{h,\beta}(0,\sigma) \subset \mathcal{N}_\gamma(\mathcal{A},\epsilon), \quad t \geq T$$

provided that $h \in (0, H)$. Since $\epsilon > 0$ is arbitrary, this yields the convergence result. □

Remarks

(i) Theorem 4.2 was first proved in [22] in a more general form where the existence of an approximate global attractor was not assumed but, rather, an approximate local atttractor constructed during the course of the proof.

(ii) Related results may be found in [33] where the weaker concept of uniformly asymptotically stable sets was studied for ordinary differential equations and in [34] where the method was generalized to partial differential equations. The paper [25] relates the approach of [33] to that of [22]. The book [44] also contains a result similar to Theorem 4.2.

(iii) In [42] the upper semicontinuity of approximate attractors was considered for (2.3) under Legendre-Galerkin approximation; in [35] such an analysis is considered for the finite element method applied to (2.3). The upper semicontinuity of attractors for the ordinary differential equations (1.1), (1.3) under approximation by Runge-Kutta methods is studied in [28]. For q-step linear multistep methods the issues are somewhat more complicated since the natural phase space for the temporal approximation is $\{X^\gamma\}^q$. This issue is considered in [24] where the existence and upper semicontinuity of attractors for strictly $A(\alpha)$-stable multistep methods is proved.

(iv) Note that, depending upon the value of γ, the error estimate in assumption (II) of Theorem 4.2 may be required for non-smooth initial data. The issue of non-smooth error estimates, and the relationship to attractor convergence, is considered in [35]. For finite difference methods the derivation of such non-smooth error estimates is considered for the Navier-Stokes equations in [45]. In [37] the use of non-smooth error estimates for finite difference approximations of the complex Ginzburg-Landau equation is avoided by deriving discrete regularity results for the approximate schemes which essentially enable β to be chosen arbitrarily large.

(v) Note that a result similar to Theorem 4.2 can be proved which incorporates the effect of time-discretization; see [22]. □

We now proceed to discuss the lower-semicontinuity of attractors under numerical approximation. As the example in section 1 shows this will not, in general, be possible unless strong hyperbolicity conditions are imposed upon the attractor. In order to make these conditions precise we need some further definitions. In the following we use dF to denote the Fréchet derivative of F.

Definition 4.3 *An* equilibrium point $\bar{u} \in X^\gamma$ *of* (2.1) *(resp.* $\bar{U} \in X^{h,\gamma}$ *of* (2.6), (2.7)*) satisfies*

$$A\bar{u} = F(\bar{u}),$$

(resp.

$$A^h\bar{U} = F^h(\bar{U}).)$$

The unstable manifold *of $\bar{u}$ (resp. $\bar{U}$) denoted by $W^u(\bar{u})$ (resp. $W_h^u(\bar{U}), W_{h,\Delta t}^u(\bar{U})$) is the set*

$$\{u_0 \in X^\gamma : (2.1) \text{ has a solution } \forall t \leq 0 \ \& \ u(t) \to \bar{u} \text{ as } t \to -\infty\},$$

(resp.

$$\{U_0 \in X^{h,\gamma} : (2.6) \text{ has a solution } \forall t \leq 0 \ \& \ U(t) \to \bar{U} \text{ as } t \to -\infty\},$$

$$\{U_0 \in X^{h,\gamma} : (2.7) \text{ has a solution } \forall n \leq 0 \ \& \ U^n \to \bar{U} \text{ as } n \to -\infty\}.)$$

The local unstable manifold *of $\bar{u}$ (resp. $\bar{U}$) denoted by $W^u(\bar{u};\epsilon)$ (resp. $W_h^u(\bar{U};\epsilon)$, $W_{h,\Delta t}^u(\bar{U};\epsilon)$) is the set*

$$\{u_0 \in W^u(\bar{u}) : u(t) \in \mathcal{B}_\gamma(\bar{u},\epsilon) \ \ \forall t \leq 0\},$$

(resp.

$$\{U_0 \in W_h^u(\bar{U}) : U(t) \in \mathcal{B}_{h,\gamma}(\bar{U},\epsilon) \ \ \forall t \leq 0\},$$
$$\{U_0 \in W_{h,\Delta t}^u(\bar{U}) : U^n \in \mathcal{B}_{h,\gamma}(\bar{U},\epsilon) \ \ \forall n \leq 0\}.)$$

An equilibrium point $\bar{u}$ of (2.1) *(resp. $\bar{U}$ of* (2.6), (2.7)*) is said to be hyperbolic if the spectrum of the linear operator $A - dF(\bar{u}) : D(A) \mapsto X$ (resp. $A^h - dF^h(\bar{U}) : \mathcal{V} \mapsto \mathcal{V}$) contains no points on the imaginary axis.*

In the following we use the notation

$$\mathcal{E} = \{v \in X^\gamma : Av = F(v)\},$$

$$\mathcal{E}^h = \{V \in \mathcal{V} : A^hV = F^h(V)\}$$

to denote the equilibria of (2.1) and of (2.6) or (2.7).

In [21] the lower semicontinuity of attractors is considered when (2.1) is in gradient form: it is supposed that a Lyapunov function $V(\bullet) : X^\gamma \mapsto \mathbb{R}$ exists for (2.1) in which case, provided the equilibria are hyperbolic, the attractor $\mathcal{A}$ is given by

$$\mathcal{A} = \overline{\bigcup_{v \in \mathcal{E}} W^u(v)}. \tag{4.7}$$

(See [20] for a precise definition of Lyapunov function in this context). Using the techniques of [21] it may then be shown, under assumptions I and II of Theorem 4.2, together with an assumption that the local unstable manifolds are lower-semicontinous in X^γ with respect to h that

$$\text{dist}_\gamma(\mathcal{A}, \mathcal{A}^h) \to 0 \ \text{ as } \ h \to 0.$$

This is a lower semicontinuity result and the same method of proof has been employed in [28] to consider the effect of time discretization on lower-semicontinuity for ordinary differential equations in gradient form. The proof of [21] explicitly uses the Morse decomposition of the attractor induced by the gradient structure. However, in [29], [30] it is shown that the assumption that the system be in gradient form is *not required* to prove lower-semicontinuity with respect to certain perturbations. Rather it is sufficient to assume that the attractor has the form (4.7) where every element of $\mathcal{E}$ is hyperbolic. Here we generalise the theorem of [29], [30] which concerns ordinary differential equations, to equation (2.1).

Theorem 4.3 Lower-Semicontinuity of Attractors *Assume that:*

(I) equations (2.1), (2.6) *have global attractors* $\mathcal{A}$ *and* $\mathcal{A}^h$ *and, furthermore, that* $\mathcal{A}$ *is given by* (4.7) *where* $\mathcal{E}$ *comprises a finite number of hyperbolic equilibria;*

(II) for any $T, \delta, r > 0 \ \exists H_1 = H_1(T, \delta, r) > 0, C = C(T, r) > 0$ *such that*

$$|S(T)u - S^h(T)U|_\gamma \leq \delta + C|u - U|_\gamma,$$

for all $u \in \mathcal{B}_\gamma(0, r)$, $U \in \mathcal{B}_{h,\gamma}(0, r)$ *and* $h \in (0, H_1)$;

(III) for every $v \in \mathcal{E}$ *and* $\delta > 0 \ \exists r = r(v, \delta) > 0, H_2 = H_2(v, \delta) > 0$ *such that*

$$dist_\gamma(W^u(v, r), \mathcal{A}^h) \leq \delta$$

for all $h \in (0, H_2)$.

Then

$$dist_\gamma(\mathcal{A}, \mathcal{A}^h) \to 0 \ \ as \ \ h \to 0.$$

Proof. It is sufficient to prove that, given any $\epsilon > 0 \ \exists H > 0$ such that for every $y \in \mathcal{A}$, $\exists y^h \in \mathcal{A}^h$ with the property that $|y - y^h|_\gamma \leq 2\epsilon$ for $h \in (0, H)$. Since $\mathcal{A}$ is given by (4.7) we need only consider $y \in \overline{W^u(v)}$ for all $v \in \mathcal{E}$. Given $v \in \mathcal{E}$ we let r be given by assumption III.

Let $\partial\mathcal{B}_\gamma(v; r)$ denote the boundary of $\mathcal{B}_\gamma(v; r)$; that is

$$\partial\mathcal{B}_\gamma(v, r) = \{u \in X^\gamma : |u - v| = r\}.$$

Now set

$$\mathcal{Q} = \overline{W^u(v; r)} \bigcap \partial\mathcal{B}_\gamma(v; r)$$

and

$$\mathcal{W} = W^u(v) \backslash \overline{W^u(v; r)}.$$

Then, for r sufficiently small,

$$\mathcal{W} = \bigcup_{t>0} S(t)\mathcal{Q}.$$

Note that for every $\tau > 0$ and every point $u \in \mathcal{Q}$ there exists a $u^- \in W^u(v; r)$ such that $u = S(\tau)u^-$. Using this fact, together with the fact that $\mathcal{W}$ is a union of trajectories all of which start on $\mathcal{Q}$, the relative compactness of (2.2) shows that $\overline{\mathcal{W}}$ is compact in X^γ. Note that $\{\mathcal{B}_\gamma(x, \epsilon) : x \in \mathcal{W}\}$ is an ϵ-cover for $\overline{\mathcal{W}}$ in X^γ and hence, since $\overline{\mathcal{W}}$ is compact in X^γ, we may extract a finite sub-cover. Denote this subcover by $\{B_i(\epsilon)\}_{i=1}^I$ and note that each $B_i(\epsilon)$ contains a point $y_i \in \mathcal{W}$, where $B_i(\epsilon) = \mathcal{B}(y_i, \epsilon)$. By construction $\exists x_i \in \mathcal{Q}$ and

$T_i > 0$ such that $S(T_i)x_i = y_i$ for each $y_i \in \mathcal{W}$. Now assumption III implies that, for any $\delta > 0, \exists x_i^h \in \mathcal{A}^h$ and $H_2^i = H_2^i(v, \delta) > 0$ such that

$$|x_i - x_i^h|_\gamma \leq \delta \quad \forall h \in (0, H_2^i],$$

and by the invariance of $\mathcal{A}^h$ it follows that $y_i^h = S^h(t)x_i^h \in \mathcal{A}^h$. It now follows from assumption II that there exists $H_1^i = H_1^i(v, \delta) > 0$ such that

$$|y_i - y_i^h|_\gamma = |S(T_i)x_i - S^h(T_i)x_i^h| \leq (1 + C)\delta \quad \forall h \in (0, \min\{H_1^i, H_2^i\}].$$

Since I is finite, we deduce that $\exists \{y_i^h\}_{i=1}^I$ and $H_3 = H_3(v, \epsilon)$ such that

$$\max_{1 \leq i \leq I} |y_i - y_i^h|_\gamma \leq \epsilon \quad \forall h \in (0, H_3].$$

Thus, for every $y \in B_i(\epsilon)$ and $i : 1 \leq i \leq I$ there exists $y_i^h \in \mathcal{A}^h$ such that

$$|y - y_i^h|_\gamma \leq 2\epsilon \quad \forall h \in (0, H_3].$$

Since the $B_i(\epsilon), i = 1, \ldots, I$ form a cover of $\overline{\mathcal{W}}$ we deduce that

$$\text{dist}_\gamma\{\overline{\mathcal{W}}, \mathcal{A}^h\} \leq 2\epsilon \quad \forall v \in (0, H_3].$$

Noting that there are only a finite number of v defining the attractor through (4.7) and using assumption III we deduce that $\exists H = H(\epsilon)$:

$$\text{dist}_\gamma\{\mathcal{A}, \mathcal{A}^h\} \leq 2\epsilon \quad \forall h \in (0, H]$$

and the result follows. □

Remarks

(i) to establish III it is sufficient to show that, for every $v \in \mathcal{E}$ $\exists r = r(v, \delta) > 0$ and $H_2 = H_2(v, \delta) > 0$ and $V \in \mathcal{E}^h, r' > 0$ such that

$$\text{dist}_\gamma(W^u(v, r), W_h^u(V, r')) \leq \delta \tag{4.8}$$

for all $h \in (0, H)$. This is since the unstable manifold of a fixed point is necessarily contained in the global attractor.

(ii) if assumption III is replaced by (4.8) then the method of proof for Theorem 4.3 shows that the closure of the unstable manifold of an equilibrium point is lower-semicontinuous with respect to numerical perturbations.

(iii) it is not difficult to generalise Theorem 4.3 to include the effect of time-discretization. □

This concludes our analysis of the upper and lower semicontinuity of attractors with respect to numerical perturbations. Two important points have been established. Firstly, lower semicontinuity will not hold in general and thus it is natural to consider the effect of numerical approximation on more robust objects; this motivates the study of inertial manifolds which contain the global attractor and perturb smoothly. Secondly, the cases in which lower semicontinuity can be established all require proving the lower semicontinuity of local unstable manifolds of hyperbolic equilibria. Thus, in the final section we consider inertial and unstable manifolds under perturbation.

5 Attractive Invariant Manifolds

Motivated by the remarks at the end of the previous section, we now study the effect of numerical approximation on attractive invariant manifolds. We start by showing how the problems of the existence of both inertial manifolds and unstable manifolds can be formulated in the same framework – a framework which requires the study of globally Lipschitz mappings in a Banach space. We then state and sketch the proof of an abstract theorem concerning perturbation of attractive invariant manifolds for such globally Lipschitz mappings.

Inertial Manifolds.
Consider (2.1) in the case where the operator A is self-adjoint. For our discussion of inertial manifolds we assume that there exists $\beta \in (0,1)$ such that the operator F in (2.1) satisfies the following conditions: $\exists \gamma \geq 0, \beta \in [0,1)$ and $E(\sigma) > 0$ such that $F : X^\gamma \to X^{\gamma-\beta}$ satisfies

$$\begin{aligned} |F(u)|_{\gamma-\beta} &\leq E(\sigma) \quad \forall u \in \mathcal{B}_\gamma(0,\sigma) \\ |F(u) - F(v)|_{\gamma-\beta} &\leq E(\sigma)|u-v|_\gamma \quad \forall u, v \in \mathcal{B}_\gamma(0,\sigma); \end{aligned} \tag{5.1}$$

We introduce the following decomposition of X :

$$X = Y \oplus Z, \quad Y = span\{\psi_1, \psi_2, \ldots, \psi_q\}, \quad Z = Y^\perp$$

where $A\varphi_i = \lambda_i \varphi_i$ (as in section 2) and the orthogonal complement is taken in X. We also denote by P and Q the projections $P : X \mapsto Y$ and $Q : X \mapsto Z$.

Now assume that (2.1) has an absorbing set $\mathcal{B}_\gamma(0,\rho)$. Then, if we are interested only in the long-time behaviour of (2.1), it is sufficient to modify F outside $\mathcal{B}_\gamma(0,\rho)$ in a smooth fashion to obtain a globally bounded and Lipschitz function. Doing this in the standard fashion (see [17]) we see that the following holds: $\exists \gamma \geq 0, \beta \in [0,1)$ and $E > 0$ such that $F : X^\gamma \to X^{\gamma-\beta}$ satisfies

$$\begin{aligned} |F(u)|_{\gamma-\beta} &\leq E \quad \forall u \in X^\gamma \\ |F(u) - F(v)|_{\gamma-\beta} &\leq E|u-v|_\gamma \quad \forall u, v \in X^\gamma \end{aligned} \tag{5.2}$$

Thus we consider equation (2.1) under (5.2). Again a semigroup $S(t) : X^\gamma \mapsto X^\gamma$ may be defined so that $u(t) = S(t)u_0$.

An inertial manifold for (2.1) under (5.2) is a set $\mathcal{M}$ defined by a graph $\Phi \in C(Y, Z)$ so that

$$\mathcal{M} = \{u \in X^\gamma : Qu = \Phi(Pu)\}$$

satisfies

$$u(0) \in \mathcal{M} \Rightarrow u(t) \in \mathcal{M} \ \forall t \in \mathbb{R}$$

and, furthermore,

$$\exists C, \mu > 0 : \ \mathrm{dist}_\gamma(S(t)u_0, \mathcal{M}) \leq Ce^{-\mu t}\mathrm{dist}_\gamma(u_0, \mathcal{M}) \quad \forall t \geq 0, \ u_0 \in X^\gamma.$$

The existence of inertial manifolds was first considered in [17] using an approach based on the Lyapunov-Perron method familiar from the construction of center manifolds. Related results concerning the existence of inertial manifolds can be found in, for example, [38] [7], [8], [14]. Here we outline a different proof for the existence of inertial manifolds which is particularly convenient for the consideration of numerical perturbations; see [32] for details.

Let $U_m = S(mT)u_0$. Then use of the variation of constants formula in (2.1) shows that

$$U_{m+1} = G(U_m) \tag{5.3}$$

where $G : X^\gamma \mapsto X^\gamma$ satisfies

$$G(u) = Lu + N(u), \quad L := e^{-AT}, \quad N(u) = \int_0^T e^{-A(T-s)} F(S(s)u)ds. \tag{5.4}$$

Furthermore, using the spectral properties of A on Y and Z, together with (5.2) and the smoothing properties of A, shows that there exist positive constants a, b, c, B such that **Assumptions G** hold:

$$|Lz|_\gamma \leq a|z|_\gamma \qquad \forall z \in Z; \tag{G1}$$

$$\exists! w \in Y : Lw = p, \ \forall p \in Y \ \ \& \ \ b|y|_\gamma \leq |Ly|_\gamma \leq c|y|_\gamma \quad \forall y \in Y; \tag{G2}$$

$$|\mathcal{R}(N(u) - N(v))|_\gamma \leq B|u - v|_\gamma \ \ \forall u, v \in X^\gamma, \quad |\mathcal{R}N(u)|_\gamma \leq B \ \ \forall u \in X^\gamma, \tag{G3}$$

where $\mathcal{R}$ equals either I, P or Q.

In particular we have

$$a = e^{-\lambda_{q+1}T}, \ b = e^{-\lambda_q T}, \ c = e^{-\lambda_1 T}, \ B \propto T^{1-\beta}. \tag{5.5}$$

If we let $p_m = Pu_m$ and $q_m = Qu_m$ then the graph Φ representing the inertial manifold must satisfy

$$p = L\xi + PN(\xi + \Phi(\xi)) \tag{5.6}$$

$$\Phi(p) = L\Phi(\xi) + QN(\xi + \Phi(\xi)) \tag{5.7}$$

together with the attractivity condition

$$|q_m - \Phi(p_m)|_\gamma \leq Ce^{-\mu mT}|q_0 - \Phi(p_0)|_\gamma. \tag{5.8}$$

To construct Φ we seek a fixed point of the mapping $\mathcal{T} : C(Y, Z) \mapsto C(Y, Z)$ defined by

$$p = L\xi + PN(\xi + \Phi(\xi)) \tag{5.9}$$

$$(\mathcal{T}\Phi)(p) = L\Phi(\xi) + QN(\xi + \Phi(\xi)). \tag{5.10}$$

Any $\Phi \in C(Y, Z)$ satisfying (5.6), (5.7), (5.8) is only an inertial manifold for the time T flow of the equation. It is necessary to show *a posteriori* that it is also invariant for the equation (2.1) for every $t > 0$; see [32] for details.

This completes our set-up of the problem of finding inertial manifolds as the graph of a function relating projections of the solution of a globally Lipsctiz mapping in a Banach space. After showing that the problem of existence for unstable manifolds may be set in the same framework, we consider the existence of Φ satisfying (5.6), (5.7) and (5.8) together with the effect of numerical perturbation. $\square$

Unstable Manifolds.
Consider the equation

$$v_t + \mathcal{C}v = g(v), \quad v(0) = v_0 \tag{5.11}$$

where $\mathcal{C}$ is a densely defined, positive-definite, sectorial operator on X. Thus we may define the Banach spaces $\mathcal{X}^\alpha := D(\mathcal{C}^\alpha)$ with norm $\|u\|_\alpha := |\mathcal{C}^\alpha u|, \alpha \in [0,\infty)$. We assume further that $g : \mathcal{X}^\gamma \mapsto X$ is locally Lipschitz for some $\gamma \in [0,1)$. Let (5.11) have an equilibrium point $\bar{v} \in \mathcal{X}^\gamma$ satisfying

$$\mathcal{C}\bar{v} = g(\bar{v}). \tag{5.12}$$

Introducing $u = v - \bar{v}$ we may write (5.11) as

$$u_t + Au = F(u), \quad u(0) = u_0 := v_0 - \bar{v} \tag{5.13}$$

where

$$\begin{aligned} A &:= \mathcal{C} - dg(\bar{v}), \\ F(u) &:= g(\bar{v}+u) - g(\bar{v}) - dg(\bar{v})u, \end{aligned} \tag{5.14}$$

We assume that the operator A has spectrum $\{\lambda_i\}_{i=1}^\infty$ satisfying

$$Re(\lambda_1) \leq \ldots \leq Re(\lambda_q) < 0 < Re(\lambda_{q+1}) \leq \ldots$$

for some integer $q \geq 1$. This ensures that $\bar{v}$ is an unstable equilibrium point of (5.11) so that the unstable manifold is non-trivial.

We assume that $dg(\bar{v})$ is a bounded linear map from $\mathcal{X}^\gamma$ to X so that $|dg(\bar{v})\mathcal{C}^{-\gamma}|$ is bounded; this implies that A is sectorial – see Corollary 1.4.5 of [23]. Thus we may construct from A the Banach spaces X^γ as in section 2. From this it follows, by application of Theorem 1.4.8 in [23], that the Banach spaces $\mathcal{X}^\alpha$ and X^α are equivalent and that $\| \bullet \|_\alpha \approx | \bullet |_\alpha$.

Finally, we assume that there exists $E(\rho)$ with $E(\rho) \to 0$ as $\rho \to 0_+$ such that, for all $u_1, u_2 \in \mathcal{B}_\gamma(0,\rho)$

$$|F(u_1) - F(u_2)| \leq E(\rho)|u_1 - u_2|_\gamma.$$

This guarantees the existence of a local solution $u(t) \in X^\gamma$ to (5.13) in the neighbourhood of $u = 0$ (equivalently for (5.11) in a neighbourhood of $v = \bar{v}$). As in the inertial manifold case we can modify the function g outside a small neighbourhood of $\bar{v}$, and hence F outside a small neighbourhood of 0, in such a way that (5.13) now satisfies the condition that there exists $E(\rho)$ with $E(\rho) \to 0$ as $\rho \to 0_+$ such that, for all $u_1, u_2 \in \mathcal{X}^\gamma$

$$|F(u_1) - F(u_2)| \leq E(\rho)|u_1 - u_2|_\gamma.$$

As for the construction of inertial manifolds we introduce a splitting of the space $X = Y \oplus Z$, where now Y and Z are found by spectral projections associated with the spectral sets $Re(\lambda) \leq Re(\lambda_q)$ and $Re(\lambda) \geq Re(\lambda_{q+1})$ respectively. As before $P : X \mapsto Y$ and $Q : X \mapsto Z$. By a similar process of using the variation of constants formula, we obtain (5.3), (5.4) where, once again, Assumptions G are satisfied. In this case, by appropriate choice of T, it may be shown that

$$a = \frac{1}{2}, \quad b = 2, \quad c \geq 2, \quad B = B(\rho) \to 0 \; as \; \rho \to 0_+. \tag{5.15}$$

Again the unstable manifold may be found as a graph $\Phi \in C(Y, Z)$ satisfying (5.6), (5.7). Thus it is possible to seek such a Φ as a fixed point of the mapping $\mathcal{T}$ defined by (5.9), (5.10).

This completes our set-up of the problem of finding unstable manifolds as the graph of a function relating projections of the solution of a globally Lipschitz mapping in a Banach space. □

A numerical approximation of (2.1) or (5.11) will yield a mapping

$$U^h_{m+1} = G^h(U^h_m) \tag{5.16}$$

where $G^h : X^{h,\gamma} \mapsto X^{h,\gamma}$ is defined by

$$G^h(u) := L^h u + N^h(u). \tag{5.17}$$

We will not be more specific about the definition of L^h and N^h since this depends upon whether inertial manifolds or unstable manifolds are being considered. We simply assume that the space $\mathcal{V}$ may be decomposed as $\mathcal{V} : Y^h \oplus Z^h$ and introduce the projections $P^h : \mathcal{V} \mapsto Y^h, \quad Q^h : \mathcal{V} \mapsto Z^h$. Here Y^h and Z^h may be considered as approximations to Y and Z. We denote by E^h the $X-$projection $E^h : X \mapsto \mathcal{V}$.

For a wide variety of numerical methods it is possible to show that the following **Assumptions** G^h are satisfied: there exist positive constants a, b, c, B, C and $C(\rho)$ such that:

$$|L^h z|_{h,\gamma} \leq a|z|_{h,\gamma} \qquad \forall z \in Z^h; \tag{G^h1}$$

$$\exists! w^h \in Y^h : L^h w^h = q^h, \forall q^h \in Y^h \;\; \& \;\; b|y|_{h,\gamma} \leq |L^h y|_{h,\gamma} \leq c|y|_{h,\gamma} \quad \forall y \in Y^h; \tag{G^h2}$$

$$|\mathcal{R}(N^h(u) - N^h(v))|_{h,\gamma} \leq B|u - v|_{h,\gamma} \;\; \forall u, v \in V^h \quad |\mathcal{R}(N^h(u))|_{h,\gamma} \leq B \;\; \forall u \in \mathcal{V} \tag{G^h3}$$

where $\mathcal{R}$ equals either I, P^h or Q^h;

$$|P - P^h|_\gamma \leq Ch; \tag{G^h4}$$

$$|G(u) - G^h(u^h)|_\gamma \leq C(\rho)(h + |u - u^h|_\gamma) \; \forall u \in \mathcal{B}_\gamma(0, \rho), \; u^h \in \mathcal{B}_\gamma(0, \rho) \bigcap \mathcal{V}; \tag{G^h5}$$

$$|E^h|_\gamma, |P|_\gamma, |P^h|_\gamma \leq C; \tag{G^h6}$$

$$C^{-1}|u|_\gamma \leq |u|_{h,\gamma} \leq C|u|_\gamma \quad \forall u \in \mathcal{V}. \tag{G^h7}$$

The choice of the constants a, b, c and B as being the same as those in Assumptions G may be acheived without loss of generality. Assumptions (G^h1) – (G^h3) are analogous to Assumptions $(G1)$ – $(G3)$ whilst Assumptions (G^h4) – (G^h7) concern the relationship between the approximate mapping and the original mapping.

As for the continuous case, an attractive invariant manifold can be represented as a graph $\Phi^h : C(Y^h, Z^h)$ satisfying

$$p = L^h\xi + PN^h(\xi + \Phi^h(\xi)) \quad (5.18)$$
$$\Phi^h(p) = L^h\Phi^h(\xi) + QN^h(\xi + \Phi^h(\xi)) \quad (5.19)$$

together with the attractivity condition

$$|q_m^h - \Phi(p_m^h)|_\gamma \leq Ce^{-\mu mT}|q_0^h - \Phi(p_0^h)|_\gamma. \quad (5.20)$$

Here $p_m = Pu_m$ and $q_m = Qu_m$.

The following theorem shows that provided certain conditions on a, b, c and B are satisfied (those yielding existence of attractive invariant manifolds for the mappings (5.3) and (5.16)), then upper and lower semicontinuity of attractive invariant manifolds may be shown under Assumptions G and G^h.

Theorem 5.1 Continuity of Invariant Manifolds *Assume that the mappings* (5.3) *and* (5.16) *satisfy Assumptions G and Assumptions G^h respectively. Assume further that there exist constants $\delta', \epsilon' \in (0, \infty)$, $\mu \in (0, 1)$ and $K \in (1, \infty)$ such that:*

$$b^{-1}B(1+\delta) \leq \mu, \quad (C1)$$

$$a\epsilon + B \leq \epsilon, \quad (C2)$$

$$\theta := a\delta + B(1+\delta) \leq \delta\phi, \quad (C3)$$

where $\phi := b - B(1+\delta) > 0$ by (C1) and

$$a + B(1+\delta) \leq \mu, \quad (C4)$$

for all $\delta \in (\delta', K\delta')$ and $\epsilon \in (\epsilon', K\epsilon')$. Then the mappings (5.3), (5.16) *both possess attractive invariant manifolds representable as graphs $\Phi : Y \mapsto Z$ and $\Phi^h : Y^h \mapsto Z^h$ respectively and satisfying* (5.6), (5.7), (5.8) *and* (5.18), (5.19), (5.20) *respectively. Furthermore if* either

$$c < 1 \;\&\; \exists r > 0 : |N(u)|_\gamma = 0 \qquad \forall u : |Pu|_\gamma \geq r \quad (5.21)$$

or

$$b > 1 \quad (5.22)$$

then:

(i) for any $p \in Y$ there exists $C(p) > 0$ such that

$$|(p + \Phi(p)) - (P^h p + \Phi^h(P^h p))|_\gamma \leq C(p)h;$$

(ii) for any $p^h \in Y^h$ there exists $C(p^h) > 0$ such that

$$|(Pp^h + \Phi(Pp^h)) - (p^h + \Phi^h(p^h))|_\gamma \leq C(p^h)h.$$

Sketch Proof The details of the proof can be found in [32]. The basic idea to establish existence is to use the contraction mapping theorem. Conditions (C1)–(C3) enable this for the mappings (5.3) and (5.16) under Assumptions $(G1)-(G3)$ and $(G^h1)-(G^h3)$ respectively. Condition (C4) yields the required exponential attraction. To prove the convergence result relating the true and approximate manifolds uses a modification of the uniform contraction principle, using (G^h5) to give the required continuity with respect to perturbations. However, since Φ and Φ^h are defined as graphs over different spaces this application of the uniform contraction principle is not entirely straightforward. The assumptions $(G^h4), (G^h6)$ and (G^h7) are used to get around this difficulty. □

Remarks

(i) The conditions (C1)–(C4) can be satisfied in the inertial manifold case provided that both λ_q and $\lambda_{q+1} - \lambda_q$ can be made sufficiently large. This is known as the *spectral gap condition* in [17] and identical conditions are derived in the existence proof sketched here; see [32]. For unstable manifolds (C1)–(C4) are satisfied since $a < 1 < b$ and B can be made arbitrarily small by choice of ρ.

(ii) The method of proof of the continuity result is a generalisation of that used by [5] to prove convergence of center-unstable manifolds in ordinary differential equations under numerical approximation. In the context of unstable manifolds for partial differential equations, a very similar *existence proof* can be found in [3].

(iii) The first proof concerning convergence of unstable manifolds under numerical approximation may be found in [4] where ordinary differential equations are considered. In [1] the effect of time-discretization on the unstable manifold of scalar reaction-diffusion equations is studied whilst in [36] the same question is considered under finite element spatial approximation. See also [21].

(iv) The convergence of inertial manifolds under spectral approximation based on the eigenfunctions of A is studied in [17], [18]. The same problem is considered for a specific time discretization in [9].

(v) The abstract framework described here used to study the existence and convergence of inertial manifolds in semi and fully discrete finite element approximations of a scalar reaction-diffusion equation and the Cahn-Hilliard equation in [32]. Furthermore, the existence and convergence of local unstable manifolds in a fully discrete reaction-diffusion equation is also analysed.

(vi) It is worth noting the different methods and assumptions employed in constructing invariant manifolds in approximation schemes. The papers by [9] and [36] are based on the Lyapunov-Perron type existence theory and, as such, require the derivation of non-standard error bounds over long-time intervals together with certain spectral approximation properties. The paper [1] is also based on a Lyapunov-Perron type approach and requires a C^1 approximation result over finite time intervals. The result of [32] described here is based on the Hadamard graph transform and requires standard C^0 error bounds on finite time intervals (G^h5) together with the closeness of certain spectral sets and their associated projections (G^h4).

References

[1] F. Alouges and A. Debussche, *On the qualitative behaviour of the orbits of a parabolic partial differential equations and its discretization in the neighbourhood of a hyperbolic fixed point*, Num. Func. Anal. and Opt. **12**(1991), 253–269.

[2] F. Armero and J. Simo, *Unconditional stability and long-term behaviour of transient algorithms for the incompressible Navier-Stokes and Euler equations.* To appear in Comp. Meth. Appl. Mech. and Eng.

[3] A. Babin and M.I. Vishik, *Attractors of evolution equations.* Studies in Mathematics and its Applications, North-Holland, Amsterdam, 1992.

[4] W.-J. Beyn, *On the numerical approximation of phase portraits near stationary points.* SIAM J. Num. Anal. **24**(1987), 1095–1113.

[5] W.-J. Beyn and J. Lorenz, *Center manifolds of dynamical systems under discretization.* Num. Func. Anal. and Opt.

[6] K. Burrage and J. Butcher, *Stability criteria for implicit Runge-Kutta processes.* SIAM J. Num. Anal., **16**(1979), 46–57.

[7] P. Constantin, C. Foias, B. Nicolaenko and R. Temam, *Integral manifolds and Inertial Manifolds for Dissipative Partial Differential Equations*, Appl. Math. Sciences, Springer Verlag, New York, 1989.

[8] P. Constantin, C. Foias, B. Nicolaenko and R. Temam, *Spectral barriers and inertial manifolds for dissiapative partial differential equations,* J. Dynamics Differential Eqs., **1**(1989), 45-73.

[9] F. Demengel and J.M. Ghidaglia, *Time-discretization and inertial manifolds*, Math. Mod. and Num. Anal, **23**(1989), 395-404.

[10] C. Devulder and M. Marion *A class of numerical algorithms for large time integration: the nonlinear Galerkin methods*, SIAM J. Numer. Anal., **29**(1990), 462-483.

[11] J. Eastwood and W. Arter, *Spurious behaviour of numerically computed fluid flow.* IMA J. Num. Anal. **7**(1987), 205–222.

[12] C.M. Elliott, *The Cahn-Hilliard model for the kinetics of phase separation.* Appears in "Mathematical models for phase change problems", edited by J.F. Rodrigues, Birkhauser, Berlin, 1989.

[13] C.M. Elliott and A.M. Stuart, *Global dynamics of discrete semilinear parabolic equations.* To appear in SIAM J. of Num. Anal.

[14] E. Fabes, M. Luskin and G. Sell, *Construction of inertial manifolds by elliptic regularization*, J. Diff. Eq., **89**(1991), 355-387.

[15] C. Foias, M.S. Jolly, I.G. Kevrekidis and E.S. Titi, *Dissipativity of numerical schemes.* Nonlinearity **4**(1991), 591–613.

[16] C. Foias, M.S. Jolly, I.G. Keverkidis and E.S. Titi, *On some dissipative fully discrete nonlinear Galerkin schemes for the Kuramoto-Sivashinsky equation*, appears in "Proc. of Athens First Interdisciplinary Olympia, November 1991, Athens, Greece." Edited by I. Babuska and L. Xanthis, (to appear).

[17] C. Foias, G. Sell and R. Temam, *Inertial manifolds for nonlinear evolutionary equations*, J. Diff. Eq., **73**(1988), 309-353.

[18] C. Foias, G. Sell and E.S. Titi, *Exponential tracking and approximation of inertial manifolds for dissipative nonlinear equations*, J. Dynamics and Diff. Eq., **1**(1989), 199-243.

[19] B. Fornberg, *On the instability of Leap-Frog and Crank-Nicolson approximations of a nonlinear partial differential equation.* Math. Comp. **27**(1973), 45–57.

[20] J.K. Hale, *Asymptotic Behaviour of Dissipative Systems.* AMS Mathematical Surveys and Monographs 25, Rhode Island, 1988.

[21] J.K. Hale and G. Raugel *Lower Semicontinuity of Attractors of Gradient Systems and Applications.* Annali di Mat. Pura. Applic. CLIV(1989), 281–326.

[22] J.K. Hale, X.-B. Lin and G. Raugel *Upper Semicontinuity of Attractors for Approximations of Semigroups and Partial Differential Equations.* Math. Comp.**50**(1988), 89–123.

[23] D. Henry, *Geometric Theory of Semilinear Parabolic Equations.* Lecture Notes in Mathematics, Springer-Verlag, New York, 1981.

[24] A.T. Hill and E. Suli. *Upper semicontinuity of attractors for linear multistep methods approximating sectorial evolution equations.* Submitted to Math. Comp. 1993.

[25] A.T. Hill and E. Suli. *Set convergence for discretizations of the attractor.* Submitted to Numer. Math. 1993.

[26] A.T. Hill, *Attractors for nonlinear convection diffusion problems and their approximation.* PhD. Thesis, Oxford University, 1992.

[27] A.T. Hill, *Global dissipativity for A−stable methods.* In preparation, 1993.

[28] A.R. Humphries and A.M. Stuart, *Runge-Kutta methods for dissipative and gradient dynamical systems*, 1992. To appear in SIAM J. Num. Anal.

[29] A.R. Humphries, *Numerical Analysis of Dynamical Systems.* University of Bath, Phd thesis, 1993.

[30] A.R. Humphries, *Approximation of attractors and invariant sets by Runge-Kutta methods.* In preparation, 1993.

[31] M.S. Jolly, I.G. Kevrekidis and E.S. Titi, *Preserving dissipation in approximate inertial forms for the Kuramoto-Sivashinsky equation,* J. Dynam. and Diff. Eq., **3**(1990), 179-197.

[32] D.A. Jones and A.M.Stuart *Attractive Invariant Manifolds Under Approximation.* In preparation, 1993.

[33] P. Kloeden and J. Lorenz. *Stable attracting sets in dynamical systems and their one-step discretizations,* SIAM J. Num. Anal. **23**(1986), 986–995.

[34] P. Kloeden and J. Lorenz. *Liapunov stability and attractors under discretization.* Appears in "Differential Equations, proceedings of the equadiff conference", edited by C.M. Dafermos, G. Ladas and G. Papanicolaou. Marcel-Dekker, New York, 1989.

[35] S. Larsson. *Non-smooth data error estimates with applications to the study of long-time behaviour of finite element solutions of semilinear parabolic problems.* Pre-print, Chalmers University, Sweden.

[36] S. Larsson and J.M. Sanz-Serna. *The behaviour of finite element solutions of semilinear parabolic problems near stationary points.* Submitted to SIAM J. Num. Anal.

[37] G.J. Lord and A.M.Stuart, *Existence and Convergence of Attractors and Inertial Manifolds for a Finite Difference Approximation of the Ginzburg-Landau equation.* In preparation, 1993.

[38] J. Mallet-Paret and G. Sell, *Inertial manifolds for reaction-diffusion equations in higher space dimension,* J. Amer. Math. Soc., **1**(1988), 805-866.

[39] Pazy *Semigroups of linear operators and applications to partial differential equations.* Springer-Verlag, New York, 1983.

[40] V. Pliss *Nonlocal problems in the theory of oscillations.* Academic Press, 1966.

[41] M. Marion and R. Temam, *Nonlinear Galerkin methods,* SIAM J. Num. Anal. **26**(1990), 1139-1157.

[42] J. Shen, *Convergence of approximate attractors for a fully discrete system for reaction-diffusion equations,* Numer. Funct. Anal. and Opt. **10**(1989), 1213–1234.

[43] A.M. Stuart and A.R. Humphries, *Model problems in numerical stability theory for initial value problems,* 1992. Submitted to SIAM Review.

[44] R. Temam, *Infinite Dimensional Dynamical Systems in Mechanics and Physics. Springer, New York, 1989.*

[45] Yin-Yan. *Attractors and error estimates for discretizations of incompressible Navier-Stokes equations.* Submitted to SIAM J. Num. Anal.

[46] T. Yoshizawa, *Stability Theory by Lyapunov's Second Method,* Publ. Math. Soc. Japan, Tokyo, 1966.

Acknowledgement
This research was supported by the NSF contract DMS-9201727 and ONR contract N00014-92-J-1876.

Anthony Humphries
Department of Mathematics
University Walk
Bristol University, BS8 1TW
UK

Don Jones
Center for Turbulence Research
Stanford University, CA94305
USA

Andrew Stuart
Program in Scientific Computing and Computational Mathematics
Division of Applied Mechanics
Durand 252
Stanford University, CA94305
USA

N K NICHOLS

Differential-algebraic equations and control system design

Abstract Techniques for the robust design of automatic feedback controllers and state estimators for systems governed by implicit linear dynamic–algebraic equations are investigated. Two computational procedures for achieving robust designs are presented: singular value and eigenstructure assignment. The procedures are based on stable decompositions of the system matrices using unitary transformations.

1. INTRODUCTION

Many control systems that arise in practice can be described by implicit dynamic–algebraic equations of the form

$$E \frac{dx}{dt} = A\, x(t) + B\, u(t) \tag{1}$$

$$y(t) = C\, x(t) \tag{2}$$

or, in the discrete–time case, of the form

$$E\, x(k+1) = A\, x(k) + B\, u(k) \tag{3}$$

$$y(k+1) = C\, x(k+1)\,. \tag{4}$$

Here the vector $x \in \mathcal{R}^n$ defines the state of the system at time t (or t_k), $u \in \mathcal{R}^m$ is the control or input to the system, and $y \in \mathcal{R}^p$ is the measured output from the system, where $m, p \leq n$. It is assumed that the matrices B, C are of full rank. The matrix E may be *singular*. Examples of such systems, called *descriptor* or *generalized state–space* systems, occur in a variety of contexts, including aircraft guidance, chemical processing, mechanical body motion, power generation, network fluid flow and many others.

The behaviour of a descriptor system is governed by the generalized eigenstructure of the matrix pencil

$$\alpha E - \beta A\,, \qquad (\alpha, \beta) \in \mathcal{C} \times \mathcal{C}\,. \tag{5}$$

The response of the system can have a complicated structure and may even contain impulsive modes. Numerical methods are needed to assess the properties of these systems and to aid system design. The main computational technique used for both system analysis and design synthesis is the reduction of the system by unitary transformation to a condensed matrix form. Such forms reveal both the structure of the system and the degrees of freedom available for the design. The synthesis problems are generally under–determined, and it is desirable to select the free parameters to give *robust* designs that are insensitive to plant disturbances and model uncertainties.

In this paper we examine two computational tools for achieving robust designs – eigenstructure assignment and singular value assignment. These methods are used in solving two basic synthesis problems: the design of automatic controllers by proportional–plus–derivative feedback and the design of observers (or state–estimators) for estimating the states of the system from measured data. In the next section we define the basic design synthesis problems. In Section 3 we examine properties of the system. In Sections 4 and 5 we develop robust design procedures based on singular value and eigenstructure assignment, respectively. Conclusions are given in Section 6.

Throughout the paper we consider continuous–time systems of form (1)–(2) only. All the results presented here also hold, however, with minor modifications, for discrete–time systems (3)–(4).

2. CONTROL SYSTEM DESIGN

The aim of a feedback controller is to ensure that the system responds automatically in a required manner to any given reference input. This is achieved by altering the sytem dynamics by 'feeding back', through the control input, information on the current state of the system, thus creating a new 'closed–loop' system.

If all the states x of the system (1)–(2) and their derivatives $\dot{x}$ can be measured (i.e. $C = I$), then a full proportional–plus–derivative state feedback can be used. The input to the system is taken to be

$$u = Fx - G\dot{x} + r\,, \tag{6}$$

where $F, G \in \mathcal{R}^{m \times n}$ are the feedback matrices to be selected and $r \in \mathcal{R}^m$ is the reference input vector. Substituting for u in (1) and rearranging gives the closed loop system equations

$$(E + BG)\dot{x} = (A + BF)x + Br\,. \tag{7}$$

The matrices F and G must be chosen to ensure that the new closed loop matrix pencil

$$\alpha(E + BG) - \beta(A + BF) \tag{8}$$

has the desired properties.

In practice all of the states of a system cannot generally be measured. In this case an auxilliary dynamical system, known as an observer, or state–estimator can be constructed to provide estimates $\hat{x}$ for all the states x of the system (1)–(2) from the measured data y:

$$E\dot{\hat{x}} = A\hat{x} + Bu + F(C\hat{x} - y) - G(C\dot{\hat{x}} - \dot{y})\,. \tag{9}$$

The observer is driven by the differences between the measured system outputs and their derivatives y and $\dot{y}$ and the estimated values $C\hat{x}$ and $C\dot{\hat{x}}$. Rearranging (9) gives the system equations for the observer

$$(E + GC)\dot{\hat{x}} = (A + FC)\hat{x} + Bu - Fy + G\dot{y} \tag{10}$$

with the system pencil

$$\alpha(E + GC) - \beta(A + FC) . \tag{11}$$

The matrices $F, G \in \mathscr{R}^{n \times p}$ must now be selected to ensure that the response of the observer $\hat{x}(t)$ converges to the system state $x(t)$ for any arbitrary starting conditions, that is, system (10) must be asymptotically *stable.* The convergence should be rapid and the converged estimate $\hat{x}$ should then track the true state x closely.

We remark that the design of an observer is equivalent to the design of an automatic controller for the *dual* system

$$\begin{aligned} E\dot{e} &= Ae + v \\ w &= Ce , \end{aligned}$$

using a feedback of the form

$$v = Fw - G\dot{w} + r .$$

The corresponding closed loop system is

$$(E + GC)\dot{e} = (A + FC)e + r$$

with system pencil given by (11), which is equivalent to the observer (10) with an appropriate choice for the reference r.

An automatic controller for the state system (1)–(2) can be obtained in the case $C \neq I$ by combining the system (1) with the observer system (10) and feeding back the estimated states $\hat{x}$ and derivatives $\dot{\hat{x}}$. A closed loop system of twice the dimension of the original system is derived. The over–all response of the closed loop system is controlled by selecting the free matrices in both the observer and the controller appropriately.

Alternatively a controller for the system (1)–(2) where $C \neq I$ can be obtained by 'feeding back' the measured output data directly. The input is taken to be

$$u = Fy - G\dot{y} + r \equiv FCx - GC\dot{x} + r , \tag{12}$$

giving the closed loop system

$$(E + BGC)\dot{x} = (A + BFC)x + Br . \tag{13}$$

The response of the closed loop system is determined by the properties of the matrix pencil

$$\alpha(E + BGC) - \beta(A + BFC) . \tag{14}$$

The results that may be achieved by selecting the feedback matrices $F,G \in \mathscr{R}^{m \times p}$ are now restricted, however, in comparison with those that may be achieved with full state feedback.

In summary, the objective of these system design problems is to choose matrices F and G such that the matrix pencil (14) has desired properties, which ensure the appropriate response of the system. The state feedback control problem, where $C = I$, and the state estimator problem, where $B = I$, are special cases. For *robust* designs it is necessary to ensure also that the properties of the pencil (14) are insensitive to perturbations in the system matrices E, A, B and C.

In the next section we examine the properties of descriptor systems and formulate design objectives.

3. PROPERTIES OF DESCRIPTOR SYSTEMS

The descriptor system (1)–(2) and its corresponding matrix pencil (5) are said to be *regular* if

$$\det(\alpha E - \beta A) \neq 0 \quad \text{for some} \quad (\alpha,\beta) \in \mathscr{C} \times \mathscr{C} \setminus \{(0,0)\} . \tag{15}$$

Regularity of the system guarantees the existence and uniqueness of classical solutions to (1) [6] [16].

For a regular system, the solutions to (1) can be characterized in terms of the eigenstructure of the pencil. The generalized eigenvalues are defined by the pairs $(\alpha_j, \beta_j) \in \mathscr{C} \times \mathscr{C} \setminus \{(0,0)\}$ such that

$$\det(\alpha_j E - \beta_j A) = 0 , \qquad j = 1, 2, 3 \ldots . \tag{16}$$

If $\beta_j \neq 0$, then $\lambda_j = \alpha_j/\beta_j$ is a *finite* eigenvalue, and if $\beta_j = 0$, then $\lambda_j \sim \infty$ is an infinite eigenvalue of the system. The right and left generalized eigenvectors and principal vectors are given by the columns of the non–singular matrices $X = [X_1, X_2]$ and $Y = [Y_1, Y_2]$ (respectively) that transform the pencil into the *Kronecker canonical form* (KCF)

$$Y^T E X = \begin{bmatrix} I_r & 0 \\ 0 & N \end{bmatrix}, \qquad Y^T A X = \begin{bmatrix} J & 0 \\ 0 & I_{n-r} \end{bmatrix}, \tag{17}$$

where J is the $r \times r$ Jordan matrix associated with the $r \leq \text{rank}(E) \equiv q$ finite eigenvalues of the pencil and N is the nilpotent Jordan matrix corresponding to the $n-r$ infinite eigenvalues [8]. The *index* of the system is defined to be equal to the degree of nilpotency of the matrix N; that is, the index is equal to k, the smallest non–negative integer such that $N^k = 0$.

For a regular system, the solution to (1) is given explicitly in terms of the KCF by

$$x(t) = X_1 z_1(t) + X_2 z_2(t) , \tag{18}$$

where

$$z_1(t) = e^{tJ} z_1(0) + \int_0^t e^{(t-s)} Y_1^T Bu(s)ds , \qquad z_1(0) \in \mathscr{R}^r ,$$

$$z_2(t) = -\sum_{i=0}^{k-1} N^i Y_2^T Bu^{(i)}(t) .$$

It is easy to see that for the solution to be continuous, the input function u must be such that $d^i[N^i Y_2^T Bu(t)]/dt^i$ exists and is continuous for all $i = 1, 2, \ldots k-1$, where k is the index of the system.

The index of a regular system is $k = 0$, by convention, if and only if rank $(E) = n$. The index k is less than or equal to one if and only if rank $[E, AS_\infty] = n$ or, equivalently, rank $[E^T, A^T T_\infty] = n$, where the columns of S_∞ and T_∞ span the null spaces of E and E^T, respectively. In this case the pencil has precisely $q \equiv$ rank (E) finite eigenvalues and $n - q$ *non–defective* infinite eigenvalues [7], [10].

An example of a regular system of index one is given by the semi–implicit equations

$$\begin{bmatrix} E_{11} & 0 \\ 0 & 0 \end{bmatrix} \begin{bmatrix} \dot{x}_1 \\ \dot{x}_2 \end{bmatrix} = \begin{bmatrix} A_{11} & A_{12} \\ A_{21} & A_{22} \end{bmatrix} \begin{bmatrix} x_1 \\ x_2 \end{bmatrix} + \begin{bmatrix} B_1 \\ B_2 \end{bmatrix} u , \tag{19}$$

where E_{11} and A_{22} are non–singular. The first block row of equations describes the dynamical behaviour of the system, while the second block row gives algebraic constraints on the states. For systems of this type, the algebraic conditions can be eliminated to give a purely dynamical *explicit* linear system. Since A_{22} is of full rank, we may write

$$x_2 = - A_{22}^{-1} (A_{21} x_1 + B_2 u) .$$

Since E_{11} is also of full rank, the system (19) then reduces to the explicit system

$$\dot{x}_1 = E_{11}^{-1} (A_{11} - A_{12} A_{22}^{-1} A_{21}) x_1 + E_{11}^{-1} (B_1 - A_{12} A_{22}^{-1} B_2) u . \tag{20}$$

We remark that the reduction to explicit form is *not* numerically reliable if E_{11}, A_{22} are ill–conditioned with respect to inversion.

Any regular system that has index at most one can, in fact, always be unitarily transformed and separated into a purely dynamical and a purely algebraic part, and the algebraic variables can be eliminated to give an explicit system of (possibly) reduced order. Higher index descriptor systems cannot be reduced to explicit systems in this way, and impulses can arise in the response if the control is not sufficiently smooth. The system can even lose causality [15] [4]. The eigenstructure of higher index systems is also necessarily less robust with respect to perturbations than systems of index at most one, since higher index systems always have *defective* multiple eigenvalues at infinity [10].

It is desirable, therefore, to design systems that are regular and of index at most one. In practice, there exist physical systems that do *not* have these properties. Such systems can, however, often be made regular and of index at most one by appropriate choice of feedback designs. Systems that *are* regular and of index at most one can, on the other hand, *lose* these properties under linear feedback. It is important, therefore, to establish conditions that ensure regularity/index ≤ 1 under feedback, and to develop numerically reliable techniques for constructing regular systems of index at most one. In the next section we give algebraic conditions that enable the regularization of a system by feedback, and describe a singular value assignment technique for designing *robust* systems that are regular and of index at most one.

Regularity and index ≤ 1 are not sufficient properties to define a satisfactory design, however. In general, we also require the system to be asymptotically stable; that is, we require the response of the system to a constant reference input to converge asymptotically to a constant state of equilibrium from any intitial state. This property holds if the finite eigenvalues of the system lie in the left–half complex plane (see [5]). In order to "shape" the response more explicitly we may wish to assign a *specific* set of (stable) eigenvalues to the system, thus guaranteeing a given modal behaviour. In Section 5 we describe methods for achieving *robust* eigenstructure assignment by feedback. The algebraic "regularizability" conditions given in Section 4 are sufficient to enable the system to be made stable as well as regular and index ≤ 1, and to permit arbitrary assignment of the finite eigenvalues of the system.

For full flexibility a combination of the singular value and eigenstructure assignment techniques can be used in practice to obtain closed loop system designs.

4. REGULARIZATION BY SINGULAR VALUE ASSIGNMENT

The aim of the system design problem is now:

> *Given real system matrices* E, A, B, C, *select real matrices* F *and* G *such that the closed loop pencil*
>
> $$\alpha(E + BGC) - \beta(A + BFC) \tag{21}$$
>
> *is regular and has index less than or equal to one.*

The pencil (21) has the required properties if and only if

$$\text{rank}\begin{bmatrix} E + BGC \\ T_\infty^T(A+BFC) \end{bmatrix} = n\,, \tag{22}$$

where the columns of T_∞ span the null space of $(E + BGC)^T$ [10]. For a *robust* solution to the problem we want the closed loop system to retain these properties under reasonable perturbations. We aim, therefore, to select matrices F and G to ensure that the matrix in (22) is as far from losing rank as possible under perturbations that preserve the range space of T_∞. Such perturbations preserve the space of admissible controls (see [5]).

It is well–known that for a matrix with full rank, the distance to the nearest matrix of lower rank is equal to its minimum singular value [9]. Hence, for robustness, we select F and G such that the pencil (21) is unitarily equivalent to a pencil of the form

$$a\begin{bmatrix} \Sigma_R & 0 \\ 0 & 0 \end{bmatrix} - \beta\begin{bmatrix} A_{11} & A_{12} \\ A_{21} & \Sigma_L \end{bmatrix}, \tag{23}$$

where the condition numers $cond(\Sigma_R)$ and $cond(\Sigma_L)$ are minimal. This choice maximises a lower bound on

$$\sigma_{\min}\left\{\begin{bmatrix} \Sigma_R & 0 \\ A_{21} & \Sigma_L \end{bmatrix}\right\} \equiv \sigma_{\min}\left\{\begin{bmatrix} E + BGC \\ T_\infty^T(A+BFC) \end{bmatrix}\right\}, \tag{24}$$

whilst retaining an upper bound on the magnitude of the gains F and G. This choice also ensures that the reduction of the closed loop descriptor system to an explicit (reduced order) system, as described in the previous section, is as well–conditioned as possible. In practice such robust systems also have improved performance characteristics (see [12] [14]).

The existence of solutions to the design problem can be established under simple conditions. The proof is based on a unitary transformation of the system to a condensed form that reveals both the structure of the system and the degrees of freedom available in the design. In the next subsection the conditions for regularizability are given, together with the main result. In the following subsections, the condensed system form is presented and a technique for selecting a robust solution to the design problem is described.

4.1 Conditions for Regularizability

Algebraic conditions that ensure regularizability of the descriptor system (1)–(2) are given by the following:

C1: $\operatorname{rank}\,[\lambda E - A, B] = n$ for all $\lambda \in \mathscr{C}$;

C2: $\operatorname{rank}\,[E, AS_\infty, B] = n$, where the columns of S_∞ span the null space of E;

01: $\operatorname{rank}\begin{bmatrix} \lambda E - A \\ C \end{bmatrix} = n$ for all $\lambda \in \mathscr{C}$;

02: $\operatorname{rank}\begin{bmatrix} E \\ T_\infty^T A \\ C \end{bmatrix} = n$, where the columns of T_∞ span the null space of E^T.

For systems that are regular, these conditions characterize the controllability and observability of the system. The conditions **C1/C2** and **01/02** guarantee that a regular system is *strongly controllable* and *strongly observable*, respectively. The conditions **C1** and **01**, together with the stronger conditions

$$\operatorname{rank}\,[E, B] = n, \quad \operatorname{rank}\,[E^T, C^T] = n \tag{25}$$

guarantee that a regular system is *completely* controllable and *completely* observable (see [16] [1] [2]). The conditions **C2** and **02** ensure 'controllability and observability at infinity' (see [15]). A regular system that has index at most one *always* satisfues **C2** and **02**.

The conditions **C1**, **C2**, **01** and **02** are all preserved under certain transformations of the system. Specifically, these conditions are all preserved under non–singular 'equivalence' transformations of the pencil and under proportional feedback. The conditions **C1** and **01** are also preserved under derivative feedback [1] [4].

The key result is given by the following:

THEOREM 1 *Given the real system matrices* E, A, B, C, *then* **C2** *and* **02** *hold if and only if there exist real matrices* F *and* G *such that*

$$\alpha(E + BGC) - \beta(A + BFC)$$

is regular with index at most one and

$$\mathrm{rank}\,(E + BGC) \;=\; r\,, \tag{26}$$

where rank(E) $\leq r \leq s \equiv$ rank(E) + t_2. (*Here* t_2 *is an integer determined by the decomposition of the system given in Theorem* 2).

If **C1** *and* **O1** *also hold, then* F *and* G *can be selected to ensure, in addition, that the corresponding closed loop system is strongly controllable and strongly observable.*

Proof. The proof follows by construction from the condensed form given in Theorem 2 of Section 4.2 [2]. □

We remark that the value of s in Theorem 1 is equal to n, the system dimension, if and only if the stronger conditions (25) hold [2]. In the special cases where $C = I$ (the state feedback control problem) and $B = I$ (the state estimator problem), the value of s is given by $s = \mathrm{rank}\,[E, B]$ and $s = \mathrm{rank}\,[E^T, C^T]$, respectively [1].

The value $r = s$ in Theorem 1 is attained in all cases by the feedback G alone, with $F = 0$. The value $r = \mathrm{rank}\,E$ is attained by feedback F alone, with $G = 0$. If **C2** and **O2** hold, solutions such that $r < \mathrm{rank}(E)$ may also exist, but the converse of the theorem does not necessarily hold [2].

4.2 A Condensed Form

The main result in Section 4.1 depends on the following.

THEOREM 2 *Given real system matrices* E, A, B, C, *there exist unitary matrices* U, V, W, Z *such that*

$$U^H E V = \begin{bmatrix} \Sigma_E & 0 \\ 0 & 0 \end{bmatrix}, \qquad U^H B W = \begin{bmatrix} B_{11} & B_{12} \\ \hat{B}_{21} & 0 \end{bmatrix}$$

$$Z^H C V = \begin{bmatrix} C_{11} & \hat{C}_{12} \\ C_{21} & 0 \end{bmatrix}, \qquad U^H A V = \begin{bmatrix} A_{11} & A_{12} \\ A_{21} & \hat{A}_{22} \end{bmatrix} \tag{27}$$

where

$$\hat{A}_{22} = \begin{bmatrix} A_{22} & A_{23} & A_{24} & 0 & 0 \\ A_{32} & A_{33} & A_{34} & \Sigma_{35} & 0 \\ A_{42} & A_{43} & \Sigma_{44} & 0 & 0 \\ 0 & \Sigma_{53} & 0 & 0 & 0 \\ 0 & 0 & 0 & 0 & 0 \end{bmatrix} \qquad \hat{B}_{21} = \begin{bmatrix} B_{21} \\ B_{31} \\ 0 \\ 0 \\ 0 \end{bmatrix}$$

$$\hat{C}_{12} = [\,C_{12} \;\; C_{13} \;\; 0 \;\; 0 \;\; 0\,] \tag{28}$$

and Σ_E, Σ_{35}, Σ_{44}, Σ_{53} *are non–singular, square diagonal matrices of dimensions* t_1, t_3, t_4 *and* t_5, *respectively.* B_{12} *and* C_{21} *are of full rank, and*

$$\begin{bmatrix} B_{21} \\ B_{31} \end{bmatrix} \quad \text{and} \quad [\, C_{12}, C_{13} \,]$$

are non–singular, square matrices of dimensions $t_2 + t_3$ *and* $s_2 + t_5$, *respectively.* (*All partitionings are compatible.*)

Proof. The proof is by construction via Algorithm 1 given in [2]. The construction uses a sequence of singular value (SVD) decompositions [9]. □

From the condensed form (27)–(28), it follows that the system pencil $\alpha E - \beta A$ is regular and of index at most one if and only if the matrix $\hat{A}_{22}$ is non–singular. Necessary and sufficient conditions for this to hold are that the last zero block rows and columns of $\hat{A}_{22}$ (and corresponding blocks of A_{21}, A_{22}) are empty (so that $t_2 = s_2$) and the square matrix A_{22} is non–singular.

From Theorem 2 it also follows that conditions **C2** and **02**, respectively, hold if and only if the last zero block rows and last zero block columns of $\hat{A}_{22}$ and $\hat{B}_{21}, \hat{C}_{12}$, respectively, are empty (together with the corresponding blocks of A_{21}, A_{22}). Furthermore, if these conditions hold, then it can be seen from the condensed form (27)–(28) that feedback matrices F and G can be constructed such that the closed loop pencil (21) is unitarily equivalent to a pencil of form (23), where Σ_R, Σ_L are square and non–singular and Σ_R is $r \times r$ with rank $\Sigma_R \leq r \leq s = t_1 + t_2$ (for details see [2]). These results, together with the fact that the conditions **C1, C2, 01** and **02** are preserved under equivalence transformations, effectively establishes Theorem 1.

In the next subsection we examine how the feedback matrices F, G can be constructed explicitly to give a *robust* solution to the regularization problem.

4.3 Robust Singular Value Assignment

To obtain a system pencil that is regular and of index at most one, the matrices F and G can be selected such that the pencil (21) is unitarily equivalent to a pencil of form (23) provided certain algebraic conditions hold, as described in the previous subsections. For a *robust* system design, we choose F and G to assign the singular values of the subsystems Σ_R, Σ_L in the equivalent system (23) to ensure that

$$cond\,(\Sigma_R) \equiv \sigma_{\max}\{\Sigma_R\}/\sigma_{\min}\{\Sigma_R\}\,, \quad cond\,(\Sigma_L) \equiv \sigma_{\max}\{\Sigma_L\}/\sigma_{\min}\{\Sigma_L\}$$

are as small as possible, and also to ensure reasonable gaps between singular values.

In general, not all singular values can be assigned arbitrarily. In the case of the state feedback controller and its dual, the state estimator problem, where $C = I$ and $B = I$, respectively, the complete singular value structure can be identified and an optimal solution can be found to the robust regularization problem. To establish the form of the optimal solution in the case $C = I$, it is convenient to reduce the form (27)–(28) still further. In this case, assuming **C2** holds so that $t_6 = 0 = s_6$ and $t_2 = s_2$, we find that $t_3 = 0 = t_4$ and, therefore, the third, fourth, and sixth block rows and fourth, fifth and sixth block columns of the form (27)–(28) are empty. We may then apply further unitary row and column operations to the first and third block rows and columns of E, A and B. It follows that there exist unitary matrices $\tilde{U}, \tilde{V}, \tilde{W}$ such that

$$\tilde{U}^H E \tilde{V} = \begin{bmatrix} \Sigma_E & 0 & 0 & 0 \\ E_{21} & E_{22} & 0 & 0 \\ 0 & 0 & 0 & 0 \\ 0 & 0 & 0 & 0 \end{bmatrix}, \qquad \tilde{U}^H B \tilde{W} = \begin{bmatrix} 0 & 0 \\ B_{21} & B_{22} \\ \Sigma_B & 0 \\ 0 & 0 \end{bmatrix},$$

$$\tilde{U}^H A \tilde{V} = \begin{bmatrix} A_{11} & A_{12} & A_{13} & A_{14} \\ A_{21} & A_{22} & A_{23} & A_{24} \\ A_{31} & A_{32} & A_{33} & A_{34} \\ A_{41} & A_{42} & 0 & \Sigma_A \end{bmatrix}, \tag{29}$$

where Σ_E, Σ_A and Σ_B are square, non–singular diagonal matrices of dimensions ℓ, $n - m - \ell$ and $m - t$ respectively, and E_{22}, B_{22} are square, non–singular matrices of dimension t. Here $\text{rank}(E) = \ell + t$ and $\text{rank}[E,B] = \ell + m$. (The form (29) can also be derived directly from the results of [1].)

From the decomposition (29) it can be seen that the matrices Σ_E and Σ_A cannot be altered by feedback. We can, however, select matrices F and G such that the system pencil $\alpha\tilde{U}^H(E + BG)\tilde{V} - \beta\tilde{U}^H(A + BF)\tilde{V}$ is equal to the pencil (23), where

$$\Sigma_R = \begin{bmatrix} \Sigma_E & 0 & 0 \\ 0 & \Sigma_1 & 0 \\ 0 & 0 & \Sigma_2 \end{bmatrix}, \qquad \Sigma_L = \begin{bmatrix} \Sigma_3 & 0 \\ 0 & \Sigma_A \end{bmatrix},$$

$\text{rank}(\Sigma_R) = r$, with $\text{rank}(E) \leq r \leq \text{rank}(E,B]$, and $\Sigma_1, \Sigma_2, \Sigma_3$ are *arbitrary*, positive diagonal matrices. Appropriate feedback matrices F and G are given by

$$\tilde{W}^H G = \begin{bmatrix} 0 & 0 & G_{13} & 0 \\ G_{21} & G_{22} & G_{23} & 0 \end{bmatrix}, \tag{30}$$

where

$$G_{13} = \Sigma_B^{-1}\begin{bmatrix} \Sigma_2 & 0 \\ 0 & 0 \end{bmatrix}, \qquad G_{22} = B_{22}^{-1}(\Sigma_1 - E_{22}),$$

$$G_{21} = -B_{22}^{-1}E_{21}, \qquad G_{23} = -B_{22}^{-1}B_{21}G_{13},$$

and

$$\tilde{W}^H F = \begin{bmatrix} 0 & 0 & F_{13} & F_{14} \\ 0 & 0 & F_{23} & F_{24} \end{bmatrix}, \tag{31}$$

where

$$F_{13} = \Sigma_B^{-1}\left(\begin{bmatrix} 0 & 0 \\ 0 & \Sigma_3 \end{bmatrix} - A_{33}\right), \qquad F_{14} = -\Sigma_B^{-1} A_{34}$$

$$F_{23} = -B_{22}^{-1} B_{21} F_{13}, \qquad F_{24} = -B_{22}^{-1} B_{21} F_{14}.$$

Some freedom in the solution remains, which has not been exploited here.

We remark that if $r = \text{rank}[E,B]$, then the dynamical part of the closed–loop system is of maximum dimension (equal to the dimension of the reachable space of the original system), and the optimal solution is obtained by feedback G alone, with $F = 0$.

The case of the state estimator problem, where $B = I$, is just the dual of the state feedback controller problem. If **02** holds, the optimal observer design can, therefore, be obtained by replacing the triple (E, A, B) by (E^T, A^T, C^T) and F by F^T in (29), (30) and (31).

For the more general output feedback problem, where $B \neq I$, $C \neq I$, the singular value structure that can be attained is more complicated. The problem of optimizing robustness can, in this case, be reduced to the problem:

Given matrix $M = \begin{bmatrix} M_{11} & M_{12} \\ M_{21} & M_{22} \end{bmatrix}$ *find* Δ *such that* $\text{cond}\left(\begin{bmatrix} M_{11}+\Delta & M_{12} \\ M_{21} & M_{22} \end{bmatrix}\right)$ *is minimal.*

This is an **open problem.** In practice an upper bound on the condition number can be minimized, using the structure in Theorem 2. Details of the procedure are described in [2] and [3] and numerical examples are also presented in [3].

In this section we have described a method for obtaining a regular closed loop system pencil of index at most one using derivative and proportional feedback. The feedback is selected to ensure that the properties of the pencil are insensitive to perturbations, using singular value assignment. It is desirable for the system design to have other additional properties, however; in particular, to be stable and, possibly, to have specified finite eigenvalues. One strategy for achieving an overall design is to use derivative feedback alone to obtain a robust, regular, index one system of maximal dimension, and then to use proportional feedback to assign the required eigenvalues to the system. In the next section we describe methods for robust eigenstructure assignment in descriptor systems using proportional feedback.

5. ROBUST EIGENSTRUCTURE ASSIGNMENT

The aim of the design problem is now to select the feedback matrices to assign specified eigenvalues to the closed loop system pencil. We consider here only the state feedback controller problem and the state estimator problem, where $C = I$ and $B = I$, respectively. The full output feedback design problem is much more difficult and is beyond the scope of this paper. We assume that if derivative feedback is available, then it is used to "pre–condition" the system pencil by singular value assignment, as described in Section 4. We therefore consider here only *proportional state* feedback designs.

The design problem is:

Given real system matrices E, A, B *and a set of* $q \equiv \text{rank}(E)$ *self–conjugate complex numbers* $\mathcal{L} = \{\lambda_1, \lambda_2, \ldots, \lambda_q\}$, *select real matrix* F *such that the closed–loop pencil*

$$\alpha E - \beta(A + BF)$$

is regular and of index at most one and has the prescribed finite eigenvalues $\lambda_j \in \mathscr{L}$, $j = 1, 2, \ldots, q$.

For a *robust* solution to the problem we want the assigned finite eigenvalues to be non–defective and we want some measure of the sensitivity of the eigenvalues (both finite and infinite) to be minimal.

If we let (a_j, β_j) denote a generalized, *simple* eigenvalue of the pencil $\alpha E - \beta A$ with right and left eigenvectors x_j and y_j satisfying

$$\alpha_j E x_j = \beta_j A x_j , \qquad a_j y_j^T E = \beta_j y_j^T A ,$$

then a measure of the sensitivity of the eigenvalue is given by the *condition number*

$$c_j = \frac{\|y_j\|_2 \; \|x_j\|_2}{(|y_j^T E x_j|^2 + |y_j^T A x_j|^2)^{\frac{1}{2}}} \tag{33}$$

(see [13]). The condition number is inversely proportional to the angles between the invariant vectors y_j and Ex_j (or y_j and Ax_j, in the case of infinite eigenvalues). The condition number c_j is, moreover, inversely proportional to the quantity $(|y_j^T E x_j| + |y_j^T A x_j|^2)^{\frac{1}{2}}$, which measures how nearly the vector x_j approximates a null vector of both E and A and, hence, how close the pencil is to losing regularity.

If the pencil $\alpha E - \beta A$ is non–defective and a perturbation of order $O(\epsilon)$ is made in E or A, then the corresponding first order perturbation in a simple eigenvalue (α_j, β_j) is of order $0(\epsilon c_j)$, where distance is measured in the chordal metric [13]. If the pencil is defective, then the corresponding perturbation in *some* eigenvalue is at least an order of magnitude worse in ϵ, and, therefore, defective systems are necessarily less robust than those that are non–defective.

The sensitivity of a multiple (non–defective) eigenvalue is proportional to the maximum of the associated condition numbers c_j, taken with respect to an *orthonormal* basis $\{x_j\}$ for the space of right eigenvectors and a corresponding set $\{y_j\}$ of left eigenvectors normalized such that $y_j^T E x_i = 0$, $y_j^T A x_i = 0$, for $i \neq j$. (See [10] for details.)

An overall measure of the sensitivity of the eigenvalues of a regular, non–defective system pencil is given by a weighted sum of all the condition numbers

$$\nu(\omega) = \left[\sum_{j=1}^{u} \omega_j^2 c_j^2 \right]^{\frac{1}{2}} , \tag{34}$$

where $\omega_j > 0$ and $\Sigma\, \omega_j^2 = 1$. A regular non–defective pencil must, by definition, be of index at most one and have precisely $q \equiv \text{rank}\,(E)$ finite eigenvalues. The robustness measure $\nu(\omega)$ can therefore be written [10]

$$\nu(\omega) = \| D_\omega [EX_q, AX_\infty]^{-1} \|_F , \tag{35}$$

where D_ω is a diagonal weighting matrix, the columns of X_∞ form an orthonormal basis for the null space of E, and $X_q = [x_1, x_2, \ldots, x_q]$ is the modal matrix of right eigenvectors associated with the finite eigenvalues, normalized such that $\|x_j\| = 1$, $j = 1, 2, \ldots, q$, and such that the vectors associated with each multiple eigenvalue form an orthonormal set. (Here $\|\cdot\|_F$ denotes the Frobenius matrix norm.) A robust feedback design can thus be achieved by selecting the eigenstructure of the closed loop system so as to minimize the measure $\nu(\omega)$ given by (35).

The existence of a solution to the eigenstructure assignment problem can be guaranteed under the algebraic conditions of Section 4.1. In the next subsection the existence results are given, together with a parameterization of the solution. In the following subsection a technique for achieving robust eigenstructure assignment using this parameterization is described.

5.1 Existence of Solutions

Necessary and sufficient conditions for the state feedback eigenvalue assignment problem to have a solution are established by the following.

THEOREM 3 *Given real system matrices* E, A, B *and any arbitrary self–conjugate set* $\mathscr{L} = \{\lambda_1, \lambda_2, \ldots, \lambda_q\}$ *of* $q \equiv \text{rank}(E)$ *complex numbers, there exists a real matrix* F *such that the matrix pencil* $\alpha E + \beta(A + BF)$ *is regular and of index at most one and has the finite eigenvalues* $\lambda_j \in \mathscr{L}, j = 1, 2, \ldots, q$, *if and only if the conditions* **C1** *and* **C2**, *defined in Section* 4.1, *hold.*

Proof. Proofs are given in [7] [10] and [1]. □

We remark that the condition **C2** ensures that the closed loop system pencil (32) can be made regular and of index at most one, and the condition **C1** guarantees that the finite poles can be assigned arbitrarily.

For a *non–defective* closed loop pencil (32) with prescribed eigenvalues $\mathscr{L} = \{\lambda_j, j = 1, 2, \ldots, q\}$, where $q = \text{rank}(E)$, we require that for some matrix $X_q \in \mathscr{C}^{n \times q}$ of full rank, the feedback matrix F satisfies

$$(A + BF)X_q = EX_q \Lambda_q , \quad \Lambda_q = \text{diag}\{\lambda_j\} , \tag{36}$$

and

$$\text{rank}\,[E, (A + BF)X_\infty] = n , \tag{37}$$

where the columns of X_∞ form an *orthonormal* basis for the null space of E and

$$\text{rank}\,[X_q, X_\infty] = n . \tag{38}$$

The condition (36) ensures that the closed loop pencil has the prescribed finite eigenvalues λ_j with a full set of independent right eigenvectors $x_j = X_q e_j$, $j = 1, 2, 3, \ldots, q$, and condition (37) ensures that the system is regular and of index at most one. The required feedback F can be parameterized in terms of the vectors x_j, $j = 1, 2, 3, \ldots, q$, and a matrix W. We have the following structure theorem.

THEOREM 4 *Given the set* $\mathscr{L} = \{\lambda_1, \lambda_2, \ldots, \lambda_q\}$ *of* distinct *self–conjugate complex numbers, where* $q = \text{rank}(E)$, *there exist vectors*

$$x_j \in \mathscr{S}_j \equiv \{x \mid (\lambda_j E - A)x \in \text{range}(B)\},\ \lambda_j \in \mathscr{L},\ j = 1, 2, \ldots, q, \tag{39}$$

such that $X_q = [x_1, x_2, \ldots, x_q]$ *satisfies* (38), *and a matrix* W *satisfying*

$$\text{rank}\,[E + AX_\infty X_\infty^T + BWX_\infty^T] = n \tag{40}$$

if and only if conditions **C1** *and* **C2** *hold. If* (38)–(40) *hold, then the matrix* F *given by*

$$F = [B^+(EX_q \Lambda_q - AX_q), W][X_q, X_\infty]^{-1} \tag{41}$$

solves the eigenvalue assignment problem, and (36) *and* (37) *are satisfied.* (*Here* B^+ *denotes the Moore–Penrose pseudo–inverse of matrix* B *and* X_∞ *denotes an orthonormal basis for the null space of* E.

Proof. The proof is established in [10]. □

If the prescribed eigenvalues are not distinct, then **C1** and **C2** are necessary but may not be sufficient to guarantee a non–defective solution to the eigenvalue assignment problem.

Theorem 4 gives a parameterization of feedback matrix F in terms of the eigenstructure of the corresponding closed loop system pencil. In the next subsection a technique is described for selecting the free parameters to give a *robust* solution to the system design problem.

5.2 A Numerical Algorithm

To construct a robust solution to the eigenvalue assignment problem, we use the parameterization of the feedback F given by (41) in Theorem 4 and select the freedom in X_q and W so as to minimize the measure of robustness $\nu(\omega)$ given by (35). We aim also to ensure that the ranks of the matrices in (38) and (40) are insensitive to perturbations; that is, we want these matrices to be well–conditioned.

In practice it is sufficient to minimize the condition numbers

$$\kappa_1 = \text{cond}_F([X_q, X_\infty]), \quad \kappa_2 = \text{cond}_2(E + AX_\infty X_\infty^T + BWX_\infty^T),$$

subject to $\|E + AX_\infty X_\infty^T + BWX_\infty^T\|_2$ remaining finite. It can be shown [10] that the measure $\nu(\omega)$ is bounded in terms of the product $\kappa_1 \kappa_2$. We have

$$\gamma_1(\kappa_1 \kappa_2)^{\frac{1}{2}} \le \nu(\omega)\, \|E + AX_\infty X_\infty^T + BWX_\infty^T\|_2 \le \gamma_2 \kappa_1 \kappa_2,$$

where γ_1 and γ_2 are fixed constants. Provided $\|E + AX_\infty X_\infty^{T^+} + BWX_\infty\|_2$ remains bounded, minimizing κ_1 and κ_2 thus minimizes an equivalent measure of the sensitivity of the assigned eigenvalues, as well as ensuring that the matrices in (38) and (40) are well–conditioned. Since the free parameters appear independently in κ_1 and κ_2, these measures can be minimized separately. Optimizing the condition numbers κ_1 and κ_2 also leads to other desirable properties of the closed loop system. In particular the transient response and the magnitude of the gains can be bounded in terms of κ_1 (see [10]), and a lower bound on the distance of the closed loop pencil to instability can be given in terms of κ_1^{-1}(see [5]).

The computational procedure for solving the robust eigenstructure assignment problem consists of four basic steps:

Step 1: Compute orthonormal bases X_∞ for kernel(E) and S_j for the subspaces $\mathscr{S}_j$, defined in (39), for $j = 1, 2, ..., q$.

Step 2: Select W to minimize $\sigma_{min}(E + AX_\infty X_\infty^T + BWX_\infty^T)$ subject to $\sigma_{max}(E + AX_\infty X_\infty^T + BWX_\infty^T) \leq \tau$, where τ is a given tolerance.

Step 3: Select vectors $x_j = S_j v_j \in \mathscr{S}_j$ with $\|x_j\|_2 = 1, j = 1, 2, ..., q,$ to minimize κ_1.

Step 4: Determine F from equation (41).

Reliable library software with procedures for computing QR, SVD and LU decompositions is used to accomplish these steps. Iterative techniques for selecting vectors from given subspaces to minimize κ_1 in *Step 3* are described in [11]. The computation of F in *Step 4* is accurate as long as κ_1 is reasonably small (relative to machine precision). A detailed description of the algorithm is given in [10]. It is not necessary that the prescribed eigenvalues be distinct. Provided that a non–defective solution to the eigenvalue assignment problem exists for the given set $\mathscr{L}$, the algorithm determines a feedback that assigns the prescribed eigenvalues.

In this section we have described a procedure for designing a robust state feedback controller with prescribed eigenvalues. The dual state estimator design problem can be solved using the same technique by replacing the system triple (E, A, B) and the feedback F in the algorithm by the triple (E^T, A^T, B^T) and the feedback F^T.

6. CONCLUSIONS

Two techniques for designing automatic feedback controllers and observers for implicit linear differential–algebraic systems are described here: singular value assignment and eigenstructure assignment. The techniques are based on stable and reliable numerical procedures for factorizing and reducing matrices to condensed forms, using unitary transformations. Measures of sensitivity for the system designs are derived and conditions for the existence of robust solutions to the synthesis problems are established. The degrees of freedom in the design are identified and computational procedures for selecting the free parameters to minimize the sensitivity measures are presented. These results all, with minor modifications, apply also to discrete–time implicit linear systems governed by dynamic–algebraic equations.

In practice a combination of the robust singular value and eigenstructure assignment procedures can be used to synthesize full state feedback controllers and observers. Applications of these combined techniques to the design of state–estimators for flow in gas networks are examined in [12] and [14]. The system dynamics are modelled by implicit, discrete–time, finite difference equations. The importance of robustness in the observer design is demonstrated by the investigations. Effects of model uncertainty and measurement noise are minimized, as far as possible, by the robust design procedures.

The synthesis of output feedback designs is more difficult and complicated than state feedback or state–estimator synthesis, and many aspects of the output design problem are still open. For systems that can be made regular and of index at most one by feedback, the condensed form presented here identifies the system structure and the available freedom for design synthesis. To exploit this freedom fully, further techniques are still needed, and extensions are required to treat higher index systems that cannot be reduced to systems of index less than or equal to one. Work on these developments is in progress.

REFERENCES

[1] A. Bunse–Gerstner, V. Mehrmann, and N.K. Nichols, Regularization of descriptor systems by derivative and proportional state feedback, *SIAM J. of Matrix Analysis and Applications*, 13, 1992, 46–67.

[2] A. Bunse–Gerstner, V. Mehrmann, and N.K. Nichols, Regularization of descriptor systems by output feedback, *IEEE Transactions on Automatic Control* (to appear).

[3] A. Bunse–Gerstner, V. Mehrmann and N.K. Nichols, Numerical methods for the regularization of descriptor systems by output feedback, IMA Preprint Series #987, Institute for Mathematics and its Applications, Minneapolis, 1992.

[4] A. Bunse–Gerstner, V. Mehrmann and N.K. Nichols, Derivative feedback for descriptor systems, FSP Mathematisierung, Universität Bielefeld, Materialien LVIII, 1989.

[5] R. Byers and N.K. Nichols, On the stability radius of a genrealized state–space system, *Linear Algebra and its Applications*, 189, 1993, 113–134.

[6] S.L. Campbell, *Singular Systems of Differential Equations*, Pitman, London, 1980.

[7] L.R. Fletcher, J. Kautsky, and N.K. Nichols, Eigenstructure assignment in descriptor systems, *IEEE Transactions on Automatic Control*, AC–31, 1986, 1138–1141.

[8] F.R. Gantmacher, *The Theory of Matrices*, Vols I, II, Chelsea, New York, 1959.

[9] G.H. Golub and C.F. Van Loan, *Matrix Computations*, John–Hopkins University Press, Baltimore, 1983.

[10] J. Kautsky, N.K. Nichols and E.K–W. Chu, Robust pole assignment in singular control systems, *Linear Algebra and its Applications*, 121, 1989, 9–37.

[11] J. Kautsky, N.K. Nichols and P. Van Dooren, Robust pole assignment in linear state feedback, *International J. of Control*, 41, 1985, 1129–1155.

[12] D.W. Pearson, M.J. Chapman and D.N. Shields, Partial singular value assignment in the design of robust obsrevers for discrete time descriptor systems, *IMA J. of Mathematical Control and Information*, 5, 1988, 203–213.

[13] G.W. Stewart, Gerschgorin theory for the generalized eigenvalue problem $Ax = \lambda Bx$, *Mathematics of Computation*, 29, 1975, 600–606.

[14] S.M. Stringer and N.K. Nichols, The state estimation of flow demands in a linear gas network from sparse pressure telemetry, *Proc. of Symposium on Postgraduate Research in Control and Information*, Institute of Measurement and Control, Nottingham, 1993.

[15] G.C. Verghese, B.C. Lévy and T. Kailath, A general state space for singular systems, *IEEE Transactions on Automatic Control*, AC–26, 1981, 811–831.

[16] E.L. Yip and R.F. Sincovec, Solvability, controllability and observability of continuous descriptor systems, *IEEE Transactions on Automatic Control*, AC–26, 1981, 702–707.

N.K. Nichols
Department of Mathematics
University of Reading
P O Box 220, Reading RG6 2AX, UK

G W STEWART

UTV decompositions

Abstract An important problem arising in a number of applications is to determine the rank of a matrix that has been contaminated with error and find an approximation to its null space. Traditionally the singular value decomposition or the pivoted QR decomposition has been used to solve this problem. However, both decompositions resist updating and parallelization. To circumvent these drawbacks, a new class of decompositions has been proposed. They factor the matrix in question into the product of an orthogonal matrix, a triangular matrix, and another orthogonal matrix. These UTV decompositions can be efficiently updated, both sequentially and in parallel. This paper is an informal introduction to UTV decompositions.

1 An Easy Problem

In order to set the background for UTV decompositions, we will first consider the traditional solution of a simple problem. Let X be an $n \times p$ matrix of rank k (for definiteness we assume that $n \geq p$). The problem is to determine k and a basis for the null space of X.

The SVD solution

The textbook solution to this problem is to invoke the singular value decomposition (SVD). Specifically [8, 4], there are orthogonal matrices

$$U = (\overset{k}{U_1} \;\; \overset{n-k}{U_2}) \quad \text{and} \quad V = (\overset{k}{V_1} \;\; \overset{p-k}{V_2})$$

such that

$$\begin{pmatrix} U_1^{\mathrm{T}} \\ U_2^{\mathrm{T}} \end{pmatrix} X(V_1 \; V_2) = \begin{pmatrix} \Sigma & 0 \\ 0 & 0 \end{pmatrix},$$

where

$$\Sigma = \operatorname{diag}(\sigma_1, \ldots, \sigma_k),$$

with

$$\sigma_1 \geq \cdots \geq \sigma_k > 0.$$

The diagonal elements σ_i are called the singular values of X. The decomposition reveals the rank of X by the fact that the last $p-k$ singular values are zero.[1]

[1] In practice rounding errors made in the computation of the SVD will cause all the singular values to be nonzero. However, unless σ_k is extremely small, it is not difficult to recognize the singular values that should be zero.

The SVD is generous in providing bases for the row and column spaces of X and X^{T}. The columns of V_1 form an orthonormal basis for the null space of X. In addition, the columns of V_2 form an orthonormal basis for the row space of X. The matrices U_2 and U_1 form orthonormal bases for the null and row spaces of X^{T}.

The QRPD solution

For all its nice properties, the SVD is expensive to compute. The QR decomposition of a matrix into the product of an orthogonal matrix and an upper triangular matrix is far cheaper. Moreover, if we allow column interchanges, the decomposition can reveal the rank of X.

Specifically [8, 4, 3], there is an orthogonal matrix

$$Q = (\overset{k}{Q_1}\ \overset{n-k}{Q_2})$$

and a permutation matrix

$$P = (\overset{k}{P_1}\ \overset{p-k}{P_2})$$

such that

$$\begin{pmatrix} Q_1^{\mathrm{T}} \\ Q_2^{\mathrm{T}} \end{pmatrix} X(P_1\ P_2) = \begin{pmatrix} R_{11} & R_{12} \\ 0 & 0 \end{pmatrix},$$

where R_{11} is a nonsingular upper triangular matrix of order k. The rank is revealed by the fact that the trailing submatrix R_{22} is zero.

The columns of Q_1 form a basis for the column space of X. Moreover, the rows of $(R_{11}\ \ R_{12})$ form a basis for the row space of X, and the columns of $(-R_{12}^{\mathrm{T}}R_{11}^{-\mathrm{T}}\ \ I)^{\mathrm{T}}$ form a basis for the null space. However, neither of these bases is orthonormal, and the second basis requires additional work to compute. In this respect the QRP decomposition is inferior to the SVD.

The permutation matrix P is necessary, as can be seen from the matrix

$$X = \begin{pmatrix} 0 & 1 \\ 0 & 1 \end{pmatrix}.$$

Here X, which is of rank one, is already in triangular form. Without a permutation we would get $Q = I$ and the $(2,2)$-element of the final triangular matrix would be nonzero. However if we interchange the two columns of X, we get a matrix whose QR decomposition has the rank-revealing upper triangular matrix

$$\begin{pmatrix} \sqrt{2} & 0 \\ 0 & 0 \end{pmatrix}.$$

2 A Harder Problem

In real life, our matrix X is likely to be contaminated with error, and our rank determination problem becomes the following. Given

$$\tilde{X} = X + E,$$

where E is small, determine the rank k of X. We would also like to determining approximations to the null space of X.

The SVD solution

When we attempt to compute the singular value decomposition, we find that the singular values $\sigma_{k+1}, \ldots, \sigma_p$ are no longer zero to working accuracy; instead they are small — in some sense proportional to E. The hard part of the problem is to find a precise relation, so that we can say when a singular value σ_k is too large to come from error alone. Obviously this cannot be done without knowing something about E itself.

One common assumption is that the elements of E are independent random variables with mean zero and known variance ϵ. It can then be shown [11] that if we ignore second order terms the expected value of $\sigma_{k+1}^2 + \cdots + \sigma_p^2$ is

$$\mathbf{E}(\sigma_{k+1}^2 + \cdots + \sigma_p^2) \cong (n-k)(p-k)\epsilon^2.$$

Thus we can choose k to be the first integer for which

$$\sigma_{k+1}^2 + \cdots + \sigma_p^2 \leq \phi(n-k)(p-k)\epsilon^2. \tag{2.1}$$

Here ϕ is a factor greater than one that insures that random fluctuations in $\sigma_{k+1}^2 + \cdots + \sigma_p^2$ will not cause the test to fail. Clearly, this test will not be effective if σ_k^2 is near $(n-k)(p-k)\epsilon^2$; but in this case the problem of determining the rank of X from $\tilde{X}$ is not well posed.

The QRPD solution

Turning now to the QRP decomposition, the rank of X is revealed by the fact that the matrix R_{22} is small, *provided that P has been properly chosen.* Here we not only have the problem of determining the meaning of the word "small", but we must find a pivoting strategy for determining P. Although there are ad-hoc strategies that work well enough in practice, a strategy that is efficient and rigorously rank revealing has eluded workers for decades [1, 2, 6]. However, Gu and Eisenstat have recently announced a promising strategy.

Some general comments

Both the SVD and the QRPD are reasonable approaches to the problem of rank determinations. The SVD is the more versatile decomposition, but it is expensive. The QRPD is not as forthcoming; but it is cheaper, and for many applications it is quite satisfactory. Both decomposition, however, have the drawback that they cannot be easily updated; that is, the decomposition cannot be cheaply recomputed when a row is added to X. Moreover, there is no really satisfactory technique for computing the SVD in parallel.

3 URV Decompositions

The problem with the SVD and the QRPD decomposition is that they are essentially unique. The requirement that Σ in the SVD be diagonal effectively determines U and

V.[2] Once the permutation P of the QRPD had been chosen, Q and R are effectively determined. All this suggests that we will gain room to build better algorithms if we work with classes of decomposition that have more degrees of freedom. UTV decompositions are a natural candidate.

For definiteness we will consider URV decompositions in which X is factored into the product of an orthogonal matrix, an upper triangular matrix, and another orthogonal matrix [9]. Specifically, there are orthogonal matrices

$$U = (\overset{k}{U_1} \;\; \overset{n-k}{U_2}) \quad \text{and} \quad V = (\overset{k}{V_1} \;\; \overset{p-k}{V_2})$$

such that

$$\begin{pmatrix} U_1^{\mathrm{T}} \\ U_2^{\mathrm{T}} \end{pmatrix} X(V_1 \; V_2) = \begin{pmatrix} R & 0 \\ 0 & 0 \end{pmatrix}$$

where R is nonsingular and upper triangular of order k. Since a diagonal matrix is also upper triangular, the SVD is an example of a URV decomposition. However, there are many more.

A URV decomposition, like the SVD, provides orthonormal bases for the row and null spaces of X and X^{T}. Specifically, the columns of V_1 and V_2 provide orthonormal bases for the null and row spaces of X. The columns of U_2 and U_1 provide orthonormal bases for the null and row spaces of X^{T}.

Now in the form described above URV decompositions have been around for a long time under the name "complete orthogonal decomposition." The novel feature is that URV decompositions can be made rank revealing in the presence of error. Given $\tilde{X} = X + E$, There are orthogonal U and V such that

$$U^{\mathrm{T}}\tilde{X}V = \begin{pmatrix} R & F \\ 0 & G \end{pmatrix},$$

where $\|G\|_{\mathrm{F}}^2 \cong \sigma_{k+1}^2 + \cdots + \sigma_p^2$ and F is negligible. (Here $\|\cdot\|_{\mathrm{F}}$ is the Frobenius norm.) Moreover this decomposition can be updated in such a way that its rank revealing character is preserved.

4 Computation of URV Decompositions

Unlike the SVD and QRPD, URV decompositions are not computed by global operations on the matrix X. Rather one starts with the zero matrix and updates the decomposition as X is brought in a row at a time.

A row is processed in three steps.

1. Updating: in which the current row of X is incorporated into the decomposition.

2. Deflation: in which an attempt is made to reduce the effective rank.

3. Refinement: in which the size of the elements of F is reduced.

[2]Subject to the proviso that only the space corresponding to multiple singular values is determined. In particular, V_2 can consist of any orthonormal basis for the null space of X.

BEFORE

```
1 2 3 4 5 6 7 8
X X X 0 0 X E E
X X̂ X X 0 E E 0
```

AFTER

```
1 2 3 4 5 6 7 8
X X X X 0 X E E
X 0 X X 0 X E E
```

Figure 4.1: Applying a Plane Rotation

A novelty of the algorithm is that we pay special attention to the small elements, represented by F and G in the decomposition and try wherever possible to keep them small. The reason we can succeed is the flexibility of our chief computational tool, the plane rotation. For this reason, we will first describe the properties of plane rotations.

Plane rotations

A plane rotations is an orthogonal transformation that combines two rows or two columns of a matrix. It can be constructed to introduce a zero at any point of the two rows, and this indeed is its chief function. However, it can have side effects which may or may not be desirable.

Figure 4.1 illustrates two rows of a matrix before and after the application of a plane rotation. The X's represent elements that are presumed nonzero. The 0's represent elements that are zero. The E's represent elements, presumed nonzero, that are small.

The following happens when a rotation is applied.

1. A zero may be introduced in a row (column 2).

2. A pair of X's remains a pair of X's (columns 1 and 3).

3. An X and an 0 are replaced by a pair of X's (column 4).

4. A pair of 0's remains a pair of 0's (column 5).

5. An X and an E are replaced by a pair of X's (column 6).

6. A pair of E's remains a pair of E's (column 7).

7. An E and a 0 are replaced by a pair of E's (column 8).

These facts, except for the first, are simply common sense. For example if you take a linear combination of a zero and a nonzero element, other things being equal, you can expect a nonzero element. The facts concerning E's are essential to preserving the rank-revealing URV structure.

$$\begin{array}{cccccc} \rightarrow & r & r & r & f & f \\ & 0 & r & r & f & f \\ & 0 & 0 & r & f & f \\ & 0 & 0 & 0 & g & g \\ & 0 & 0 & 0 & 0 & g \\ \rightarrow & \hat{y} & y & y & z & z \end{array} \Longrightarrow \begin{array}{cccccc} & r & r & r & f & f \\ \rightarrow & 0 & r & r & f & f \\ & 0 & 0 & r & f & f \\ & 0 & 0 & 0 & g & g \\ & 0 & 0 & 0 & 0 & g \\ \rightarrow & 0 & \hat{y} & y & z & z \end{array} \Longrightarrow$$

$$\begin{array}{cccccc} & r & r & r & f & f \\ & 0 & r & r & f & f \\ \rightarrow & 0 & 0 & r & f & f \\ & 0 & 0 & 0 & g & g \\ & 0 & 0 & 0 & 0 & g \\ \rightarrow & 0 & 0 & \hat{y} & z & z \end{array} \Longrightarrow \begin{array}{cccccc} & r & r & r & f & f \\ & 0 & r & r & f & f \\ & 0 & 0 & r & f & f \\ \rightarrow & 0 & 0 & 0 & g & g \\ & 0 & 0 & 0 & 0 & g \\ \rightarrow & 0 & 0 & 0 & \hat{z} & z \end{array} \Longrightarrow$$

$$\begin{array}{cccccc} & r & r & r & f & f \\ & 0 & r & r & f & f \\ & 0 & 0 & r & f & f \\ & 0 & 0 & 0 & g & g \\ \rightarrow & 0 & 0 & 0 & 0 & g \\ \rightarrow & 0 & 0 & 0 & 0 & \hat{z} \end{array} \Longrightarrow \begin{array}{ccccc} r & r & r & f & f \\ 0 & r & r & f & f \\ 0 & 0 & r & f & f \\ 0 & 0 & 0 & g & g \\ 0 & 0 & 0 & 0 & g \\ 0 & 0 & 0 & 0 & 0 \end{array}$$

Figure 4.2: Simple Updating

Updating

The updating step folds a row x^{T} of X into the current decomposition. The first step is to transform x^{T} into the coordinate system of the decomposition by replacing it with $x^{\mathrm{T}}V$. The decomposition then takes the form illustrated below for $p = 5$ and $k = 3$

$$\begin{array}{ccccc} r & r & r & f & f \\ 0 & r & r & f & f \\ 0 & 0 & r & f & f \\ 0 & 0 & 0 & g & g \\ 0 & 0 & 0 & 0 & g \\ y & y & y & z & z \end{array}$$

The next step depends on whether the elements denoted by z are small enough to be folded into the part of the decomposition corresponding to F and G. If they are, the reduction proceeds as in Figure 4.2. The arrows to the right of each array indicate which two rows are being combined, and the element with a hat is the one that is annihilated. Because the z's are sufficiently small, they do not destroy the rank-revealing structure as they combine with the f's and g's (Rule 7 in the discussion of plane rotations above).

$$\begin{array}{ccccc} & & & \downarrow & \downarrow \\ r & r & r & f & f \\ 0 & r & r & f & f \\ 0 & 0 & r & f & f \\ 0 & 0 & 0 & g & g \\ 0 & 0 & 0 & 0 & g \\ y & y & y & z & \hat{z} \end{array} \Longrightarrow \begin{array}{cccccc} & r & r & r & f & f \\ & 0 & r & r & f & f \\ & 0 & 0 & r & f & f \\ \rightarrow & 0 & 0 & 0 & g & g \\ \rightarrow & 0 & 0 & 0 & \hat{g} & g \\ & y & y & y & z & 0 \end{array} \Longrightarrow$$

$$\begin{array}{ccccc} r & r & r & f & f \\ 0 & r & r & f & f \\ 0 & 0 & r & f & f \\ 0 & 0 & 0 & g & g \\ 0 & 0 & 0 & 0 & g \\ y & y & y & z & 0 \end{array}$$

Figure 4.3: Redution of z

If the z's are too large, the above reduction will make the f's and g's large and destroy the rank-revealing structure. This, in some sense, is to be expected. For if the z's are large enough, the addition of the row x^{T} increases the rank. But the addition of a single row should not increase the rank by more than one; yet the reduction completely overwhelms F and G. The cure for this problem is to reduce all but the first z to zero before performing the reduction. This means that only the first columns of F and G become large, which corresponds to an increase in rank of one.

The technique for reducing the trailing z's is illustrated in Figure 4.3. First a plane rotation is applied to the columns of the matrix to annihilate the last of the z's. This does not make any element of F or G large, but it does put a small element on the subdiagonal of G. This element is in turn annihilated by a rotation applied to the rows of the matrix. If there are more than two z's, all but the first can be annihilated by an analogous procedure.

Deflation

An apparent increase in rank in the updating procedure can turn out to be spurious. We therefore need a means of detecting when R is rank degeneracy and computing a corresponding URV decomposition for it.

To detect a rank degeneracy, we use a condition estimator [5] to find a vector w of norm one such that

$$\|Rw\| \cong \inf(R),$$

where $\inf(R)$ is the smallest singular value of R. If $\|Rw\|$ is less than a prescribed tolerance, R is judged to be degenerate.

If R is not degenerate, we return to the update step. Otherwise we proceed as follows.

$$
\begin{array}{cccc} \downarrow & \downarrow & & \\ r & r & r & r \\ 0 & r & r & r \\ 0 & 0 & r & r \\ 0 & 0 & 0 & r \end{array}
\Longrightarrow
\begin{array}{ccccc} \rightarrow & r & r & r & r \\ \rightarrow & \hat{r} & r & r & r \\ & 0 & 0 & r & r \\ & 0 & 0 & 0 & r \end{array}
\Longrightarrow
\begin{array}{cccc} & \downarrow & \downarrow & \\ r & r & r & r \\ 0 & r & r & r \\ 0 & 0 & r & r \\ 0 & 0 & 0 & r \end{array}
\Longrightarrow
$$

$$
\begin{array}{ccccc} & r & r & r & r \\ \rightarrow & 0 & r & r & r \\ \rightarrow & 0 & \hat{r} & r & r \\ & 0 & 0 & 0 & r \end{array}
\Longrightarrow
\begin{array}{cccc} & & \downarrow & \downarrow \\ r & r & r & r \\ 0 & r & r & r \\ 0 & 0 & r & r \\ 0 & 0 & 0 & r \end{array}
\Longrightarrow
\begin{array}{ccccc} & r & r & r & r \\ & 0 & r & r & r \\ \rightarrow & 0 & 0 & r & r \\ \rightarrow & 0 & 0 & \hat{r} & r \end{array}
\Longrightarrow
$$

$$
\begin{array}{cccc} r & r & r & e \\ 0 & r & r & e \\ 0 & 0 & r & e \\ 0 & 0 & 0 & e \end{array}
$$

Figure 4.4: Transformation of R

First we use plane rotations as shown below to reduce all the components of w but the last to zero.

$$
\begin{array}{cc} \rightarrow & \hat{w} \\ \rightarrow & w \\ & w \\ & w \end{array}
\Longrightarrow
\begin{array}{cc} & 0 \\ \rightarrow & \hat{w} \\ \rightarrow & w \\ & w \end{array}
\Longrightarrow
\begin{array}{cc} & 0 \\ & 0 \\ \rightarrow & \hat{w} \\ \rightarrow & w \end{array}
\Longrightarrow
\begin{array}{c} 0 \\ 0 \\ 0 \\ 1 \end{array}
$$

Let Q be the product of the rotations. We then determine a product P of rotations so that PRQ^{T} is triangular. The technique is illustrated in Figure 4.4. As a rotation from Q is applied to two columns of R it generates a subdiagonal element, which is annihilate by a rotation applied to two rows of R. The matrix P is just the product of these row rotations.

The last column of the final matrix has been filled with e's to indicate that it is small. In fact, its norm is approximately equal to $\inf(R)$. To see this, note that

$$
\|Rw\| = \|PRQ^{\mathrm{H}}Qw\| = \|PRQ^{\mathrm{H}}\mathbf{e}_p\|,
$$

where $\mathbf{e}_p$ is the vector whose last component is one and whose other components are zero. Hence the last column of PRQ^{H} has norm $\|Rw\| \cong \inf(R)$. This means that PRQ is the triangular part of a rank-revealing URV decomposition of the original R.

Refinement

It has been found useful to apply a refinement step after deflation to reduce the elements above the diagonal in the last column of the newly deflated matrix. The deflation step is illustrated below.

$$\begin{array}{cccc} r & r & r & e \\ 0 & r & r & e \\ 0 & 0 & r & e \\ 0 & 0 & 0 & e \end{array} \Longrightarrow \begin{array}{cccc} r & r & r & 0 \\ 0 & r & r & 0 \\ 0 & 0 & r & 0 \\ e^2 & e^2 & e^2 & e \end{array} \Longrightarrow \begin{array}{cccc} r & r & r & e^3 \\ 0 & r & r & e^3 \\ 0 & 0 & r & e^3 \\ 0 & 0 & 0 & e \end{array}$$

First plane rotations are applied to the columns of R to annihilate the elements in the last column. This reduction places elements in the last row, which are in turn annihilated by rotations applied to the rows. In most cases the off-diagonal in the last column elements will be considerably reduced by this process, which is related of the QR algorithm. For example, if $\inf(R) \cong 1$, the final e's will usually be proportional to the cube of the original e's, as indicated in the figure. As we shall see, this reduction improves the quality of the approximate null space produced by the URV decomposition.

5 Perturbation theory

It is now time to examine how good the decomposition is. We will take as our standard the SVD, and ask how well the approximate null spaces produced by a URVD compare with the approximate null spaces produced by the SVD. We will also ask how the singular values of G are related to those of X.

Quality of the null space

Recall that in the URVD

$$\begin{pmatrix} U_1^{\mathrm{T}} \\ U_2 \end{pmatrix} \tilde{X}(V_1\ V_2) = \begin{pmatrix} R & F \\ 0 & G \end{pmatrix}.$$

it is the matrix V_2 that furnishes the orthonormal basis for the approximate null space. If we let $\hat{V}_2$ be the corresponding matrix for the SVD, then we want to compare the subspaces spanned by the V_2 and $\hat{V}_2$.

It turns out that a good measure of the difference is the size of the matrix $\hat{V}_1^{\mathrm{T}}V_2$, where $\hat{V}_1$ is also from the SVD. The rationale is that since $\hat{V}_1^{\mathrm{T}}\hat{V}_2 = 0$, the matrix $\hat{V}_1^{\mathrm{T}}V_2$ will be small whenever the space spanned by V_2 is near the space spanned by $\hat{V}_2$. In fact the singular values of $\hat{V}_1^{\mathrm{T}}V_2$ are just the signs of the canonical angles between the spaces spanned by $\hat{V}_2$ and V_2 (for details see [13]).

It can be shown that [7]

$$\|\hat{V}_1^{\mathrm{T}}V_2\| \leq \frac{\|F\|}{(1-\rho^2)\inf(R)}, \tag{5.1}$$

where

$$\rho = \frac{\|G\|}{\inf(R)}$$

is assumed to be less than one. Thus the quality of the approximate null space depends on the quantity $\|F\|$. This furnishes a justification for the refinement step, since its goal is to reduce the size of F.

Singular values

It can also be shown [7] that if $\rho < 1$, then the singular values of G are $O(\|F\|^2)$ approximations to the smallest $n-k$ singular values of $\tilde{X}$. In particular,

$$\|G\|_{\mathrm{F}}^2 \cong \sigma_{k+1}^2 + \cdots + \sigma_p^2.$$

Thus we can use $\|G\|_{\mathrm{F}}^2$ in the test (2.1) to determine rank during the computation of a URV decomposition.

6 Variants and Generalizations

Parallelization

The algorithm sketched above can be parallelized to run in $O(p)$ time on a linear array of p processors. For details see [10].

The ULV decomposition

The ULV decomposition is a lower triangular version of the URV decomposition [12]. Specifically, we can find orthogonal matrices U and V such that

$$\begin{pmatrix} U_1^{\mathrm{T}} \\ U_2^{\mathrm{T}} \end{pmatrix} \tilde{X}(V_1\ V_2) = \begin{pmatrix} L & 0 \\ F & G \end{pmatrix}$$

where L is nonsingular and lower triangular of order k and G is lower triangular with $\|G\|_{\mathrm{F}}^2 = \sigma_{k+1}^2 + \cdots \sigma_p^2$. At first glance, the ULV decomposition would appear to be a trivial variant of the URV decomposition. Actually the two are quite different.

In the first place, a perturbation analysis of the kind presented in the last section shows that for the ULV decomposition

$$\|\hat{V}_1^{\mathrm{T}} V_2\| \leq \frac{\rho\|F\|}{(1-\rho^2)\inf(R)}.$$

The difference between this bound and (5.1) is the factor ρ multiplying $\|F\|$. Since ρ is less than one — in some applications much less that one — the ULV furnishes better approximation to the space spanned by the small singular values of $\tilde{X}$.

The price to be paid is that the algorithm for computing a ULV decomposition is more complicated than the one for computing a URV decomposition. And it is richer in rotations that are applied to the columns of the matrix. Since these rotations must be accumulated in V, the algorithm is more expensive for applications, such as signal processing, where V but not U is required.

Generalized ULV decompositions

In some applications, we need the space spanned by the eigenvectors of the pencil

$$X^{\mathrm{T}}X + \lambda Y^{\mathrm{T}}Y$$

corresponding to the smallest eigenvalues. Luk has shown how to find orthogonal matrices U_X, U_Y, and V such that

$$U_X^{\mathrm{T}} X V = DL \quad \text{and} \quad U_Y^{\mathrm{T}} Y V = L,$$

where L is lower triangular and D is diagonal. The space required is revealed by small values of the diagonal elements of D. This decomposition, like the ordinary ULV decomposition can be updated.

References

[1] C. H. Bischof. A parallel QR factorization algorithm with controlled local pivoting. *SIAM Journal on Scientific and Statistical Computing*, 12:36–57, 1991. Citation communicated by Per Christian Hansen.

[2] T. F. Chan. Rank revealing QR factorizations. *Linear Algebra and Its Applications*, 88/89:67–82, 1987.

[3] G. H. Golub. Numerical methods for solving least squares problems. *Numerische Mathematik*, 7:206–216, 1965.

[4] G. H. Golub and C. F. Van Loan. *Matrix Computations*. Johns Hopkins University Press, Baltimore, Maryland, 2nd edition, 1989.

[5] N. J. Higham. A survey of condition number estimation for triangular matrices. *SIAM Review*, 29:575–596, 1987.

[6] Y. P. Hong and C.-T. Pan. Rank-revealing QR factorizations and the singular value decomposition. *Mathematics of Computation*, 58:213–232, 1992.

[7] R. Mathias and G. W. Stewart. A block qr algorithm and the singular value decomposition. *Linear Algebra and Its Applications*, 182:91–100, 1992.

[8] G. W. Stewart. *Introduction to Matrix Computations*. Academic Press, New York, 1973.

[9] G. W. Stewart. An updating algorithm for subspace tracking. *IEEE Transactions on Signal Processing*, 40:1535–1541, 1992.

[10] G. W. Stewart. Updating URV decompositions in parallel. Technical Report CS-TR-2880, Department of Computer Science, University of Maryland, College Park, 1992. To appear in *Parallel Computing*.

[11] G. W. Stewart. Determining rank in the presence of error. In M. S. Moonen, G. H. Golub, and B. L. R. DeMoor, editors, *Linear Algebra for Large Scale and Real-Time Applications*, pages 275–292, Dordrecht, 1993. Kluwer Academic Publishers.

[12] G. W. Stewart. Updating a rank-revealing ULV decomposition. *SIAM Journal on Matrix Analysis and Applications*, 14:494–499, 1993.

[13] G. W. Stewart and J.-G. Sun. *Matrix Perturbation Theory*. Academic Press, Boston, 1990.

G. W. Stewart
Department of Computer Science and
Institute for Advanced Computer Studies
University of Maryland
College Park, MD 20742
USA
stewart@cs.umd.edu

M J TODD

A lower bound on the number of iterations of an interior-point algorithm for linear programming

Abstract We describe some elements of interior-point methods for linear programming. In contrast to Dantzig's well-known simplex method, these algorithms generate a sequence of points in the relative interior of the feasible region. They have some remarkable properties: certain of these methods have very attractive theoretical upper bounds on the number of iterations required ($O(\sqrt{n}t)$ iterations for problems with n inequalities to attain a precision of t additional digits), while others can be highly effective in solving large-scale problems. We analyze in detail one of these algorithms (the primal-dual affine-scaling method) that is very close to what is implemented in practice, and show that it may take at least $n^{1/3}$ iterations to improve the initial duality gap by a factor of twenty. We also discuss how far this analysis can be extended to other primal-dual interior-point methods.

One unusual feature of our approach is that we do not construct bad examples explicitly. Instead, our viewpoint is more like that of information-based complexity in nonlinear programming; we reveal to the algorithm at each iteration a bad pair of search directions, which may depend on the previous iterations, but we show that all our directions are consistent with some initial data for the linear programming problem.

1 Introduction

For over forty years, Dantzig's simplex method has been the standard solution algorithm for linear programming problems, which seek to minimise a linear function of several variables subject to linear equations and inequalities [5]. In practice, its performance remains excellent – the number of iterations required is usually a small number of the number of equality constraints in a standard-form problem, even though there are examples (see for instance Klee and Minty [10]) needing an exponential number of iterations for many particular simplex pivoting rules. This exponential gap has been explained to some extent by theoretical analyses showing that the expected number of iterations required by some simplex variants on random problems generated from certain probability distributions is polynomial in the dimensions of the problem instance; for a survey, see Borgwardt [4].

In the last eight years, there has been enormous activity in optimisation in the field of interior-point methods for linear programming and extensions – the bibliography of Kranich [13] lists 1303 items. This explosion of research was instigated by the work of Karmarkar [9], who provided a polynomial-time algorithm whose extensions and variants have proved to be very efficient in solving large-scale linear programming problems; see, e.g., Bixby et al. [3] and Lustig et al. [15]. For an overview of such methods the reader is referred to [6, 7, 27].

The most effective interior-point methods computationally are primal-dual methods, and these are variants of polynomial-time algorithms having the best complexity theoretically also. The latter methods, either path-following methods (see, e.g., Kojima et al. [11], Monteiro and Adler [19], and Gonzaga [7]) or potential-reduction methods (see Kojima et al. [12]), require $O(\sqrt{n}t)$ iterations to attain an additional t digits of accuracy in a problem with n inequality constraints, given a sufficiently "centered" pair of initial primal and dual strictly feasible points. On the other hand, computational experience with sophisticated primal-dual interior-point codes suggests that the number of iterations necessary grows much more slowly with the dimension n. Early papers cited an almost constant number of iterations to solve a range of small to reasonably large problems, while the results of Lustig, Marsten, and Shanno on problems with n up to two million suggested that the growth was logarithmic in n [14]. Once again there is an exponential gap between observed performance and theoretical bounds, even though the latter are now polynomial. We seek to investigate this gap.

There have been attempts to study the "expected" number of iterations theoretically. These analyses are not rigorous as in the case of the simplex method; instead of assuming a random problem held fixed throughout the iterations, they make a probabilistic assumption about the data at a particular iteration, analyze the performance at that iteration, and hence make heuristic estimates of the "typical" behaviour of interior-point algorithms. Nevertheless, these studies indicate behaviour closer to what is observed in practice. Nemirovsky [21] for a Karmarkar-like method and Mizuno et al. [17] for a primal-dual wide-neighbourhood method derive "anticipated" bounds growing only logarithmically with n.

Instead, we seek here to understand whether the theoretical upper bounds are close to tight; perhaps a better analysis would yield worst-case bounds nearer to what is observed in practice. Thus we look for *lower* bounds on the number of iterations required. Such bounds have been investigated before, mainly for Karmarkar's original projective-scaling method. Thus Anstreicher [1] showed that $\Omega(\ln(n))$ iterations might be necessary to obtain a fixed improvement in the objective function value. Ji and Ye [8] improved Anstreicher's analysis and obtained a bound of $\Omega(n)$ iterations from a starting point quite close to the boundary. Powell [24] (see also [23]) also derived a lower bound of $\Omega(n)$ iterations for a discretisation of a semi-infinite problem, again using a starting point quite close to the boundary. These results suggest that Karmarkar's result (an upper bound of $O(nt)$ iterations) may be essentially tight.

Very recently, Bertsimas and Luo [2] considered algorithms reducing the Tanabe-Todd-Ye primal-dual potential function, and showed that the $O(\sqrt{n}t)$- iteration bound is tight by proving a similar lower bound. The algorithms they consider are "primal-or-dual" methods that are not symmetric between the primal and dual, and update an iterate in just one of these problems at each iteration.

All these papers construct a specific problem on which the algorithm performs particularly poorly; none addresses the currently popular methods used in implementations – symmetric primal-dual algorithms.

By contrast, Sonnevend et al. [26] discuss a particular problem for which the primal-dual central trajectory has large "total curvature," which shows that primal-dual algorithms that follow this trajectory closely will require $\Omega(n^{1/3})$ iterations. However, most practical algorithms use much longer step sizes, and their iterates do not stay close to the central path.

In this paper we also obtain a bound of at least $n^{1/3}$ iterations to obtain a constant factor decrease in the duality gap. The algorithm we study is the primal-dual affine-scaling algorithm, which is very close to the methods used in practical implementations. We also allow almost any reasonable step size rule, such as going 99.5% of the way to the boundary of the feasible region, again as used in practical codes; such step size rules definitely do not lead to iterates lying close to the central trajectory. However, to give the algorithms every benefit, we start at points on the central path. Note also that, since our results are lower bounds, they apply also to more general methods, e.g. those that allow infeasible iterates, and more general problems, e.g. convex quadratic programming, as long as the algorithms reduce to the studied method for feasible starts and linear programming instances.

We also discuss how far our results extend to other primal-dual interior-point methods that use directions including some centering component. Many practical algorithms include such a component to a small degree to keep the iterates from approaching the boundary too closely prematurely. Our discussion indicates that the lower bounds we obtain for the affine-scaling method can often be expected to hold for other algorithms also.

In contrast to previous constructions of lower bounds, we do not give explicit problems that cause the algorithm to take many iterations. Instead, our approach is much closer to that used in analyses of the informational complexity of nonlinear optimisation (see Nemirovsky and Yudin [22]), where an oracle is assumed to generate information about the problem instance at each iteration. In this view, the process can be viewed as a game between the oracle, which tries to generate information about the problem instance as unhelpfully as possible, and the algorithm, which uses this information as efficiently as possible and tries to ask the oracle questions which will severely limit the set of possible problem instances. (Think of the bisection algorithm to determine a zero of a continuous function of a single variable.) This is an unnatural view to take of linear programming, where the problem instance is determined by the finitely many real numbers in the data. But interior-point methods, in determining their steps at each iteration, use remarkably little of this information; the search directions are just certain projections of appropriate vectors, and a wide range of problem instances will lead to the same pair of search directions. Thus our approach is to generate, for a long sequence of iterations, a pair of singularly unhelpful search directions which lead to little improvement. Then we show that there is a problem instance that will give rise to exactly these search directions. (Of course it is necessary that no other part of the algorithm "look at" the data; hence we require that the step size rule depend only on the current iterates and search directions.)

The paper is organised as follows. In the next section, we describe the class of algorithms under discussion, show how the duality gap is reduced at each iteration, and consider possible step size rules. One particularly important algorithm in this class is the so-called primal-dual affine-scaling method, to which we confine ourselves for most of the paper. The section ends with a theorem giving sufficient conditions for a sequence of pairs of search directions to arise from some problem instance.

Section 3 provides the main tool, showing inductively that a sequence of particularly unpleasant search directions might occur. In Section 4, we demonstrate that these search directions do lead to very slight reduction of the duality gap, and hence obtain our lower bound for the primal-dual affine-scaling algorithm.

In Section 5, we argue that, as long as the iterates satisfy a certain condition, the lower bound remains valid for other primal-dual methods with a centering component in the search directions. We verify that this condition seems to hold for many such algorithms by making some computational tests in MATLAB [18].

We must mention that these results are mainly theoretical. For n less than a few billion, there is little difference between logarithmic growth in n and growth like $n^{1/3}$. Indeed, a good fit to the iteration counts in [14] can be obtained using $n^{1/3}$ instead of $\ln(n)$. The correlation coefficient in [14] using a logarithmic fit was $R^2 = 0.979$; using a fit to $n^{1/3}$ gives instead $R^2 = 0.952$.

2 Primal-Dual Interior-Point Algorithms

We consider the primal linear programming problem in standard form

$$(P) \qquad \begin{array}{rl} \min_x & c^\top x \\ & Ax = b \\ & x \geq 0, \end{array}$$

with dual problem

$$(D) \qquad \begin{array}{rl} \max_{y,s} & b^\top y \\ & A^\top y + s = c \\ & s \geq 0, \end{array}$$

where $A \in I\!R^{m \times n}$, $b \in I\!R^m$, and $c \in I\!R^n$ are the data, and $x, s \in I\!R^n$, and $y \in I\!R^m$ the variables. For any x feasible in (P) and (y, s) feasible in (D), it is easy to see that the duality gap is

$$c^\top x - b^\top y = x^\top s \geq 0; \tag{2.1}$$

the strong duality theorem of linear programming states that x and (y, s) are optimal if and only if $x^\top s = 0$.

As long as (P) is feasible, (2.1) shows that the objective function value of (y, s) is determined from s alone, and we will view our algorithms as generating a sequence of pairs (x^k, s^k) lying in

$$\begin{aligned} \mathcal{F}^0 := \{(x, s) \in I\!R^{2n} : Ax = b, A^\top y + s = c \\ \text{for some } y, x > 0, s > 0\}; \end{aligned} \tag{2.2}$$

the strict inequalities are the reason for the name "interior-point algorithms." It is easy to update y^k along with s^k so that (y^k, s^k) remains feasible in (D), but we will suppress this for simplicity.

We assume that $\mathcal{F}^0$ is nonempty, which implies that both (P) and (D) have bounded sets of optimal solutions, and that we have an initial $(x^0, s^0) \in \mathcal{F}^0$.

The pair (x^k, s^k) will be updated at each iteration by taking a damped perturbed Newton step for the optimality conditions

$$\begin{aligned} Ax &= b, && (2.3)\\ A^\top y + s &= c, && (2.4)\\ XSe &= 0, && (2.5)\\ x \geq 0, &\quad s \geq 0, && (2.6)\end{aligned}$$

where e denotes $(1, ..., 1)^\top \in I\!R^n$ and X and S the diagonal matrices with $Xe = x$, $Se = s$. The damping is carried out to maintain positivity in $\{(x^k, s^k)\}$. Frequently, in order to encourage the possibility of a full step, while maintaining positivity, a damped Newton step is taken for the perturbed system

$$\begin{aligned} Ax &= b, && (2.7)\\ A^\top y + s &= c, && (2.8)\\ XSe &= \gamma\mu e, && (2.9)\\ x \geq 0, &\quad s \geq 0, && (2.10)\end{aligned}$$

where $\mu = (x^k)^\top s^k/n$ and $0 \leq \gamma < 1$. The set of solutions to (2.7)–(2.10), for $0 < \gamma < \infty$, forms the so-called central trajectory for (P) and (D); path-following algorithms are based on approximately following this path as γ decreases to 0.

The Newton direction for (2.7)–(2.9) at (x^k, s^k) is the solution to

$$\begin{aligned} A\xi &= 0, && (2.11)\\ A^\top \eta + \sigma &= 0, && (2.12)\\ X^k\sigma + S^k\xi &= \gamma\mu e - X^kS^ke. && (2.13)\end{aligned}$$

(Superscripts are used throughout for indices; nonnegative integer powers are indicated by enclosing their arguments in parentheses.) The form of the solution to (2.11)–(2.13) can be made more apparent by the following scaling. Let

$$\begin{aligned} D^k &:= (X^k)^{1/2}(S^k)^{-1/2}, \quad V^k := (X^k)^{1/2}(S^k)^{1/2}, && (2.14)\\ v &:= v^k := V^ke, \quad w := v - \gamma\mu(V^k)^{-1}e, && (2.15)\\ \tilde{A} &:= AD^k, \quad \tilde{\xi} := (D^k)^{-1}\xi, \quad \tilde{\sigma} := D^k\sigma. && (2.16)\end{aligned}$$

Then (2.11)–(2.13) is equivalent to

$$\tilde{A}\tilde{\xi} = 0, \tag{2.17}$$

$$\tilde{A}^{\mathsf{T}}\eta + \tilde{\sigma} = 0, \tag{2.18}$$

$$\tilde{\sigma} + \tilde{\xi} = -w, \tag{2.19}$$

whose solution can be written as

$$\tilde{\xi} = -P_{\tilde{A}}w, \quad \tilde{\sigma} = -(I - P_{\tilde{A}})w, \tag{2.20}$$

where P_M denotes the matrix that projects a vector orthogonally onto the null space of M. For future reference, we note that

$$P_M = I - \bar{M}^{\mathsf{T}}(\bar{M}\bar{M}^{\mathsf{T}})^{-1}\bar{M}, \tag{2.21}$$

where the rows of $\bar{M}$ form a basis for the row space of M.

The scaling above corresponds to the change of variables taking x to $(D^k)^{-1}x$ and s to $D^k s$. Note that both x^k and s^k are thus transformed to v. If $\gamma = 0$, i.e., if we do not perturb (2.5) to (2.9), then $w = v$ and our steps in scaled space are then $\tilde{\xi} = -P_{\tilde{A}}v$, $\tilde{\sigma} = -(I - P_{\tilde{A}})v$.

We can now state our generic primal-dual interior-point algorithm:

Algorithm

Choose $(x^0, s^0) \in \mathcal{F}^0$ and set $k = 0$.
While $(x^k)^{\mathsf{T}}s^k > \tau$ do

begin

choose $0 \leq \gamma^k \leq 1$;

set

$$\mu^k := (x^k)^{\mathsf{T}}s^k/n; \tag{2.22}$$

$$D := D^k := (X^k)^{1/2}(S^k)^{-1/2}, \quad \tilde{A} := AD; \tag{2.23}$$

$$V := V^k := (X^k)^{1/2}(S^k)^{1/2}; \tag{2.24}$$

$$v := v^k := Ve, \quad w := w^k := v - \gamma^k\mu^k V^{-1}e; \tag{2.25}$$

$$\tilde{x} := \tilde{x}^k := D^{-1}x^k = v, \quad \tilde{s} := \tilde{s}^k := Ds^k = v; \tag{2.26}$$

compute

$$\tilde{\xi} := \tilde{\xi}^k := -P_{\tilde{A}}w, \quad \tilde{\sigma} := \tilde{\sigma}^k := -(I - P_{\tilde{A}})w; \tag{2.27}$$

$$\xi^k := D\tilde{\xi}, \quad \sigma^k = D^{-1}\tilde{\sigma}^k; \tag{2.28}$$

choose $\rho_P^k > 0$ and $\rho_D^k > 0$ so that

$$x^{k+1} := x^k + \rho_P^k\xi^k > 0, \quad s^{k+1} := s^k + \rho_D^k\sigma^k > 0; \tag{2.29}$$

and set $k := k + 1$.

end

This general framework includes most primal-dual methods. For instance, short-step path-following algorithms (e.g., [11, 19]) arise by taking $\gamma^k = 1 - \kappa/\sqrt{n}$ for some fixed κ and $\rho_P^k = \rho_D^k = 1$ for all k; the version of OB1 described in [15] chooses $\gamma^k = 1/n$ or $1/\sqrt{n}$ and ρ_P^k and ρ_D^k equal to .995 of their maximum values. Of particular interest to us are the primal-dual affine-scaling methods, which choose $\gamma^k = 0$ at each iteration, with various choices for the step sizes ρ_P^k and ρ_D^k [20, 16, 28]. The parameter τ in the algorithm is the stopping criterion; for theoretical purposes, we assume it is zero (or sufficiently small) in the analysis.

Proposition 2.1 *If $\rho_P^k = \rho_D^k = \rho^k$, then*

$$(x^{k+1})^\top s^{k+1} = \{1 - \rho^k(1 - \gamma^k)\}(x^k)^\top s^k. \tag{2.30}$$

Proof. Note that $\tilde{\xi}^\top \tilde{\sigma} = 0$ ($\tilde{\xi}$ lies in the null space of $\tilde{A}$, while $\tilde{\sigma}$ lies in its row space). It is convenient to write

$$\tilde{x}^+ := (D^k)^{-1} x^{k+1} = \tilde{x} + \rho_P^k \tilde{\xi} \tag{2.31}$$

and

$$\tilde{s}^+ := D^k s^{k+1} = \tilde{s} + \rho_D^k \tilde{\sigma}. \tag{2.32}$$

Then

$$\begin{aligned}
(x^{k+1})^\top s^{k+1} &= (\tilde{x}^+)^\top \tilde{s}^+ \\
&= (\tilde{x} + \rho_P^k \tilde{\xi})^\top (\tilde{s} + \rho_D^k \tilde{\sigma}) \\
&= \tilde{x}^\top \tilde{s} + \rho^k(\tilde{x}^\top \tilde{\sigma} + \tilde{\xi}^\top \tilde{s}) + (\rho^k)^2 \tilde{\xi}^\top \tilde{\sigma} \\
&= \tilde{x}^\top \tilde{s} + \rho^k(v^\top(\tilde{\sigma} + \tilde{\xi})) \\
&= \tilde{x}^\top \tilde{s} + \rho^k(-v^\top w) \\
&= \tilde{x}^\top \tilde{s} + \rho^k(-v^\top v + \gamma^k \mu^k e^\top e) \\
&= \tilde{x}^\top \tilde{s} + \rho^k(-1 + \gamma^k)\tilde{x}^\top \tilde{s} \\
&= \{1 - \rho^k(1 - \gamma^k)\}(x^k)^\top s^k.
\end{aligned}$$

Here we have used $\mu^k e^\top e = \mu^k n = (x^k)^\top s^k = \tilde{x}^\top \tilde{s}$. □

Hence to achieve a large decrease in the duality gap, we would like ρ^k to be large and γ^k to be small. Most of the results of this paper concern the affine-scaling algorithm, for which $\gamma^k = 0$ for all k. Then the algorithm will perform poorly if ρ^k is small. We will show that for a long sequence of iterations we can ensure that the directions are such that maintaining feasibility forces small values of ρ^k.

It appears that the restriction $\rho_P^k = \rho_D^k$ in Proposition 2.1 is unduly limiting on the algorithm we consider; indeed, practical algorithms of this type allow different step sizes in the primal and dual problems. However, under a very natural symmetry property, the examples we construct will have the property that $\rho_P^k = \rho_D^k$. Specifically, we assume that

$$\rho_P^k = \rho^{(k)}(x^k, \xi^k) \quad \text{and} \quad \rho_D^k = \rho^{(k)}(s^k, \sigma^k), \tag{2.33}$$

where the function $\rho^{(k)}$ satisfies the property

$$\rho^{(k)}(\Pi x^k, \Pi \xi^k) = \rho^{(k)}(x^k, \xi^k) \tag{2.34}$$

for every x^k, ξ^k and every $n \times n$ permutation matrix Π. Thus the step sizes depend only on the current iterate and search direction (and possibly the iteration number) and are symmetric between primal and dual and between different components. Hence if $s^k = \Pi x^k$ and $\sigma^k = \Pi \xi^k$ for some permutation Π, (2.33) and (2.34) imply

$$\rho_P^k = \rho_D^k. \tag{2.35}$$

We assume until Section 5 that

$$\gamma^k = 0 \quad \text{for all} \quad k, \tag{2.36}$$

so that we are considering an affine-scaling algorithm.

We describe below a particular step size rule satisfying (2.33) and (2.34). Suppose at some iteration that $\tilde{\xi} = 0$, so that $\tilde{\sigma} = -v$. Then $\rho_P^k = \rho_D^k = 1$ leads to $x^{k+1} = x^k$, $s^{k+1} = 0$, with $(x^{k+1})^\mathsf{T} s^{k+1} = 0$. It follows that x^{k+1} and s^{k+1} are optimal in (P) and (D) and the algorithm stops. But if $\tilde{\xi} = 0$, v lies in the row space of $\tilde{A}$, whence s^k and so c lie in the row space of A. In this case, at the initial iteration we would find v in the row space of $\tilde{A}$, so the algorithm would terminate immediately. Similarly, $\tilde{\sigma} = 0$ at some iteration implies $\tilde{\xi} = -v$, and then $\rho_P^k = \rho_D^k = 1$ leads to $x^{k+1} = 0$, $s^{k+1} = s^k$, and x^{k+1} and s^{k+1} are then again optimal in (P) and (D). If this happens, v lies in the null space of $\tilde{A}$, so $b = \tilde{A}v = 0$. Again, the initial iteration would then find $\tilde{\sigma} = 0$ and the algorithm would terminate immediately.

Let us therefore assume that at each iteration $\tilde{\xi} \neq 0$ and $\tilde{\sigma} \neq 0$. Then $v^\mathsf{T}\tilde{\xi} = -v^\mathsf{T} P_{\tilde{A}} v = -\|\tilde{\xi}\|^2 < 0$ shows that $\tilde{\xi}$ has a negative component, and similarly so does $\tilde{\sigma}$. Then we can write

$$\rho_P^k = \lambda_P^k \min_{j:\xi_j^k < 0} \left\{ \frac{x_j^k}{-\xi_j^k} \right\}, \quad \rho_D^k = \lambda_D^k \min_{j:\sigma_j^k < 0} \left\{ \frac{s_j^k}{-\sigma_j^k} \right\}, \tag{2.37}$$

for positive λ's, and (2.29) holds iff $\lambda_P^k, \lambda_D^k < 1$. The λ's represent the proportion of the way to the boundary of the feasible region taken by the step. Note that (2.33) and (2.34) hold if $\lambda_P^k = \lambda_D^k = \lambda^k$ depends only on the iteration k. (The code OB1 [15] chooses $\lambda^k = .995$ for all k.)

The only obvious restriction on the directions $\tilde{\xi}$ and $\tilde{\sigma}$ is that they be orthogonal and sum to $-v$. The following example from [17] shows that at least one of ρ_P^k, ρ_D^k could be forced to be very small:

$$\begin{aligned}\tilde{x} &:= \tilde{s} := v := e,\\ \tilde{\xi} &:= -(1+\sqrt{n},1,\ldots,1)^\top/2,\\ \tilde{\sigma} &:= -(1-\sqrt{n},1,\ldots,1)^\top/2.\end{aligned}$$

Then ρ_P^k must be less than $2(1+\sqrt{n})^{-1}$ in order that $\tilde{x}^+$ be positive. In order to force both ρ_P^k and ρ_D^k to be small, we modify the example slightly:

$$\tilde{x} := \tilde{s} := v := e, \tag{2.38}$$

$$\tilde{\xi} := -(1+\sqrt{n/2},1-\sqrt{n/2},1,\ldots,1)^\top/2, \tag{2.39}$$

$$\tilde{\sigma} := -(1-\sqrt{n/2},1+\sqrt{n/2},1,\ldots,1)^\top/2. \tag{2.40}$$

Then any step size rule satisfying (2.33) and (2.34) must yield $\rho_P^k = \rho_D^k =: \rho^k$, and feasibility demands $\rho^k < 2(1+\sqrt{n/2})^{-1}$. Proposition 2.1 then gives

$$(x^{k+1})^\top s^{k+1} > \left(1 - \frac{2}{1+\sqrt{n/2}}\right)(x^k)^\top s^k. \tag{2.41}$$

Suppose that in fact this happens at the initial iteration, with $x^0 = s^0 = e$. It is not hard to construct a matrix A so that (2.39)–(2.40) hold, so that (2.41) implies a very small decrease in the duality gap. Can we continue to choose "bad" directions for a long sequence of iterations? We will show in the next two sections that this is indeed possible. To conclude this section we present a result giving sufficient conditions for a sequence of directions to be produced by the affine-scaling algorithm.

Suppose at the kth iteration we have $\tilde{\xi} + \tilde{\sigma} = -v$. A sufficient condition for $\tilde{\xi}$ to be $-P_{\tilde{A}}v$ and $\tilde{\sigma}$ to be $-(I-P_{\tilde{A}})v$ is then that $\tilde{\xi}$ lies in the null space and $\tilde{\sigma}$ in the row space of $\tilde{A}$. Removing the scaling, we want ξ^k and σ^k to lie in the null space and row space of A respectively.

Theorem 2.1 *Let $(\xi^j, \sigma^j) \in \mathbb{R}^{2n}$ be given for $0 \le j \le k$. A sufficient condition that there exists $A \in \mathbb{R}^{m\times n}$ and $\eta^j \in \mathbb{R}^m$, $0 \le j \le k$, such that*

$$A\xi^j = 0,\ A^\top \eta^j + \sigma^j = 0,\ 0 \le j \le k, \tag{2.42}$$

is that

(i) $k+1 \le \min\{m, n-m\}$; and

(ii) $\Xi := [\xi^0, \ldots, \xi^k]$ and $\Sigma := [\sigma^0, \ldots, \sigma^k]$ satisfy $\Xi^\top \Sigma = 0$.

Proof. Let the columns of $\hat{\Xi}$ and $\hat{\Sigma}$ form bases for the column spaces of Ξ and Σ respectively, and suppose they have $r_\xi \le k+1$ and $r_\sigma \le k+1$ columns. By (ii), the orthogonal complement of the column space of Ξ (which is the same as that of $\hat{\Xi}$) is an $(n-r_\xi)$-dimensional subspace of $\mathbb{R}^n$ containing the r_σ linearly independent columns of $\hat{\Sigma}$ (which span those of Σ). Extend these r_σ vectors to a basis for the subspace, and let the m rows of A consist of m of these basis vectors including the r_σ columns of $\hat{\Sigma}$ ($r_\sigma \le k+1 \le m \le n-k-1 \le n-r_\xi$ from (i)). Then (2.42) clearly holds. □

3 The Inductive Step

In this section and the next we demonstrate how the bad behaviour of (2.39)–(2.41) can continue for several iterations. The 0th iteration will "contaminate" the first pair of components of x and s; similarly the $(k-1)$st will "contaminate" the kth pair of components, but all subsequent components of x and s will still be equal. It is convenient therefore to index the components in pairs. We assume that $n =: 2p$ is even, and index the components $1, -1, 2, -2, \ldots, p, -p$.

In addition, we will preserve symmetry in the first $2k$ components of x^k and s^k. Specifically, x_1^k will equal s_{-1}^k, x_{-1}^k will equal s_1^k, and so on. To describe these properties conveniently, we let $\bar{\Pi}$ denote the permutation matrix that switches the jth and $(-j)$th components of each vector in $I\!R^n$, $1 \leq j \leq p$. We also let S_n^k denote the set of permutation matrices that leave fixed the first k pairs of components of vectors in $I\!R^n$.

We suppose that $(e, e) \in \mathcal{F}^0$, and let the initial iterates be

$$x^0 = s^0 = e. \tag{3.1}$$

We assume that at the beginning of the kth iteration we have $(x^k, s^k) \in \mathcal{F}^0$ satisfying symmetry of the pairs,

$$\bar{\Pi} x^k = s^k, \tag{3.2}$$

and equality of the final components,

$$\Pi x^k = x^k,\ \Pi s^k = s^k, \text{ for all } \Pi \in S_n^k. \tag{3.3}$$

We make similar assumptions about the previous search directions. Let $\Xi^k := [\xi^0, \ldots, \xi^{k-1}]$ and $\Sigma^k := [\sigma^0, ..., \sigma^{k-1}]$ be the matrices of previous primal and dual search directions. We assume

$$(\Xi^k)^\mathsf{T} \Sigma^k = 0, \tag{3.4}$$

$$\bar{\Pi} \Xi^k = \Sigma^k, \tag{3.5}$$

and, for each $0 \leq j \leq k$,

$$\Pi \xi^{j-1} = \xi^{j-1},\ \Pi \sigma^{j-1} = \sigma^{j-1}, \text{ for all } \Pi \in S_n^j. \tag{3.6}$$

Here, (3.4) ensures that the previous search directions are consistent with some matrix A (see Theorem 2.1), while (3.5) maintains the symmetry between pairs of components. Finally, (3.6) shows that the search directions ξ^{j-1} and σ^{j-1} treat all components after the first j pairs equally.

Note that, by setting $x^0 = s^0 = e$, (3.2) and (3.3) hold for $k = 0$, while (3.4)–(3.6) hold vacuously. Also, note that, if Ξ^k and Σ^k contain the single columns (2.39) and (2.40), (3.4)–(3.6) hold for $k = 1$.

We next examine the effect of these assumptions on X^k, S^k, D^k and $v = v^k$. From (3.2) and (3.3),

$$\bar{\Pi} X^k \bar{\Pi} = S^k, \quad \text{and}$$
$$\Pi X^k \Pi^\top = X^k, \quad \Pi S^k \Pi^\top = S^k, \text{ for all } \Pi \in S_n^k.$$

We then find

$$\begin{aligned} \bar{\Pi} D^k \bar{\Pi} &= \bar{\Pi}(X^k)^{1/2}\bar{\Pi}\ \bar{\Pi}(S^k)^{-1/2}\bar{\Pi} \\ &= (S^k)^{1/2}(X^k)^{-1/2} = (D^k)^{-1}, \end{aligned} \tag{3.7}$$

and similarly

$$\Pi D^k \Pi^\top = D^k \text{ for all } \Pi \in S_n^k. \tag{3.8}$$

In the same way,

$$\begin{aligned} \bar{\Pi} v^k &= \bar{\Pi}(X^k)^{1/2}\bar{\Pi}\ \bar{\Pi}(S^k)^{1/2}\bar{\Pi}\ \bar{\Pi} e \\ &= (S^k)^{1/2}(X^k)^{1/2} e = v^k, \end{aligned} \tag{3.9}$$

and

$$\Pi v^k = v^k, \text{ for all } \Pi \in S_n^k. \tag{3.10}$$

Our final assumption is

$$\max\{v_j^k, v_{-j}^k : 1 \le j \le k\} \le v_{k+1}^k = v_{-k-1}^k = \ldots = v_p^k = v_{-p}^k. \tag{3.11}$$

(The equalities follow from (3.10).) We shall see the importance of this assumption later. We have thus made the

Inductive Hypothesis

At the beginning of the kth iteration, the iterates (x^k, s^k) and the previous search directions Ξ^k and Σ^k satisfy (3.2)–(3.6) and (3.11).

We now make the scaling D^k to primal and dual iterates and directions:

$$\begin{aligned} \tilde{x}^k &:= (D^k)^{-1} x^k = v^k, & (3.12)\\ \tilde{s}^k &:= D^k s^k = v^k, & (3.13)\\ \tilde{\Xi}^k &:= (D^k)^{-1}\Xi^k, \quad \text{and} & (3.14)\\ \tilde{\Sigma}^k &:= D^k \Sigma^k. & (3.15) \end{aligned}$$

We note that $\tilde{\Xi}^k$ and $\tilde{\Sigma}^k$ satisfy

$$(\tilde{\Xi}^k)^\top \tilde{\Sigma}^k = 0 \tag{3.16}$$

from (3.4), and (3.5) and (3.7) show that they satisfy

$$\bar{\Pi}\tilde{\Xi}^k = \tilde{\Sigma}^k. \tag{3.17}$$

From (3.6) we deduce $\Pi\Xi^k = \Xi^k$ and $\Pi\Sigma^k = \Sigma^k$ for all $\Pi \in S_n^k$, so (3.8) shows

$$\Pi\tilde{\Xi}^k = \tilde{\Xi}^k, \quad \Pi\tilde{\Sigma}^k = \tilde{\Sigma}^k, \tag{3.18}$$

for all such Π's.

The purpose of this section is to show that we can choose the next directions in the form

$$\tilde{\xi}^k = -\frac{1}{2}(v^k + (\epsilon_1, \epsilon_{-1}, \ldots, \epsilon_{k+1}, \epsilon_{-k-1}, 0, \ldots, 0)^\top), \tag{3.19}$$

$$\tilde{\sigma}^k = -\frac{1}{2}(v^k - (\epsilon_1, \epsilon_{-1}, \ldots, \epsilon_{k+1}, \epsilon_{-k-1}, 0, \ldots, 0)^\top), \tag{3.20}$$

where $\epsilon_j + \epsilon_{-j} = 0$, $1 \leq j \leq k+1$. Note that $\bar{\Pi}\tilde{\xi}^k = \tilde{\sigma}^k$ and $\Pi\tilde{\xi}^k = \tilde{\xi}^k$, $\Pi\tilde{\sigma}^k = \tilde{\sigma}^k$, for all $\Pi \in S_n^{k+1}$.

Theorem 3.1 *As long as $k+1 < p = n/2$ and the inductive hypothesis holds, we can choose $\tilde{\xi}^k$ and $\tilde{\sigma}^k$ as in (3.19)–(3.20) so that $\Xi := [\tilde{\Xi}^k, \tilde{\xi}^k]$ and $\Sigma := [\tilde{\Sigma}^k, \tilde{\sigma}^k]$ satisfy $\Xi^\top\Sigma = 0$.*

Proof. For simplicity, we write P_ξ for $P_{(\tilde{\Xi}^k)^\top}$ and P_σ for $P_{(\tilde{\Sigma}^k)^\top}$. Set

$$\begin{aligned} \tilde{\xi}^{(1)} &:= -P_\sigma v^k, \quad \tilde{\xi}^{(2)} := -(I - P_\xi)v^k, \\ \tilde{\sigma}^{(1)} &:= -P_\xi v^k, \quad \tilde{\sigma}^{(2)} := -(I - P_\sigma)v^k. \end{aligned}$$

Then $(\tilde{\Sigma}^k)^\top\tilde{\xi}^{(1)} = 0$ and, since $\tilde{\xi}^{(2)}$ lies in the column space of $\tilde{\Xi}^k$, (3.4) shows that $(\tilde{\Sigma}^k)^\top\tilde{\xi}^{(2)} = 0$. Similarly, $(\tilde{\Xi}^k)^\top\tilde{\sigma}^{(1)} = (\tilde{\Xi}^k)^\top\tilde{\sigma}^{(2)} = 0$. Thus, if we set $\tilde{\xi}^{(3)} := \frac{1}{2}(\tilde{\xi}^{(1)} + \tilde{\xi}^{(2)})$ and $\tilde{\sigma}^{(3)} := \frac{1}{2}(\tilde{\sigma}^{(1)} + \tilde{\sigma}^{(2)})$, we have

$$(\tilde{\Sigma}^k)^\top\tilde{\xi}^{(3)} = (\tilde{\Xi}^k)^\top\tilde{\sigma}^{(3)} = 0. \tag{3.21}$$

Also, from the definitions,

$$\tilde{\xi}^{(3)} + \tilde{\sigma}^{(3)} = -v^k. \tag{3.22}$$

Let $\hat{\Xi}^k$ have as columns a column basis for $\tilde{\Xi}^k$. Then from (3.17) we deduce that $\hat{\Sigma}^k = \bar{\Pi}\hat{\Xi}^k$ has as columns a column basis for $\tilde{\Xi}^k$. Using (2.21) we find

$$\begin{aligned} \bar{\Pi}\tilde{\xi}^{(1)} &= -\bar{\Pi}v^k + \bar{\Pi}\hat{\Sigma}^k((\hat{\Sigma}^k)^\top\hat{\Sigma}^k)^{-1}(\hat{\Sigma}^k)^\top v^k \\ &= -\bar{\Pi}v^k + (\bar{\Pi}\hat{\Sigma}^k)((\bar{\Pi}\hat{\Sigma}^k)^\top(\bar{\Pi}\hat{\Sigma}^k))^{-1}(\bar{\Pi}\hat{\Sigma}^k)^\top\bar{\Pi}v^k. \end{aligned}$$

But $\bar{\Pi}v^k = v^k$ by (3.9) and $\bar{\Pi}\hat{\Sigma}^k = \hat{\Xi}^k$ so $\bar{\Pi}\tilde{\xi}^{(1)} = -P_\xi v^k = \tilde{\sigma}^{(1)}$. Similarly, $\bar{\Pi}\tilde{\xi}^{(2)} = \tilde{\sigma}^{(2)}$, and thus

$$\bar{\Pi}\tilde{\xi}^{(3)} = \tilde{\sigma}^{(3)}. \tag{3.23}$$

An analogous argument using (3.10) and (3.18) shows that $\Pi\tilde{\xi}^{(1)} = \tilde{\xi}^{(1)}$, and similarly $\Pi\tilde{\xi}^{(2)} = \tilde{\xi}^{(2)}$ and $\Pi\tilde{\sigma}^{(i)} = \tilde{\sigma}^{(i)}$, $i = 1, 2$, for all $\Pi \in S_n^k$, so that

$$\Pi\tilde{\xi}^{(3)} = \tilde{\xi}^{(3)}, \quad \Pi\tilde{\sigma}^{(3)} = \tilde{\sigma}^{(3)} \text{ for all } \Pi \in S_n^k. \tag{3.24}$$

We can now almost choose $\tilde{\xi}^k = \tilde{\xi}^{(3)}$, $\tilde{\sigma}^k = \tilde{\sigma}^{(3)}$. Indeed, (3.22), (3.23) and (3.24) show that they have the correct form (3.19) and (3.20) (with $\epsilon_{k+1} = \epsilon_{-k-1} = 0$), while (3.4) and (3.21) show that we only need $(\tilde{\xi}^{(3)})^\top\tilde{\sigma}^{(3)} = 0$ to satisfy the conclusions of the theorem. Now by their definitions,

$$\begin{aligned}
(\tilde{\xi}^{(1)})^\top\tilde{\sigma}^{(1)} &= (v^k)^\top P_\sigma P_\xi v^k, \\
(\tilde{\xi}^{(1)})^\top\tilde{\sigma}^{(2)} &= (v^k)^\top P_\sigma(I - P_\sigma)v^k = 0, \\
(\tilde{\xi}^{(2)})^\top\tilde{\sigma}^{(1)} &= (v^k)^\top(I - P_\xi)P_\xi v^k = 0, \\
(\tilde{\xi}^{(2)})^\top\tilde{\sigma}^{(2)} &= (v^k)^\top(I - P_\xi)(I - P_\sigma)v^k.
\end{aligned}$$

Since $I - P_\xi = R(\tilde{\Xi}^k)^\top$ for some R and $I - P_\sigma = \tilde{\Sigma}^k S$ for some S, (3.4) shows that $(\tilde{\xi}^{(2)})^\top\tilde{\sigma}^{(2)} = 0$ also. Now, since $(\tilde{\Xi}^k)^\top\tilde{\Sigma}^k = 0$,

$$P_\sigma P_\xi = P_M =: P_{\xi\sigma} \quad \text{for} \quad M = [\tilde{\Xi}^k, \tilde{\Sigma}^k]^\top,$$

and

$$\begin{aligned}
(\tilde{\xi}^{(3)})^\top\tilde{\sigma}^{(3)} &= \frac{1}{4}(\tilde{\xi}^{(1)})^\top\tilde{\sigma}^{(1)} \\
&= \frac{1}{4}(v^k)^\top P_{\xi\sigma} v^k \\
&= \frac{1}{4}\|P_{\xi\sigma} v^k\|^2 \geq 0.
\end{aligned} \tag{3.25}$$

We then define

$$\begin{aligned}
\tilde{\xi}^k &:= \tilde{\xi}^{(3)} + (0, \ldots, 0, -\epsilon, +\epsilon, 0, \ldots, 0)^\top, \\
\tilde{\sigma}^k &:= \tilde{\sigma}^{(3)} + (0, \ldots, 0, +\epsilon, -\epsilon, 0, \ldots, 0)^\top,
\end{aligned}$$

where the epsilons occur in the $(k+1)$st pair of components and where $\epsilon \geq 0$ is chosen so that $(\tilde{\xi}^k)^\top\tilde{\sigma}^k = 0$. (Thus, $\epsilon = (\frac{1}{8})^{1/2}\|P_{\xi\sigma}v^k\|$.) Then $\tilde{\xi}^k$ and $\tilde{\sigma}^k$ are of the correct form (3.19), (3.20) (with $\epsilon_{k+1} \geq 0$), and defining Ξ and Σ as in the theorem, we easily check that $\Xi^\top\Sigma = 0$. (Note that each column of $\tilde{\Xi}^k$ has equal entries in the $(k+1)$st pair of components, so $(\tilde{\Xi}^k)^\top\tilde{\sigma}^k = (\tilde{\Xi}^k)^\top\tilde{\sigma}^{(3)} = 0$, and similarly $(\tilde{\Sigma}^k)^\top\tilde{\xi}^k = 0$.) □

Theorem 3.1 establishes almost all of the inductive hypothesis with k replaced by $k+1$. Indeed, the new matrices of past search directions are

$$\Xi^{k+1} = D^k\Xi, \quad \Sigma^{k+1} = (D^k)^{-1}\Sigma,$$

where Ξ and Σ are defined in the theorem. Then (3.4)–(3.6) hold for $k+1$ by the conclusions of the theorem, the inductive hypothesis for k, and the form (3.19)–(3.20) of the new directions.

It only remains to show that (3.2), (3.3), and (3.11) hold for $k+1$. If we assume that (2.33) and (2.34) hold, then (3.2) for k and $\bar{\Pi}\xi^k = \sigma^k$ show that $\rho_P^k = \rho_D^k$, and hence that $\bar{\Pi}x^{k+1} = s^{k+1}$. Similarly, (3.3) and $\Pi\xi^k = \xi^k$, $\Pi\sigma^k = \sigma^k$ for all $\Pi \in S_n^{k+1}$ show that (3.3) holds for $k+1$.

We now turn to (3.11). We see that

$$\begin{aligned} v_j^{k+1} &= \{x_j^{k+1}s_j^{k+1}\}^{1/2} = \{([D^k]^{-1}x^{k+1})_j(D^k s^{k+1})_j\}^{1/2} \\ &= \{(v_j^k + \rho^k\tilde{\xi}_j^k)(v_j^k + \rho^k\tilde{\sigma}_j^k)\}^{1/2} \end{aligned}$$

for each j, $1 \le |j| \le p$, where $\rho^k := \rho_P^k = \rho_D^k$. Let $\bar{\rho}^k := \rho^k/2$. Then, from (3.19) and (3.20),

$$\begin{aligned} v_j^{k+1} &= \{([1-\bar{\rho}^k]v_j^k + \bar{\rho}^k\epsilon_j)([1-\bar{\rho}^k]v_j^k - \bar{\rho}^k\epsilon_j)\}^{1/2} \\ &= \{(1-\bar{\rho}^k)^2(v_j^k)^2 - (\bar{\rho}^k\epsilon_j)^2\}^{1/2} \le (1-\bar{\rho}^k)v_j^k \end{aligned}$$

for $1 \le |j| \le k+1$, while

$$v_j^{k+1} = \{([1-\bar{\rho}^k]v_j^k)([1-\bar{\rho}^k]v_j^k)\}^{1/2} = (1-\bar{\rho}^k)v_j^k$$

for $k+1 < |j| \le p$. Thus (3.11) for k establishes that it remains true for $k+1$.

We therefore have

Theorem 3.2 *Suppose the inductive hypothesis holds for $k < p$, and that at the kth iteration of the primal-dual affine-scaling algorithm, the step sizes are chosen by some rule satisfying (2.33) and (2.34). Then search directions can be chosen such that the inductive hypothesis remains true for $k+1$.* □

4 The Main Result

By Theorem 3.2, we can continue generating search directions satisfying the inductive hypothesis for $p = n/2$ iterations. Moreover, for $m < n$, the first $\min\{m, n-m\}$ of these iterations will be consistent with some $m \times n$ matrix A by Theorem 2.1, and hence, if we choose $b = Ae$ and $c = A^\top y + e$ for any y, will be the directions obtained in the affine-scaling algorithm applied to (P) starting at $x^0 = s^0 = e$.

We now examine how the duality gap changes in the kth iteration, assuming that the inductive hypothesis holds and the search directions are given by (3.19) and (3.20). We suppose the step sizes are chosen by (2.33) and (2.34) so that $\rho_P^k = \rho_D^k =: \rho^k =: 2\bar{\rho}^k$. We show that $\bar{\rho}^k$ must be very small, and apply Proposition 2.1.

Let β denote $v_{k+1}^k = v_{-k-1}^k = \ldots = v_p^k = v_{-p}^k$. Then

$$(v^k)^\top v^k \geq 2(p-k)(\beta)^2.$$

Now $\tilde{\xi}^k$ and $\tilde{\sigma}^k$ (and hence $2\tilde{\xi}^k$ and $2\tilde{\sigma}^k$) are orthogonal, and using (3.19) and (3.20) this yields

$$\sum_{1\leq|j|\leq k+1} (\epsilon_j)^2 = (v^k)^\top v^k \geq 2(p-k)(\beta)^2.$$

Hence some $|\epsilon_j|$ is at least $\sqrt{\frac{p-k}{k+1}}\beta$. Recalling that $\epsilon_j + \epsilon_{-j} = 0$, we have

$$\epsilon_j \geq \sqrt{\frac{n-2k}{2k+2}}\beta \tag{4.1}$$

for some $1 \leq |j| \leq k+1$. Since

$$\tilde{x}_j^+ = (1-\bar{\rho}^k)v_j^k - \bar{\rho}^k\epsilon_j > 0,$$

we have

$$\bar{\rho}^k < \frac{v_j^k}{\epsilon_j + v_j^k} \leq \frac{\beta}{\epsilon_j + \beta} \leq \frac{1}{1+\sqrt{\frac{n-2k}{2k+2}}}, \tag{4.2}$$

where we have used $v_j^k \leq \beta$ from (3.11).

For $k = 0$, the bound (4.2) is like that which led to (2.41), giving a reduction of only a factor of $1 - O(n^{-1/2})$ in the duality gap. As k increases, the bound in (4.2) increases. However, suppose K is the first iteration for which

$$(x^K)^\top s^K \leq \exp(-t)(x^0)^\top s^0.$$

Then, using Proposition 2.1,

$$\prod_{k=0}^{K-1}(1-2\bar{\rho}^k) \leq \exp(-t),$$

whence

$$\sum_{k=0}^{K-1} \ln(1+\frac{2\bar{\rho}^k}{1-2\bar{\rho}^k}) = -\sum_{k=0}^{K-1}\ln(1-2\bar{\rho}^k) \geq t,$$

and, using $\ln(1+\theta) \leq \theta$, we obtain

$$\sum_{k=0}^{K-1}\left(\frac{2}{1/\bar{\rho}^k - 2}\right) \geq t.$$

Next, substituting the bound (4.2) on $\bar{\rho}^k$, we reach

$$\sum_{k=0}^{K-1} \left(\frac{2\sqrt{2}\sqrt{k+1}}{\sqrt{n-2k}-\sqrt{2k+2}} \right) \geq t. \tag{4.3}$$

Now let us assume that

$$K \leq n/50,$$

so that, for $k \leq K-1$, $\sqrt{2k+2} \leq \sqrt{2K} \leq \sqrt{n}/5$ and $\sqrt{n-2k} \geq \sqrt{n-2K} \geq \sqrt{24n/25} \geq 19\sqrt{n}/20$. Thus the denominators in (4.3) are always at least $3\sqrt{n}/4$, and we deduce that

$$\sum_{k=1}^{K} \sqrt{k} \geq \frac{1}{2\sqrt{2}} \cdot \frac{3}{4} \cdot \sqrt{n} t.$$

But

$$\sum_{k=1}^{K} \sqrt{k} \leq \int_{1}^{K+1} \sqrt{\theta} d\theta \leq \frac{2}{3}(K+1)^{3/2},$$

whence

$$K+1 \geq (\frac{3}{2} \cdot \frac{1}{2\sqrt{2}} \cdot \frac{3}{4} \cdot \sqrt{n} t)^{2/3} \geq .54 n^{1/3} t^{2/3}. \tag{4.4}$$

Combining this analysis with Theorems 2.1 and 3.2, we arrive at

Theorem 4.1 *Consider a primal-dual affine-scaling algorithm that uses a step size rule satisfying (2.33) and (2.34) at each iteration. Also, suppose $n/50 \leq m \leq 49n/50$. Then there is an instance of (P), with $A \in I\!R^{m \times n}$, $b = Ae \in I\!R^m$, and $c = A^\top y + e \in I\!R^n$ for any $y \in I\!R^m$, such that to decrease the duality gap by a factor of $\exp(t)$, starting with $x^0 = e$ and $s^0 = e$, the algorithm requires at least*

$$\min\{n/50, .54 n^{1/3} t^{2/3} - 1\}$$

iterations.

(Observe that for large n and moderate t, the second term attains the minimum. Hence for n at least 1000, it takes no less than $n^{1/3} - 1$ iterations to achieve the modest reduction of a factor of $\exp(3)$ – about 20 – and at least $2n^{1/3} - 1$ iterations to achieve a reduction of a factor of $\exp(8)$ – about 3000.)

We conclude this section with a few remarks. Note that we have to be given the algorithm's step size rule before we can construct the bad example, and this must satisfy (2.33) and (2.34). Thus the rule cannot use any information about the problem instance besides the current iterate and search direction. (It could use some properties of previous iterates and search directions, as long as it is symmetrical enough to yield $\rho_P^k = \rho_D^k$.) But it cannot inspect A globally, because A is not yet known – only after constructing bad search directions for $\Omega(n^{1/3})$ iterations do we provide the matrix A that produced them. Subject to this minor restriction, any primal-dual affine-scaling algorithm requires $n^{1/3}$ iterations to achieve even a modest constant factor reduction in the duality gap.

5 Extensions and Discussion

In this section we investigate how the analysis of Sections 3 and 4 can be extended to more general primal-dual interior-point algorithm using a centering component. Thus, at some or all iterations, γ^k in (2.25) is nonzero, and $\tilde{\xi}^k$ and $\tilde{\sigma}^k$ are projections of $w^k \neq v^k$, where we recall that

$$w^k := v^k - \gamma^k \mu^k V^{-1} e = v^k - \gamma^k ((v^k)^\top v^k / n) V^{-1} e. \tag{5.1}$$

Like the step sizes, we allow γ^k to depend on k and the current iterates x^k and s^k, but not on any other information in A, b, or c.

We again assume that the algorithm is initiated with $x^0 = s^0 = e$, and that at iteration k, the current iterate $(x^k, s^k) \in \mathcal{F}^0$ satisfies (3.2) and (3.3) and the past search directions Ξ^k and Σ^k satisfy (3.4)–(3.6). As in Section 3, these assumptions imply that D^k and v^k satisfy (3.7)–(3.10), and we deduce from (5.1) that

$$\bar{\Pi} w^k = w^k \quad \text{and} \quad \Pi w^k = w^k \quad \text{for all } \Pi \in S_n^k. \tag{5.2}$$

It may be unrealistic to assume (3.11) when there is a centering component in the search direction; let us instead assume that

$$\alpha \max\{v_j^k, v_{-j}^k : 1 \leq j \leq k\} \;\leq\; v_{k+1}^k = v_{-k-1}^k = \ldots = v_p^k = v_{-p}^k, \tag{5.3}$$

where $0 < \alpha \leq 1$.

With this new inductive hypothesis, we can check that Theorem 3.1 remains true where now, in (3.19) and (3.20) v^k is replaced with w^k. We call the resulting expressions (3.19)$'$ and (3.20)$'$. Moreover, Theorem 3.2 is still valid, except that the inequality in (5.3) may not be true for $k+1$.

We now turn to the effect of the kth iteration on the duality gap as in the previous section. Our tool is again Proposition 2.1. To simplify the notation, we suppress the dependence on k. If $\gamma = 1$ (a so-called *centering* or *constant-cost centering* step), then the proposition shows that the duality gap is unchanged.

If $\gamma^k = 1 - \kappa/\sqrt{n}$ for some fixed κ and $\rho^k = 1$, as in short-step path-following methods, we see

$$(x^{k+1})^\top s^{k+1} = (1 - \kappa/\sqrt{n})(x^k)^\top s^k,$$

and hence $\Omega(n^{1/2})$ steps will be necessary to achieve a constant factor reduction in the duality gap. Our interest is in algorithms using small but possibly nonzero values of γ, perhaps interspersed with centering steps. Thus we assume that

$$0 \leq \gamma \leq \bar{\gamma} < \frac{1}{2}, \tag{5.4}$$

where $\bar{\gamma}$ does not depend on the iteration number.

Again, let $\rho_P^k = \rho_D^k =: \rho^k =: 2\bar{\rho}^k$. We wish to show that $\bar{\rho}^k$ is small, so that the reduction in the duality gap (no better than the factor $1 - 2\bar{\rho}^k$) is also small. As in Section 4, we find, for $1 \leq |j| \leq k+1$,

$$\begin{aligned} \tilde{x}_j^+ &= \tilde{x}_j + 2\bar{\rho}^k \tilde{\xi}_j \\ &= v_j + \bar{\rho}^k(-v_j + \gamma\mu v_j^{-1} - \epsilon_j) \end{aligned}$$

using (5.1) and (3.19)$'$. Hence, if there is such a j with $(v_j)^2 + \epsilon_j v_j - \gamma\mu > 0$, we have

$$\bar{\rho}^k < \frac{(v_j)^2}{(v_j)^2 + \epsilon_j v_j - \gamma\mu}. \tag{5.5}$$

Thus we wish to show that there exists a j such that the denominator of (5.5) is large.

Since $\tilde{\xi}^k$ and $\tilde{\sigma}^k$ are orthogonal, we find as before

$$\sum_{1\leq|j|\leq k+1} (\epsilon_j)^2 = w^\top w = (1-2\gamma)v^\top v + (\gamma\mu)^2 \sum_{1\leq|i|\leq p} v_i^{-2},$$

where the last equation follows from the definition (5.1) of w. Hence

$$\sum_{1\leq|j|\leq k+1} ((\epsilon_j)^2 - (\gamma\mu)^2 v_j^{-2}) \geq (1-2\gamma)n\mu,$$

and there is therefore some j, $1 \leq |j| \leq k+1$, with

$$(\epsilon_j)^2 \geq (\gamma\mu)^2 v_j^{-2} + \frac{(1-2\gamma)n}{2k+2}\mu$$

or

$$v_j|\epsilon_j| \geq \gamma\mu\left(1 + \frac{1-2\gamma}{(\gamma)^2} \cdot \frac{(v_j)^2}{\mu} \cdot \frac{n}{2k+2}\right)^{1/2}. \tag{5.6}$$

Since $\epsilon_j + \epsilon_{-j} = 0$, we may assume that $\epsilon_j > 0$, and hence the denominator in (5.5) is positive.

We now use (5.3) to get a bound on $(v_j)^2/\mu$. Indeed

$$n\mu = v^\top v \geq (n-2k)(v_p)^2 \geq (n-2k)(\alpha)^2(v_j)^2,$$

so that

$$\delta := \frac{(v_j)^2}{\mu} \leq \frac{n}{n-2k}\alpha^{-2} \leq 2\alpha^{-2} \tag{5.7}$$

as long as $k \leq n^{1/3} - 1$ so that $n - 2k \geq n/2$. Combining (5.5) and (5.6), and using δ as in (5.7), we obtain

$$\bar{\rho}^k < \frac{\delta}{\delta + \gamma\left[\left(1 + \frac{1-2\gamma}{(\gamma)^2} \cdot \frac{n}{2k+2} \cdot \delta\right)^{1/2} - 1\right]} = \frac{\delta}{\delta + \gamma[(1+\nu\delta)^{1/2} - 1]}$$
$$=: \frac{1}{f(\delta)},$$

where $\nu := \frac{1-2\gamma}{(\gamma)^2} \cdot \frac{n}{2k+2} > 0$ using (5.4).

It is easy to show that $f(\delta)$ is monotonically decreasing for $\delta > 0$, and thus we can obtain a valid bound on $\bar{\rho}^k$ by substituting the bound on δ in (5.7). Hence we find, for $k \leq n^{1/3} - 1$,

$$\begin{aligned}
\bar{\rho}^k &< \frac{2}{2 + \gamma(\alpha)^2\left[\left(1 + \frac{2(1-2\gamma)}{(\alpha\gamma)^2} \cdot \frac{n}{2k+2}\right)^{1/2} - 1\right]} \\
&\leq \frac{2}{\gamma(\alpha)^2\left(1 + \frac{2(1-2\gamma)}{(\alpha\gamma)^2} \cdot \frac{n}{2k+2}\right)^{1/2}} \\
&\leq \frac{2}{\gamma(\alpha)^2\left(\frac{2(1-2\gamma)}{(\alpha\gamma)^2} \cdot \frac{n}{2k+2}\right)^{1/2}} \\
&= \frac{2}{(1-2\gamma)^{1/2}\alpha}\sqrt{\frac{k+1}{n}} \\
&\leq \frac{2}{(1-2\bar{\gamma})^{1/2}\alpha}\sqrt{\frac{k+1}{n}},
\end{aligned} \tag{5.8}$$

where we have used the uniform bound (5.4) on γ.

We therefore have

Theorem 5.1 *Suppose at the kth iteration, with $k \leq n^{1/3} - 1$, the inductive hypothesis holds, with condition (5.3) replacing (3.11). Suppose that the search directions are given by (3.19)′ and (3.20)′. Then, if the step size rule satisfies (2.33) and (2.34), and if either $\gamma^k = 1$ or $0 \leq \gamma^k \leq \bar{\gamma} < \frac{1}{2}$, then*

$$(x^{k+1})^\top s^{k+1} \geq \left(1 - \frac{4}{(1-2\bar{\gamma})^{1/2}\alpha} n^{-1/3}\right)(x^k)^\top s^k.$$

Moreover, if (5.3) holds for all such iterations, then $\Omega(n^{1/3})$ iterations are required to obtain a constant factor decrease in the duality gap. □

The weakness of this result, in contrast to Theorem 4.1, is that we must *assume* that the iterates satisfy (5.3) at each iteration, whereas this could be *proved* by induction (with $\alpha = 1$) in the case of the affine-scaling algorithm. One way to assure (5.3) is to require that all iterates satisfy the fairly weak centering condition that they lie in

$$\{(x,s) \in \mathcal{F}^0 : \|XSe - \mu e\|_\infty \leq \chi\mu, \text{ where } \mu = x^T s/n\}$$

for some $0 < \chi < 1$ (see [17]). Then all $x_j s_j'$s lie within a factor of $(1+\chi)/(1-\chi) \leq (1-\chi)^{-2}$ of each other, and (5.3) holds for $\alpha = 1 - \chi$. However, this centering condition is not imposed for most practical implementations. In order to see whether this assumption is reasonable in practice, we have made a number of computational runs with $n = 10^6$ using MATLAB [18].

At each iteration, we do not compute the search directions by projections. Instead, we use the form (3.19)′–(3.20)′ of the directions, computing $\epsilon_1^k \ldots \epsilon_k^k$ by the conditions $(\tilde{\Sigma}^k)^\top \tilde{\xi}^k = 0$, and then ϵ_{k+1}^k so that $(\tilde{\xi}^k)^\top \tilde{\sigma}^k = 0$. In theory, $\epsilon_{k+1}^k = (\frac{1}{2}(\tilde{\xi}^{(3)})^\top \tilde{\sigma}^{(3)})^{1/2} = \frac{1}{8^{1/2}} \| P_{\xi\sigma} w^k \|$ is real (see the proof of Theorem 3.1), but we found many times that $(\tilde{\xi}^{(3)})^\top \tilde{\sigma}^{(3)}$ was numerically negative and these runs were aborted. (Again, $\tilde{\xi}^{(3)}$ and $\tilde{\sigma}^{(3)}$ were not computed by projections; they are the vectors of the form (3.19)′ and (3.20)′, with $\epsilon_{k+1}^k = 0$, satisfying $(\tilde{\Sigma}^k)^\top \tilde{\xi}^{(3)} = 0$, $(\tilde{\Xi}^k)^\top \tilde{\sigma}^{(3)} = 0$.) These numerical difficulties were caused by near linear dependence in the columns of $\tilde{\Sigma}^k$ and $\tilde{\Xi}^k$. However the remaining runs (about a fifth of the total) successfully completed 150 iterations ($n^{1/3} = 100$) with the anticipated results.

We tried keeping γ^k fixed at the level $0, .1, \ldots, .9$ (not all satisfying (5.4)), with step sizes chosen by (2.37) with $\lambda_p^k = \lambda_D^k = \lambda$ equal to .1, .2, ..., .9, .95, .99. In these runs, (5.3) was always satisfied with $\alpha = 1$, i.e., (3.11) remained true. We next tried the same values for λ with γ^k fixed at .95, .97, .99 and .995. Even with such a large centering component, (5.3) remained true for $\alpha = .98$.

Finally, we tried predictor-corrector algorithms as in [17] – see also [25, 26]. Here every other iteration used $\gamma^k = 1$ (a centering step), while the remaining iterations used a fixed γ^k with the values mentioned in the previous paragraph. For the noncentering steps, we used the step size rule of the previous paragraph, with the values of λ listed there. For the centering steps, we chose ρ to be the smaller of 1 and that given by strategy above; $\rho = 1$ corresponds to a Newton step for a centering problem, and was chosen in all but two runs. Once again, (5.3) was true for all runs if we set $\alpha = .75$.

It thus appears that the poor behaviour established rigorously for the primal-dual affine-scaling algorithm might also be exhibited by most primal-dual algorithms of the form given in Section 2. We must stress again, however, that this is a theoretical bound on the asymptotic behaviour of such algorithms. For problems with a million variables, $n^{1/3}$ is only 100, and this is not an unreasonable number of iterations, and seems to be in line with what has been observed in practice for such large problems – see Lustig, Marsten and Shanno [15].

References

[1] K. M. Anstreicher, "On the performance of Karmarkar's algorithm over a sequence of iterations," SIAM Journal on Optimization **1**, 22-29 (1991).

[2] D. Bertsimas and X. Luo, "On the worst case complexity of potential reduction algorithms for linear programming." Working Paper 3558-93, Sloan School of Management, MIT, Cambridge, MA 02139, USA (1993).

[3] R. E. Bixby, J. W. Gregory, I. J. Lustig, R. E. Marsten, and D. F. Shanno, "Very large-scale linear programming: A case study in combining interior point and simplex methods," Operations Research **40**, 885-897 (1992).

[4] K. H. Borgwardt, "The Simplex Method, a Probabilistic Analysis." Springer-Verlag, Berlin, 1987.

[5] G. B. Dantzig, "Linear Programming and Extensions." Princeton University Press, Princeton, NJ, 1960.

[6] D. Goldfarb and M. J. Todd, "Linear Programming." In: G. L. Nemhauser, A. H. G. Rinnooy Kan, and M. J. Todd, eds., "Optimization," volume 1 of Handbooks in Operations Research and Management Science, pp. 73-170, North Holland, Amsterdam, The Netherlands, 1989.

[7] C. C. Gonzaga, "Path following methods for linear programming," SIAM Review **34**, 167-227 (1992).

[8] J. Ji and Y. Ye, "A complexity analysis for interior-point algorithms based on Karmarkar's potential function." Working Paper, Dept. of Mathematics, The University of Iowa, Iowa City, IA 52242, USA, revised (1991).

[9] N. K. Karmarkar, "A new polynomial-time algorithm for linear programming," Combinatorica **4**, 373-395 (1984).

[10] V. Klee and G. J. Minty, "How good is the simplex algorithm." In: O. Shisha, ed., Inequalities III, pp. 159-175, Academic Press, New York, 1972.

[11] M. Kojima, S. Mizuno, and A. Yoshise, "A polynomial-time algorithm for a class of linear complementarity problems," Mathematical Programming **44**, 1-26 (1989).

[12] M. Kojima, S. Mizuno, and A. Yoshise, "An $O(\sqrt{n}L)$ iteration potential reduction algorithm for linear complementarity problems," Mathematical Programming **50**, 331-342 (1991).

[13] E. Kranich, "Interior point methods for mathematical programming: A bibliography." Discussion Paper 171, Institute of Economy and Operations Research, FernUniversität Hagen, P.O. Box 940, D-5800 Hagen 1, West-Germany (1991). The (actual) bibliography can be accessed electronically by sending e-mail to 'netlib@research.att.com' with the message 'send intbib from bib.'

[14] I. J. Lustig, R. E. Marsten, and D. F. Shanno, "The primal-dual interior point method on the Cray supercomputer." In: T. F. Coleman and Y. Li, eds., Large-Scale Numerical Optimization, volume 46 of SIAM Proceedings in Applied Mathematics, pp. 70-80, Society of Industrial and Applied Mathematics (SIAM), Philadelphia, PA, USA, 1990.

[15] I. J. Lustig, R. E. Marsten, and D. F. Shanno, "Computational experience with a primal-dual interior point method for linear programming," Linear Algebra and Its Applications **152**, 191-222 (1991).

[16] S. Mizuno and A. Nagasawa, "A primal-dual affine scaling potential reduction algorithm for linear programming." Research Memorandum 427, The Institute of Statistical Mathematics, Tokyo, Japan (1992).

[17] S. Mizuno, M. J. Todd, and Y. Ye, "On adaptive step primal-dual interior-point algorithms for linear programming." Technical Report 944, School of Operations Research and Industrial Engineering, Cornell University, Ithaca, NY 14853-3801, USA (1990). To appear in Mathematics of Operations Research.

[18] C. B. Moler, J. Little, S. Bangert and S. Kleiman, "Pro-Matlab User's Guide." MathWorks, Sherborn, MA, 1987.

[19] R. D. C. Monteiro and I. Adler, "Interior path following primal-dual algorithms: Part I: Linear programming," Mathematical Programming **44**, 27-41 (1989).

[20] R. D. C. Monteiro, I. Adler, and M. G. C. Resende, "A polynomial-time primal-dual affine scaling algorithm for linear and convex quadratic programming and its power series extension," Mathematics of Operations Research **15**, 191-214 (1990).

[21] A. S. Nemirovsky, "An algorithm of the Karmarkar type," Tekhnicheskaya Kibernetika **1**, 105-118 (1987). Translated in: Soviet Journal on Computers and System Sciences **25**, 61-74 (1987).

[22] A.S. Nemirovsky and D.B. Yudin, "Problem Complexity and Method Efficiency in Optimization." John Wiley and Sons, Chichester, 1983.

[23] M. J. D. Powell, "The complexity of Karmarkar's algorithm for linear programming." In: D. F. Griffiths and G. A. Watson, eds., Numerical Analysis 1991, volume 260 of Pitman Research Notes in Mathematics, pp. 142-163, Longman, Burnt Hill, UK, 1992.

[24] M. J. D. Powell, "On the number of iterations of Karmarkar's algorithm for linear programming." Technical Report DAMTP 1991/NA23, Dept. of Applied Mathematics and Theoretical Physics, University of Cambridge, Silver Street, Cambridge CB3 9EW, UK (1991).

[25] G. Sonnevend, J. Stoer, and G. Zhao, "On the complexity of following the central path by linear extrapolation in linear programming," Methods of Operations Research **62**, 19-31 (1990).

[26] G. Sonnevend, J. Stoer, and G. Zhao, "On the complexity of following the central path of linear programs by linear extrapolation II," Mathematical Programming **52**, 527-553 (1991).

[27] M. J. Todd, "Recent developments and new directions in linear programming." In: M. Iri and K. Tanabe, eds., Mathematical Programming: Recent Developments and Applications, pp. 109-157, Kluwer Academic Press, Dordrecht, The Netherlands, 1989.

[28] L. Tunçel, "A note on the primal-dual affine scaling algorithms." Technical Report 92-8, Cornell Computational Optimization Project, Center for Applied Mathematics, Cornell University, Ithaca, NY 14853-3801, USA (1992).

Acknowledgement

This research was supported in part by NSF, AFOSR, and ONR through NSF Grant DMS-8920550.

Dr. M.J. Todd
School of Operations Research and Industrial Engineering
Cornell University
Ithaca, New York 14853-3801
USA

P TOWNSEND AND M F WEBSTER

Computational analysis in rheological flow problems

1 Introduction

Numerical techniques have been well established as essential tools in the study of classical fluid dynamics for quite some time. For the study of the flow of rheologically complex fluids the picture is rather different. It is only relatively recently that one has seen the emergence of commercially available software packages specifically designed for this area. Even these have their limitations and there are many unresolved problems to tax the abilities of the numerical analyst.

In this paper a brief survey is given of some of these problems in the context of particular applications in the process industry. A finite element technique, which may be used to address such problems, is described, and some results presented to illustrate the effectiveness of the technique.

2 Non-Newtonian Fluids

A Newtonian fluid is said to be one that satisfies Newton's law:-

> *'The resistance which arises from the lack of slipperiness of the parts of the liquid, other things being equal, is proportional to the velocity with which the parts of the liquid are separated from one another'.*

Such a law is remarkably accurate for some fluids, and in particular, water, and as such, of course, gives rise to the Navier Stokes equations, which may be used for modelling a wide range of applications. It is the case, however, that the vast majority of fluids do not satisfy such a law, and we refer to those fluids as non-Newtonian fluids. In particular, there are many fluids for which

(a) the viscosity is a function of shear rate,

(b) the viscosity is a function of elongation rate,

(c) the fluid exhibits elastic properties (a viscoelastic liquid).

To emphasise that there are many fluids which exhibit one or more of these properties we list the following:-

polymer solutions	polymer melts
liquid detergents	multigrade oils
non-drip paints	contact adhesives
foodstuffs	blood and other body fluids

It is the case that one wishes to exploit the fluid properties to give different flow behaviour. If we take paint as an example, one requires a low viscosity when the paint is brushed onto a surface, but then, when the brushing stress is removed, a high viscosity is needed to prevent the paint flowing (running) off the surface. For further information on the rheology of a range of materials, the reader is directed to Barnes et al. [1989].

The properties described under (a) – (c) above give rise to many unexpected flow phenomena. Often the fluid behaviour is totally different from that you would expect on the basis of the Navier–Stokes equations. A pictorial record of many experimental observations of non-Newtonian liquids is to be found in Boger and Walters [1993]. Such flow phenomena provide a significant challenge to those who wish to model such flows. One can identify three major tasks. Firstly one must construct appropriate constitutive models to describe the fluid behaviour. Secondly, it is necessary to design experiments to measure the parameters associated with the models, and finally one needs advanced numerical procedures to solve the relevant equations for specific flows.

In this paper we concentrate on the last of these tasks, and give consideration to numerical techniques for the solution of the partial differential equations derived from the Phan-Thien/Tanner (PTT) model, Phan-Thien and Tanner [1977]. Such a model, which is popular at the present time, is able to represent all of the properties (a) – (c) above, and in particular, viscoelastic behaviour.

3 The equations

The PTT constitutive equations are given by

$$\mathbf{T} = \tau + 2\mu_2\mathbf{D} \tag{3.1}$$

$$\left\{1 + \frac{\varepsilon\lambda_1}{\mu_1}\mathrm{trace}(\tau)\right\}\tau + \lambda_1\mathcal{D}\tau = 2\mu_1\mathbf{D} \tag{3.2}$$

$$\mathcal{D}\tau = (1-\xi/2)\overset{\nabla}{\tau} + (\xi/2)\overset{\Delta}{\tau} \tag{3.3}$$

where in equation (3.1) the extra stress tensor $\mathbf{T}$ has been split into a conventional Newtonian contribution $2\mu_2\mathbf{D}$ and an elastic contribution τ. τ satisfies the differential equation (3.2) where $\mathcal{D}$ is a material derivative comprised of a linear combination of lower and upper convected Oldroyd derivatives(cf. Oldroyd [1950]), and ε, λ_1, and ξ are material constants. For a polymer solution for example, μ_1 and μ_2 may be thought of respectively as the viscosities of the polymer and its solvent.

Coupled with equations (3.1) and (3.2) one must also satisfy the equations for the conservation of mass and momentum, namely

$$\nabla\cdot\mathbf{u} = 0 \tag{3.4}$$

$$\rho\mathbf{u}_t + \rho\mathbf{u}\cdot\nabla\mathbf{u} = \nabla\cdot\mathbf{T} - \nabla p \tag{3.5}$$

where subscript t denotes time differentiation, $\mathbf{u}$ is the velocity vector, ρ is the fluid density and p is the pressure.

The system represented by equations (3.1) - (3.5) is notoriously difficult to solve. For the majority of practical flows of importance in the processing industry, viscosities are high and consequently, Reynolds number are low so that the non-linear convective terms in the momentum equations do not dominate. However, comparable terms in the constitutive equations, whose magnitude depend essentially on the 'elastic' time constant, λ_1, give rise to major problems for highly elastic materials, and for many years it was not possible to obtain solutions for high values of λ_1. This problem became known as the 'high Weissenberg number problem' and it is only relatively recently that progress has been made by taking great care in the construction of appropriate numerical solution techniques, see for example Marchal and Crochet [1987].

Apart from the above, the modelling of non-Newtonian flows is subject to a variety of other problems which we may summarise as follows:

- The fluid viscosity can be very sensitive to shear and extension.
- Very severe gradients occur in some flow variables.
- There are stress singularities at re-entrant corners – nature unknown.
- Problems are computationally expensive.
- Geometries can be complex with moving boundaries, for example, in injection moulding applications.

Another dimension is added to one's problems if one also takes into account non-isothermal behaviour. In many processing operations temperatures can vary either through imposed cooling, or through viscous heating within the body of the fluid, or both. Small changes in temperature can lead to very large changes in fluid viscosity and result again in the occurrence of extremely sharp boundary layers.

Perhaps the most significant challenge to the numerical analyst is to construct robust algorithms which are efficient. To resolve sharp boundary layers requires algorithms capable of retaining great spatial accuracy in specific regions of the solution domain. In the finite element context this undoubtedly points to adaptive meshing, although it is not yet clear how one might construct such a scheme in the viscoelastic context. Whatever, it is the case that large numbers of elements are needed to achieve appropriate levels of accuracy, even in two spatial dimensions. Add to this the requirement to model complex three-dimensional geometries and to incorporate transient behaviour, that is, a time-stepping procedure, and one can see that efficiency is a high priority, as is the need to exploit the computational power of modern parallel computer architectures. In the next section we outline an algorithm which has been designed to meet some of these points.

4 A numerical algorithm for modelling viscoelastic flow

Computational rheology has attracted the attention of numerical analysts for many years and in the literature one can find examples of work with finite differences, finite elements, boundary elements, finite volume algorithms and spectral methods.

Here we concentrate on a so-called Taylor-Galerkin finite element algorithm which couples a second order Taylor series expansion in time with a Galerkin spatial projection (see Donea [1984]). A pressure correction method is used to satisfy the incompressibility constraint and to solve for the pressure (see Van Kan [1986]). To illustrate the basic elements of the algorithm we give the formulation for the Navier Stokes equations. The comparable form for a PTT model follows in a straightforward manner. Further detail and associated references are to be found in Carew et al. [1993]

From the Navier Stokes equations

$$\rho \mathbf{u}_t = \mu \Delta \mathbf{u} - \rho \mathbf{u} \cdot \nabla \mathbf{u} - \nabla p \tag{4.1}$$

we discretize in time as follows :

$$\frac{\rho}{\Delta t}(\mathbf{u}^{n+1} - \mathbf{u}^n) = \frac{1}{2}([\mu \Delta \mathbf{u}]^{n+1} + [\mu \Delta \mathbf{u}]^n) - [\rho \mathbf{u} \cdot \nabla \mathbf{u}]^{n+1/2} - \frac{1}{2}(\nabla p^{n+1} + \nabla p^n). \tag{4.2}$$

We now perform a Galerkin spatial projection, at the same time introducing a number of discrete steps to achieve the second order time accuracy:

$$\text{Step 1a}: \quad (\frac{2\rho}{\Delta t}M + \frac{\mu}{2}S)(U^{n+1/2} - U^n) = \frac{1}{2}(F^n + F^{n+1/2}) + \left\{-[\mu S + \rho N(U)]U + L^T P\right\}^n \tag{4.3}$$

$$\text{Step 1b}: \quad (\frac{\rho}{\Delta t}M + \frac{\mu}{2}S)(U^* - U^n) = \frac{1}{2}(F^{n+1} + F^n) + \left\{-\mu S U + L^T P\right\}^n - [\rho N(U)]U]^{n+1/2} \tag{4.4}$$

$$\text{Step 2}: \quad KQ^{n+1} = -\frac{2\rho}{\Delta t}LU^*, \qquad \text{where } Q^{n+1} = P^{n+1} - P^n, \tag{4.5}$$

$$\text{Step 3}: \quad \frac{2\rho}{\Delta t}M(U^{n+1} - U^*) = L^T Q^{n+1}. \tag{4.6}$$

Here M is a consistent mass matrix, N is an advection matrix, S is a diffusion matrix, L is an incompressibility matrix and K is a pressure stiffness matrix.

In the two half steps of step 1 a nonsolenoidal velocity field U^* is computed which is then used to compute the pressure from the Poisson equation in step 2. In practice one computes the pressure difference Q over a time step and then updates the pressure. Finally, U^* is corrected at step 3 to produce a divergence-free velocity field U^{n+1}. The algorithm is semi-implicit in the sense that diffusion terms appear on the left hand side of equations (4.3) and (4.4) to greatly enhance stability properties. Steps 1 and 3, which involve the mass matrices, are solved iteratively with only a few iterations, typically less than ten, whereas the Poisson equation is solved directly using Choleski decomposition.

Close examination of the structure of the algorithm reveals that it is well-suited to a parallel implementation and thus offers promise for the task of modelling transient flows in complex three-dimensional geometries. The algorithm has been tried and tested for many different flow problems involving both two and three space dimensions, Newtonian, shear-thinning, strain-thickening and fully viscoelastic fluids, and isothermal and non-isothermal conditions. For the latter a coupled energy equation is solved for the temperature distribution.

5 A typical flow calculation: a contraction flow

The computation of flow into two-dimensional or axi-symmetric three-dimensional contractions such as illustrated in figure 1, is important for two reasons. Firstly, these are geometries which occur frequently in processing operations. Secondly, the attention paid to such problems, both experimentally and numerically, gives rise to much data which may be used to validate computer codes. Of particular interest is the development of recirculating regions in such flows.

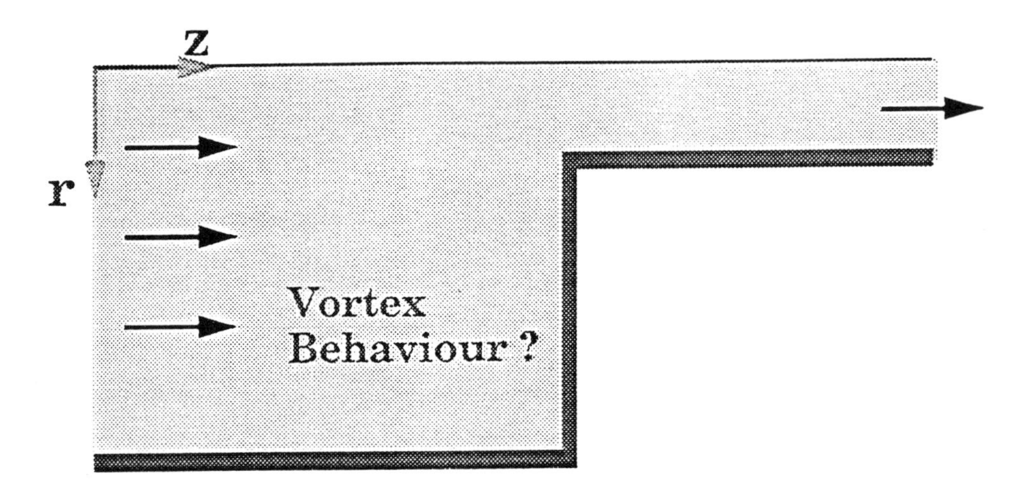

Figure 1.

In non-dimensionalising the governing equations for the flow of a PTT liquid through such a contraction, in addition to a conventional Reynolds number, another non-dimensional group, the Weissenberg number, $\mathbf{We} = \lambda_1 U/R$ emerges, where U and R are characteristic velocity and length, respectively. **We** is thus directly related to the elasticity of the fluid. Initially we turn attention to a special case of the PTT model in which $\varepsilon = \xi = 0$. Such a model is called an Oldroyd B model.

We mentioned above the so-called high Weissenberg number problem and we can illustrate this here. It should be pointed out that for some processing operations on highly elastic materials, such as polymer melts, Weissenberg numbers in excess of 10^2 are common. It is with some consternation, therefore, that when one applies the algorithm set out in section 4 for an Oldroyd B fluid, all is well for very small **We** but at a value of unity numerical inaccuracies are encountered. This can be seen in figure 2a where we plot the steady state shear stress field, τ_{xy}. A non-dimensional time step of 0.01 is used in the calculation. One would expect smooth contours here, and although this solution is stationary, that is, does not change with further time integration, it is physically unacceptable. It proves to be the case that the spatial projection is at fault, and to achieve the smooth solution shown in figure 2b, one must apply a streamline upwinding, Petrov Galerkin projection. The details are to be found in Carew et al. [1993].

One now sets out with some confidence to compute solutions for large **We**, only to be brought to a halt at **We**=2. The solution appears to settle to a steady state but then experiences a very rapid divergence. This proves to be a time stepping instability which may be controlled by halving the time step. Further increments in **We** lead eventually to a limit point at about **We**=5. Beyond that point convergence would not seem to be possible with this algorithm. The limit point is, to some extent, mesh dependent, but whatever one tries it seems unlikely that one can achieve the levels of **We** appropriate for industrial processing.

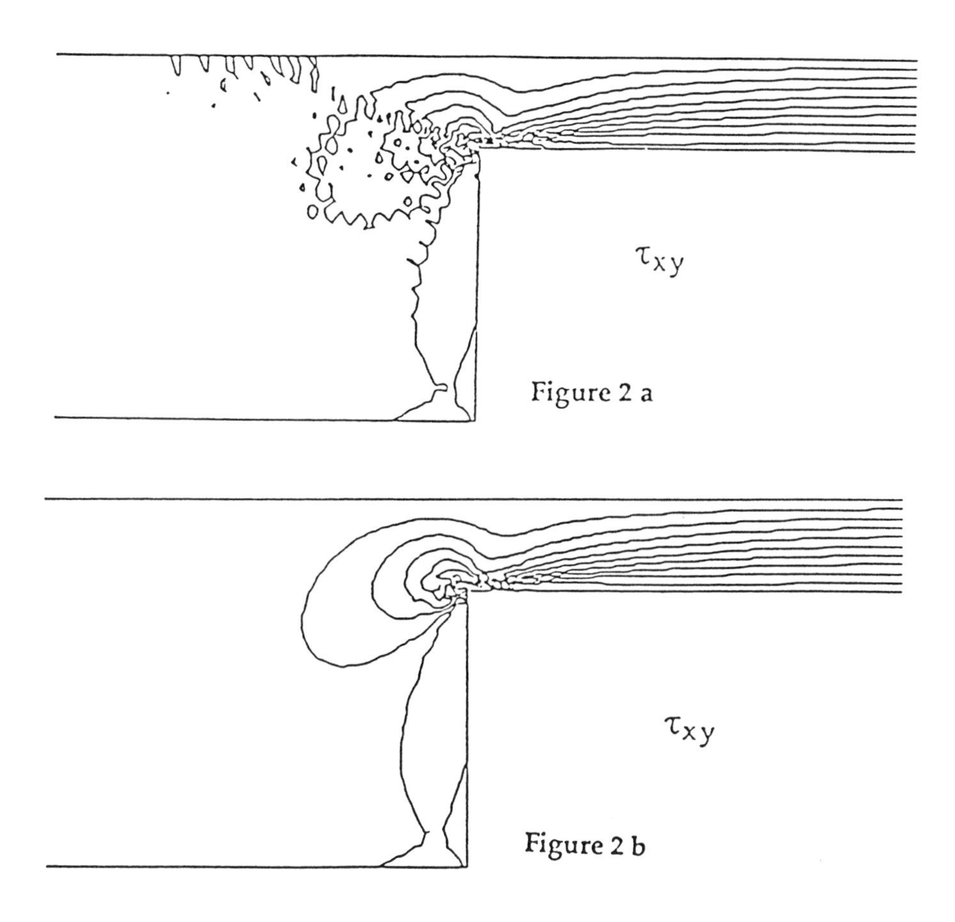

Computed shear stress for the contraction flow of an Oldroyd viscoelastic fluid at **We** = 1 for (a) the Taylor–Galerkin algorithm and (b) SUPG.

Unlikely, that is, if one perseveres with this constitutive model. Careful consideration of the behaviour of an Oldroyd model reveals that it fails to represent appropriately the extensional behaviour of the fluid. In this contraction flow, the fluid experiences rapid extension as it accelerates through the contraction. Essentially the fluid behaves as if its viscosity is greatly increased, and in the Oldroyd case this increase becomes infinite. Taking appropriate values for ε and ξ controls the 'extensional viscosity' and allows one to compute solutions for apparently unlimited values of **We**.

The above illustrates the care the computational rheologist must take to ensure one has both an appropriate mathematical model of the non-Newtonian fluid and a robust numerical algorithm to achieve a solution to the governing equations.

6 Conclusions

Rheological fluid dynamics is a well established field with many important industrial applications. Numerical methods have been applied to rheological flows for many years but it is only recently that major success has been achieved in the qualitative prediction of flows. Challenges still remain, notably the need for computationally efficient algorithms to attack difficult three-dimensional viscoelastic flows. An algorithm has been presented here which might allow some progress towards this goal.

References:

Barnes, H.A., Hutton, J.F. and Walters, K, *An introduction to Rheology*, Elsevier, 1989.

Boger, D.V. and Walters, K, *Rheological Phenomena in Focus*, Elsevier, 1993.

Carew, E.O.A., Townsend, P. and Webster, M.F., *A Taylor-Galerkin algorithm for viscoelastic flows*, to appear in Int. J. non-Newtonian Fluid Mech., 1993.

Donea, J., *A Taylor-Galerkin method for convective transport problems*, Int. J. Num. Meth. Eng., **20**, 101, 1984.

Marchal, J.M. and Crochet, M.J.,*A new mixed finite element for calculating viscoelastic flows*, Int. J. non-Newtonian Fluid Mech., **26**, 77, 1987.

Oldroyd, J.G.,*On the formulation of rheological equation of state*, Proc. Roy. Soc., **A200**, 523, 1950.

Phan-Thien, N and Tanner, R.I.,*A new constitutive equation derived from network theory*, Int. J. non-Newtonian Fluid Mech., **2**, 353, 1977.

Van Kan, J.,*A second-order accurate pressure-correction scheme for viscous incompressible flow*, SIAM J. Sci. Stat. Comput., **7**, 3, 870, 1986.

Professor P. Townsend & Dr M.F.Webster,
Department of Computer Science,
University College,
Swansea,
U.K.
EMAIL: p.townsend@swansea.ac.uk

Contributed Papers[1]

Reducing the Degree of Bézier Curves
M.L. Baart (Potchefstroom University, Dept of Mathematics and Computer Science, PO Box 1174, Vanderbijlpark 1900, South Africa.) & M.A. Coetzee.

Adaptive Grid Generation and Monotonic Adaptive Solutions of Time Dependent PDE's
M. J. Baines (University of Reading, Mathematics Dept, Whiteknights, PO BOX 220, Reading, RG6 2AX Berks, UK.)

On the stability of linear multistep formulae adapted for Volterra functional equations
Christopher T H Baker (University of Manchester, Dept of Mathematics, Oxford Road, Manchester M13 9PL, UK.)

The structured total least squares problem
Sven G. Bartels (Universität Karlsruhe, Inst Fur Statistik und Math, Postfach 6380, 7500 Karlsruhe 1, Germany)

Orthogonality-preserving difference equations and an application to the analytic singular value decomposition
S. Bell (University of Reading, Mathematics Dept, Whiteknights, PO Box 220, Reading, RG6 2AX Berks, UK)

A finite element estimate with trigonometric hat functions for Sturm-Liouville eigenvalues
G. Vanden Berghe (University of Gent, RUG Laboratorium voor Numerieke Wiskunde en Informatica, Krijgslaan 281-S9, B-9000 Gent, Belgium.) & H. De Meyer.

Lebesgue constant minimizing linear rational interpolation of continuous functions over the interval
Jean–Paul Berrut (Université de Fribourg, Dept de Mathematiques, Perolles, CH-1700 Fribourg/Perolles, Switzerland)

The numerical analysis of a model for phase separation of a multi-component mixture
J.F. Blowey (University of Durham, Dept of Mathematical Sciences, Science Laboratories, South Road, Durham DH1 3LE, UK.) M.I.M. Copetti, & C.M. Elliott.

Discrete least squares approximation and prewavelets from radial function spaces
M.D. Buhmann (University of Cambridge, DAMPT, Silver Street, Cambridge, CB3 9EW.)

[1]The addresses given are those of the first named author, the presenter of each paper.

The Heat Balance Integral Method for cylindrical and spherical problems
J. Caldwell (City Polytechnic of Hong Kong, Dept of Applied Mathematics, 83 Tat Chee Avenue, Kowloon, Hong Kong.)

Some recent results on symplectic integration
M. P. Calvo (Departmento Matemática Aplicada y Computación, Universidad de Valladolid, Facultad de Ciencias, Valladolid, Spain.)

The Finite Section Method for integral equations on the half-line in weighted spaces
Simon N Chandler–Wilde (University of Bradford, Dept of Civil Engineering, Bradford BD7 1DP, UK.)

Quality assessment of adaptive quadrilateral meshes by function interpolations
Ke Chen (University of Liverpool, Dept of Stats and Comp Math, Victoria Building, Brownlow Hill, PO Box 147, Liverpool L69 3BX, UK.)

Limit loads computed as the minimum of a sum of norms
Edmund Christiansen (Odense University , Matematisk Institut, Campusvej 55, DK-5230 Odense M, Denmark)

The Faber polynomials for annular sectors
J.P. Coleman (University of Durham, Dept of Mathematical Sciences, South Road, Durham DH1 3LE, UK.) & N.J. Myers.

An adaptive scheme for the derivation of harmonic impedance contours
I. D. Coope (University of Canterbury, Dept of Mathematics, Christchurch 1, New Zealand.) M. Dominguez, J. Arrillaga, & N.R. Watson.

A one-dimensional quasi-static contact problem in linear thermoelasticity
M. I. M. Copetti (Universidade Federal de Santa Maria, Departmento de Matematica, 97 119-900 Santa Maria, RS, Brazil.) & C.M.Elliott.

Discretisation and multigrid solution of elliptic equations with strongly discontinuous coefficients
P. I. Crumpton (University of Oxford, Computing Lab, Wolfson Building, Parks Road, Oxford OX1 3QD, UK.) G. J. Shaw, & A. F. Ware.

Continuation and the detection of bifurcations in large systems of differential algebraic equations using the recursive projection method
Bryan D. Davidson (University of Bristol, School of Mathematics, University Walk, Bristol BS8 1TW, UK.)

Accurate & stable numerical schemes for scattering problems
Penny J. Davies (University of Dundee, Department of Mathematics and Computer

Science, Dundee DD1 4HN,UK.) & Dugald B. Duncan.

Multiplicative Schwartz Factorization for Preconditioning Reservoir Simulation Applications
J. C. Díaz (University of Tulsa, Dept of Comp and Math Sciences, 600 S College, Tulsa, Oklahoma 74104, USA) L. Chu, M. Komara & A. C. Reynolds,

Stability of fully discrete spectral-collocation methods in the numerical solution of convection-diffusion problems
J.L.M. van Dorsselaer (University of Leiden, Department of Mathematics, PO Box 9512, 2300 RA Leiden, The Netherlands.)

Approximation of lattice equations
D. B. Duncan (Heriot Watt University, Mathematics Dept, Riccarton, Edinburgh, EH14 4AS, UK.)

Inexact and preconditioned Uzawa algorithms for saddle point problems
Howard C. Elman (University of Maryland, Dept of Computer Science, College Park, MD 20742, USA.)

Fast numerical solution of the exterior Helmholtz problem by imbedding
Oliver Ernst (Stanford University, SCCM Program MJH 312, Stanford CA 94305, USA.)

DECUHR: An algorithm for automatic integration of singular functions over a hyperrectangular region
Terje O. Espelid (University of Bergen, Department of Informatics, Hoyteknologisteret, N-5020 Bergen, Norway) & Alan Genz.

Convergence rates for polynomial interpolation of analytic functions
B Fornberg (Exxon Research and Eng. Co., Clinton Township, Route 22 East, Annandale, New Jersey, 08801, USA.)

Parallel globally adaptive quadrature on the KSR-1
T. L. Freeman (University of Manchester, Dept of Mathematics, Manchester M13 9PL,UK.)

Some practical experience with the time integration of the nonlinear Galerkin method
B. García-Archilla (Departmento Matemática Aplicada y Computación, Universidad de Valladolid, Facultad de Ciencias, Valladolid, Spain.)

Numerical solution of non-convex Ginzburg-Landau variational problems
A R Gardiner (University of Sussex, Mathematical and Physics Building, Falmer, Brighton BN1 9QH, UK.)

A non-linear second-order method for solving ODEs
R Garrapa (via A Muolo 12, 70043 Monopoli BA, Italy.)

Iterative methods for the detection of Hopf bifurcations in finite element discretisations of incompressible flow problems
T. J. Garratt (Aspentech UK Ltd, Sheraton House, Castle Park, Cambridge CB3 0AX, UK.) K. A. Cliffe, & A. Spence.

Optimization of an inverse siesmic problem
Susana Gomez (INRIA Rocquencourt, France & University of Mexico)

Is SOR being revived?
P.R. Graves-Morris (University of Bradford, Department of Mathematics, Bradford, West Yorkshire BD7 1DP, UK.)

Krylov subspace methods for the incompressible Navier-Stokes equations
Richard Hanby (UMIST, Dept of Mathematics, PO Box 88, Manchester M60 1QD, UK.)

The evaluation of wavelet integrals
David Handscomb (University of Oxford, Computing Laboratory, Wolfson Building, Parks Road, Oxford, OX1 3QD, UK.)

The matrix sign decomposition and its relation to the polar decomposition
Nicholas J. Higham (University of Manchester, Dept of Mathematics, Manchester M13 9PL, UK.)

Nonlinear stability theory for ODEs: T-stability, An asymptotic approach
A.R. Humphries (University of Bristol, School of Mathematics, University Walk, Bristol BS8 1TW, UK.)

The numerical splitting error in advection-reaction equations
Willem Hundsdorfer (Centre for Mathematics & Comp Science, Kruislaan 413, PO Box 4079, 1009 AB Amsterdam, The Netherlands.)

Three-dimensional fluid flow in a tube using finite and pseudospectral discretisations
R. Hunt (University of Strathclyde, Department of Mathematics, Livingstone Tower, 26 Richmond Street, Glasgow G1 1XH, UK.)

The numerical computation of zeros of Bessel function $J_\nu(z)$ regarded as a function of ν for any given real or complex z
Yasuhiko Ikebe (University of Tsukuba, College of Information Sciences , Tsukuba-City, Ibaraki-Ken, Japan 305.) Nobuyoshi Asai, Yoshinori Miyazaki, DongSheng Cai, & Issei Fujishiro.

Solution of functional equations by Dirichlet series
A. Iserles (University of Cambridge, DAMTP, Silver Street, Cambridge CB3 9EW, UK.)

Symplectic partitioned Runge-Kutta methods for constrained Hamiltonian systems
Laurent Jay (Université de Genève, Section de Mathematiques, 2-4 rue du Lievre, Case postale 240, 1211 Genève 24, Switzerland.)

On IOM(q): Incomplete orthogonalization methods for large unsymmetric linear systems
Zhongxiao Jia (Universität Bielefeld, Department of Mathematics, 4800 Bielefield 1,Germany)

On mesh adaptivity for time-dependent problems
Peter K. Jimack (University of Leeds, School of Computer Studies, Leeds LS2 9JT, UK.)

Experiments with a Hessian-free method for sparse minimax problems
Kristjan Jonasson (Technical University of Denmark, Institute for Numerical Analysis, 2800 Lyngby, Denmark.)

A Volterra integral type method for solving initial-boundary value problems with nonlinear boundary conditions
B.Jumarhon (University of Strathclyde, Dept of Mathematics, Livingstone Tower, 26 Richmond Stret, Glasgow G1 1XH, UK.) & S.McKee.

Numerical solution of Hamiltonian normal forms
B. Karasözen (Middle East Technical University, Department of Mathematics, 06531 Ankara, Turkey.)

Implicit Runge-Kutta methods for singularly perturbed integrodifferential equations
Jean-Paul Kauthen (University of Fribourg, Institut de Mathematiques, CH-1700 Fribourg, Switzerland.)

Solutions to large Sylvester equations using the Arnoldi and GMRES methods
Ebrahim M. Kasenally (Imperial College, Interdisciplinary Research Centre for Process Systems Engineering, Exhibition Road, London SW7 2BY, UK.) & Imad M. Jaimoukha.

On chaotic behaviour of some difference equations
S Keras (University of Cambridge, DAMTP, Silver Street, Cambridge CB39EW, UK.)

Matrix powers in finite precision arithmetic
Philip A. Knight (University of Manchester, Dept of Mathematics, Manchester M13 9PL.)

Anti-Gaussian quadrature formulas
Dirk P. Laurie (Potchefstroom University, Dept of Maths, PO Box 1174, Vanderbijlpark 1900, South Africa.)

Local stability of translates of polyharmnic splines in even space dimension
Jeremy Levesley (University of Leicester, Department of Mathematics & Computer Science, Leicester LE1 7RH,UK.)

Improving eigenstructure assignment by a minimisation technique
D M Littleboy (University of Reading, Mathematics Dept, Whiteknights, PO Box 220, Reading, RG6 2AX Berks,UK.)

Stability analysis of the θ-methods for neutral differential equations
Yunkang Liu (University of Cambridge, DAMTP, Silver Street, Cambridge CB3 9EW,UK.)

A unitary integrator for the matrix differential equation $\Theta' = i\Theta\Omega(x, \Theta)$
M Marletta (University of Leicester, Dept of Mathematics and Comp Sci, Leicester LE1 7RH,UK.)

Polynomial splines of even degree: Best approximation by γ-splines
M.F.G. Martins (Universidade Nove de Lisboa, Departmento de Mathematica, Quinta da Torre, 2825 Monte da Caparica, Portugal.)

On applications of GENSMAC: a multiple free surface code for Newtonian and non-Newtonian flow
S. McKee (University of Strathclyde, Dept of Mathematics, 26 Richmond Street, Glasgow, G1 1XH, UK.) & M. Tomé.

Expert systems to select numerical solvers for second order initial value problems
G.V. Mitsou (Agricultural University of Greece, Informatics Laboratory, Iera Odos 75, Athens 118 55, Greece.) & T.E. Simos.

An expert system for the numerical solution of the radial Schrödinger equation
G.V. Mitsou (Agricultural University of Greece, Informatics Laboratory, Iera Odos 75, Athens 118 55, Greece.) & T.E. Simos.

Computing the global structures in dynamical systems
G. Moore (Imperial College, Dept of Mathematics, Huxley Building, Queen's Gate, London SW7 2BZ,UK.)

Non-symmetry in the modelling of contaminant transport in porous media
K.J.Neylon (University of Reading, Dept of Mathematics, Whiteknights, PO Box 220, Reading, RG6 2AX,UK)

Pseudospectral Chebyshev methods for the solution of the KDV equation domain-decomposition techniques
F.Z. Nouri (University of Annaba, cité des 88 logts., Bloc 2 Appts 12 Bd d'Afrique, Annaba 23000, Algeria.)

A generalized orthogonality relation for multidimensional wavelets
Gerhard Opfer (University of Hamburg, Inst fur Angewandte Math, Bundesstr. 55, D 2000 Hamburg 13, Germany.)

On Runge–Kutta discretizations for time-dependent parabolic problems
Alexander Ostermann (Universitaet Innsbruck, Institut fur Mathematik, Technikerstrasse 13, A-6020 Innsbruck, Austria.)

Full discretizations of parabolic initial value problems
C. Palencia (Departmento Matemática Aplicada y Computación, Universidad de Valladolid, Facultad de Ciencias, Valladolid, Spain.)

Bounded homotopies to solve systems of algebraic nonlinear equations
Jorge R. Paloschi (Imperial College, Centre for Process Systems Engineering, London SW7 2BY,UK)

Parallel multiple shooting by partial updating
A.Papini (Universitá di Firenze,Dipartmento di Energetica, Via Cesare Lombroso 6/17, 50134 Firenze,Italy.) S. Bellavia, M.G.Gasparo, M.Macconi,

High order approximations to the matrix exponential via operator splitting
Dimpy Pathria (Princeton University, Program in Applied and Comp Maths, Princeton, New Jersey 08544-1000, USA) & J. Bronski.

Numerical solution of second kind integral equations on infinite cylindrical surfaces
Andrew Peplow (University of Southampton, Structural Dynamics Group, ISVR, Southampton, UK.) & Simon Chandler-Wilde.

Iterative application of FFT methods for the numerical solution of partial differential equations
W.M. Pickering (University of Sheffield, PO Box 597, Dept of Applied & Comp. Mathematics, Sheffield S10 2UN,UK.) & P.J. Harley.

Iterative solution techniques for the Navier-Stokes equations
Alison Ramage (University of Strathclyde, Dept of Mathematics, 26 Richmond Street, Glasgow G1 1XH,UK.) & Andy Wathen.

Numerical Solutions of Hamiltonian Systems
G. Ramaswami (University of Cambridge, DAMTP, Silver Street, Cambridge CB3 9EW,UK.)

High precision integration of stiff O.D.E's by multistep Radau methods
Stefan Schneider (Université de Genève, Section de Mathematiques, 2-4 rue de Lievre, C.P. 240, 1211 Genève 24,Switzerland)

Inverse and implicit function theorems for piecewise differentiable functions
Stefan Scholtes (Universität Karlsruhe, Inst Fur Statistik und Math, Postfach 6380, 7500 Karlsruhe 1,Germany.)

Multiple scales analysis for a spectral method
S.W. Schoombie (University of the Orange Free State, Dept of Applied Mathematics, PO Box 339, Bloemfontein 9300, South Africa.) & E. Maré.

MA42 - A new frontal code for solving sparse unsymmetric systems
J.A. Scott (Rutherford Appleton Laboratory, Atlas Centre, Chilton, Didcot, Oxon OX11 0QX,UK.) & I.S. Duff.

Convergence rates and classification for irregular finite element meshes
P.M. Selwood (University of Bristol, School of Mathematics, University Walk, Bristol BS8 1TW,UK.) & A.J. Wathen.

Multigrid preconditioning of incompressible flow equations
D.J. Silvester (UMIST, Dept of Mathematics, PO Box 88, Manchester M60 1QD,UK.)

The oscillatory and pulsatile flow of a Newtonian fluid through square corrugated tubes
Z. Skalak (Czechoslovak Academy of Sciences, Institute of Hydrodynamics, Podbabska 13, 166 12 Praha 6, Czech Republic.) & P. Kucera.

Pseudospectral solution of a nonlinear, dissipative system
D. M. Sloan (University of Strathclyde, Dept of Mathematics, 26 Richmond Street, Glasgow, G1 1XH, UK.) & R. Wallace.

Order conditions for symplectic Runge-Kutta schemes
M. Sofroniou (Loughborough University, Dept of Mathematical Sciences, Loughborough, Leicestershire LE11 3TU,UK.) & W. Oevel.

Wavelets, filters and dilation equations
Gilbert Strang (M.I.T., ROOM 2-240, Mathematics Dept, Cambridge, Massachussetts 02139, USA.)

Analysis of a cell vertex finite volume method for convection-diffusion in two dimensions
Martin Stynes (University College Cork, Dept of Mathematics, Cork, Ireland.) K.W.Morton & E.Süli,

A posteriori error analysis for Galerkin approximations of Friedrichs systems
Endre Süli (University of Oxford, Computing Laboratory, Wolfson Building, Parks Road, Oxford OX1 3QD,UK.)

A block Markowitz strategy for the solution of sparse linear equations
V Swift (University of Portsmouth, School of Mathematical Studies, Mercantile House, Hampshire Terrace, Portsmouth,UK.)

Implicit Newton's method in the update of matrix factorizations
Marek Szularz (Kingston University, School of Mathematics, Kingston-upon-Thames, Surrey KT1 2EE,UK.) & Svetlana Lyle.

On singular integro-differential equations
Ezio Venturino (Citta Universitaria, Dipartmento di Matematica, viale A doria 6, 95100 Catania, Italy.)

An asymptotic expansion for integration over a disk
P. Verlinden (Katholieke Universiteit Leuven, Dept of Computing Science, Celestijnenlaan 200A, B-3001 Leuven, Belgium.)

A Shadowing Lemma Approach to Global Error Analysis for Initial Value ODEs
Erik S. Van Vleck (Simon Fraser University, Dept of Maths & Statistics, Burnaby, BC V5A 1S6 , Canada.)

Approximations to shallow water flows using variational principles
S. L. Wakelin (University of Reading, Mathematics Dept, Whiteknights, PO Box 220, Reading, RG6 2AX Berks,UK.)

Package PARASPAR for general sparse matrices
Jerzy Waśniewski (The Danish Computer Centre for Research and Education (UNI-C), DTH, Bgn. 305, 2800 Lyngby, Denmark.)

Approximate Greens function preconditioning for the semiconductor device equations
Andy Wathen (University of Bristol, School of Mathematics, University Walk, Bristol BS8 1TW,UK.) & Steve Chandler.

Solving generalizations of orthogonal Procrustes problems
G A Watson (University of Dundee, Dept of Mathematical Sciences, Dundee DD1 4HN, Scotland,UK.)

Computation of Hopf bifurcation with bordered matrices
Bodo Werner (University of Hamburg, Institut fur Angewandte Mathematik, Bundesstrasse 55, D-2000 Hamburg 13, Germany.)

ODEs with singular points and singular collocation matrices
K. Wright (University of Newcastle, University Computing LAB, Claremont Tower, Claremont Road, Newcastle upon Tyne NE1 7RU,UK.)